NONPARAMETRIC REGRESSION AND SPLINE SMOOTHING

STATISTICS: Textbooks and Monographs

A Series Edited by

1. The Generalized Jackknife Statistic, *H. L. Gray and W. R. Schucany*
2. Multivariate Analysis, *Anant M. Kshirsagar*
3. Statistics and Society, *Walter T. Federer*
4. Multivariate Analysis: A Selected and Abstracted Bibliography, 1957–1972, *Kocherlakota Subrahmaniam and Kathleen Subrahmaniam*
5. Design of Experiments: A Realistic Approach, *Virgil L. Anderson and Robert A. McLean*
6. Statistical and Mathematical Aspects of Pollution Problems, *John W. Pratt*
7. Introduction to Probability and Statistics (in two parts), Part I: Probability; Part II: Statistics, *Narayan C. Giri*
8. Statistical Theory of the Analysis of Experimental Designs, *J. Ogawa*
9. Statistical Techniques in Simulation (in two parts), *Jack P. C. Kleijnen*
10. Data Quality Control and Editing, *Joseph I. Naus*
11. Cost of Living Index Numbers: Practice, Precision, and Theory, *Kali S. Banerjee*
12. Weighing Designs: For Chemistry, Medicine, Economics, Operations Research, Statistics, *Kali S. Banerjee*
13. The Search for Oil: Some Statistical Methods and Techniques, *edited by D. B. Owen*
14. Sample Size Choice: Charts for Experiments with Linear Models, *Robert E. Odeh and Martin Fox*
15. Statistical Methods for Engineers and Scientists, *Robert M. Bethea, Benjamin S. Duran, and Thomas L. Boullion*
16. Statistical Quality Control Methods, *Irving W. Burr*
17. On the History of Statistics and Probability, *edited by D. B. Owen*
18. Econometrics, *Peter Schmidt*
19. Sufficient Statistics: Selected Contributions, *Vasant S. Huzurbazar (edited by Anant M. Kshirsagar)*

20. Handbook of Statistical Distributions, *Jagdish K. Patel, C. H. Kapadia, and D. B. Owen*
21. Case Studies in Sample Design, *A. C. Rosander*
22. Pocket Book of Statistical Tables, *compiled by R. E. Odeh, D. B. Owen, Z. W. Birnbaum, and L. Fisher*
23. The Information in Contingency Tables, *D. V. Gokhale and Solomon Kullback*
24. Statistical Analysis of Reliability and Life-Testing Models: Theory and Methods, *Lee J. Bain*
25. Elementary Statistical Quality Control, *Irving W. Burr*
26. An Introduction to Probability and Statistics Using BASIC, *Richard A. Groeneveld*
27. Basic Applied Statistics, *B. L. Raktoe and J. J. Hubert*
28. A Primer in Probability, *Kathleen Subrahmaniam*
29. Random Processes: A First Look, *R. Syski*
30. Regression Methods: A Tool for Data Analysis, *Rudolf J. Freund and Paul D. Minton*
31. Randomization Tests, *Eugene S. Edgington*
32. Tables for Normal Tolerance Limits, Sampling Plans and Screening, *Robert E. Odeh and D. B. Owen*
33. Statistical Computing, *William J. Kennedy, Jr., and James E. Gentle*
34. Regression Analysis and Its Application: A Data-Oriented Approach, *Richard F. Gunst and Robert L. Mason*
35. Scientific Strategies to Save Your Life, *I. D. J. Bross*
36. Statistics in the Pharmaceutical Industry, *edited by C. Ralph Buncher and Jia-Yeong Tsay*
37. Sampling from a Finite Population, *J. Hajek*
38. Statistical Modeling Techniques, *S. S. Shapiro and A. J. Gross*
39. Statistical Theory and Inference in Research, *T. A. Bancroft and C.-P. Han*
40. Handbook of the Normal Distribution, *Jagdish K. Patel and Campbell B. Read*
41. Recent Advances in Regression Methods, *Hrishikesh D. Vinod and Aman Ullah*
42. Acceptance Sampling in Quality Control, *Edward G. Schilling*
43. The Randomized Clinical Trial and Therapeutic Decisions, *edited by Niels Tygstrup, John M Lachin, and Erik Juhl*
44. Regression Analysis of Survival Data in Cancer Chemotherapy, *Walter H. Carter, Jr., Galen L. Wampler, and Donald M. Stablein*
45. A Course in Linear Models, *Anant M. Kshirsagar*
46. Clinical Trials: Issues and Approaches, *edited by Stanley H. Shapiro and Thomas H. Louis*
47. Statistical Analysis of DNA Sequence Data, *edited by B. S. Weir*
48. Nonlinear Regression Modeling: A Unified Practical Approach, *David A. Ratkowsky*
49. Attribute Sampling Plans, Tables of Tests and Confidence Limits for Proportions, *Robert E. Odeh and D. B. Owen*
50. Experimental Design, Statistical Models, and Genetic Statistics, *edited by Klaus Hinkelmann*
51. Statistical Methods for Cancer Studies, *edited by Richard G. Cornell*
52. Practical Statistical Sampling for Auditors, *Arthur J. Wilburn*
53. Statistical Methods for Cancer Studies, *edited by Edward J. Wegman and James G. Smith*
54. Self-Organizing Methods in Modeling: GMDH Type Algorithms, *edited by Stanley J. Farlow*
55. Applied Factorial and Fractional Designs, *Robert A. McLean and Virgil L. Anderson*
56. Design of Experiments: Ranking and Selection, *edited by Thomas J. Santner and Ajit C. Tamhane*
57. Statistical Methods for Engineers and Scientists: Second Edition, Revised and Expanded, *Robert M. Bethea, Benjamin S. Duran, and Thomas L. Boullion*

58. Ensemble Modeling: Inference from Small-Scale Properties to Large-Scale Systems, *Alan E. Gelfand and Crayton C. Walker*
59. Computer Modeling for Business and Industry, *Bruce L. Bowerman and Richard T. O'Connell*
60. Bayesian Analysis of Linear Models, *Lyle D. Broemeling*
61. Methodological Issues for Health Care Surveys, *Brenda Cox and Steven Cohen*
62. Applied Regression Analysis and Experimental Design, *Richard J. Brook and Gregory C. Arnold*
63. Statpal: A Statistical Package for Microcomputers—PC-DOS Version for the IBM PC and Compatibles, *Bruce J. Chalmer and David G. Whitmore*
64. Statpal: A Statistical Package for Microcomputers—Apple Version for the II, II+, and IIe, *David G. Whitmore and Bruce J. Chalmer*
65. Nonparametric Statistical Inference: Second Edition, Revised and Expanded, *Jean Dickinson Gibbons*
66. Design and Analysis of Experiments, *Roger G. Petersen*
67. Statistical Methods for Pharmaceutical Research Planning, *Sten W. Bergman and John C. Gittins*
68. Goodness-of-Fit Techniques, *edited by Ralph B. D'Agostino and Michael A. Stephens*
69. Statistical Methods in Discrimination Litigation, *edited by D. H. Kaye and Mikel Aickin*
70. Truncated and Censored Samples from Normal Populations, *Helmut Schneider*
71. Robust Inference, *M. L. Tiku, W. Y. Tan, and N. Balakrishnan*
72. Statistical Image Processing and Graphics, *edited by Edward J. Wegman and Douglas J. DePriest*
73. Assignment Methods in Combinatorial Data Analysis, *Lawrence J. Hubert*
74. Econometrics and Structural Change, *Lyle D. Broemeling and Hiroki Tsurumi*
75. Multivariate Interpretation of Clinical Laboratory Data, *Adelin Albert and Eugene K. Harris*
76. Statistical Tools for Simulation Practitioners, *Jack P. C. Kleijnen*
77. Randomization Tests: Second Edition, *Eugene S. Edgington*
78. A Folio of Distributions: A Collection of Theoretical Quantile-Quantile Plots, *Edward B. Fowlkes*
79. Applied Categorical Data Analysis, *Daniel H. Freeman, Jr.*
80. Seemingly Unrelated Regression Equations Models: Estimation and Inference, *Virendra K. Srivastava and David E. A. Giles*
81. Response Surfaces: Designs and Analyses, *Andre I. Khuri and John A. Cornell*
82. Nonlinear Parameter Estimation: An Integrated System in BASIC, *John C. Nash and Mary Walker-Smith*
83. Cancer Modeling, *edited by James R. Thompson and Barry W. Brown*
84. Mixture Models: Inference and Applications to Clustering, *Geoffrey J. McLachlan and Kaye E. Basford*
85. Randomized Response: Theory and Techniques, *Arijit Chaudhuri and Rahul Mukerjee*
86. Biopharmaceutical Statistics for Drug Development, *edited by Karl E. Peace*
87. Parts per Million Values for Estimating Quality Levels, *Robert E. Odeh and D. B. Owen*
88. Lognormal Distributions: Theory and Applications, *edited by Edwin L. Crow and Kunio Shimizu*
89. Properties of Estimators for the Gamma Distribution, *K. O. Bowman and L. R. Shenton*
90. Spline Smoothing and Nonparametric Regression, *Randall L. Eubank*
91. Linear Least Squares Computations, *R. W. Farebrother*
92. Exploring Statistics, *Damaraju Raghavarao*

93. Applied Time Series Analysis for Business and Economic Forecasting, *Sufi M. Nazem*
94. Bayesian Analysis of Time Series and Dynamic Models, *edited by James C. Spall*
95. The Inverse Gaussian Distribution: Theory, Methodology, and Applications, *Raj S. Chhikara and J. Leroy Folks*
96. Parameter Estimation in Reliability and Life Span Models, *A. Clifford Cohen and Betty Jones Whitten*
97. Pooled Cross-Sectional and Time Series Data Analysis, *Terry E. Dielman*
98. Random Processes: A First Look, Second Edition, Revised and Expanded, *R. Syski*
99. Generalized Poisson Distributions: Properties and Applications, *P. C. Consul*
100. Nonlinear L_p-Norm Estimation, *Rene Gonin and Arthur H. Money*
101. Model Discrimination for Nonlinear Regression Models, *Dale S. Borowiak*
102. Applied Regression Analysis in Econometrics, *Howard E. Doran*
103. Continued Fractions in Statistical Applications, *K. O. Bowman and L. R. Shenton*
104. Statistical Methodology in the Pharmaceutical Sciences, *Donald A. Berry*
105. Experimental Design in Biotechnology, *Perry D. Haaland*
106. Statistical Issues in Drug Research and Development, *edited by Karl E. Peace*
107. Handbook of Nonlinear Regression Models, *David A. Ratkowsky*
108. Robust Regression: Analysis and Applications, *edited by Kenneth D. Lawrence and Jeffrey L. Arthur*
109. Statistical Design and Analysis of Industrial Experiments, *edited by Subir Ghosh*
110. *U*-Statistics: Theory and Practice, *A. J. Lee*
111. A Primer in Probability: Second Edition, Revised and Expanded, *Kathleen Subrahmaniam*
112. Data Quality Control: Theory and Pragmatics, *edited by Gunar E. Liepins and V. R. R. Uppuluri*
113. Engineering Quality by Design: Interpreting the Taguchi Approach, *Thomas B. Barker*
114. Survivorship Analysis for Clinical Studies, *Eugene K. Harris and Adelin Albert*
115. Statistical Analysis of Reliability and Life-Testing Models: Second Edition, *Lee J. Bain and Max Engelhardt*
116. Stochastic Models of Carcinogenesis, *Wai-Yuan Tan*
117. Statistics and Society: Data Collection and Interpretation, Second Edition, Revised and Expanded, *Walter T. Federer*
118. Handbook of Sequential Analysis, *B. K. Ghosh and P. K. Sen*
119. Truncated and Censored Samples: Theory and Applications, *A. Clifford Cohen*
120. Survey Sampling Principles, *E. K. Foreman*
121. Applied Engineering Statistics, *Robert M. Bethea and R. Russell Rhinehart*
122. Sample Size Choice: Charts for Experiments with Linear Models: Second Edition, *Robert E. Odeh and Martin Fox*
123. Handbook of the Logistic Distribution, *edited by N. Balakrishnan*
124. Fundamentals of Biostatistical Inference, *Chap T. Le*
125. Correspondence Analysis Handbook, *J.-P. Benzécri*
126. Quadratic Forms in Random Variables: Theory and Applications, *A. M. Mathai and Serge B. Provost*
127. Confidence Intervals on Variance Components, *Richard K. Burdick and Franklin A. Graybill*
128. Biopharmaceutical Sequential Statistical Applications, *edited by Karl E. Peace*
129. Item Response Theory: Parameter Estimation Techniques, *Frank B. Baker*
130. Survey Sampling: Theory and Methods, *Arijit Chaudhuri and Horst Stenger*
131. Nonparametric Statistical Inference: Third Edition, Revised and Expanded, *Jean Dickinson Gibbons and Subhabrata Chakraborti*
132. Bivariate Discrete Distribution, *Subrahmaniam Kocherlakota and Kathleen Kocherlakota*

133. Design and Analysis of Bioavailability and Bioequivalence Studies, *Shein-Chung Chow and Jen-pei Liu*
134. Multiple Comparisons, Selection, and Applications in Biometry, *edited by Fred M. Hoppe*
135. Cross-Over Experiments: Design, Analysis, and Application, *David A. Ratkowsky, Marc A. Evans, and J. Richard Alldredge*
136. Introduction to Probability and Statistics: Second Edition, Revised and Expanded, *Narayan C. Giri*
137. Applied Analysis of Variance in Behavioral Science, *edited by Lynne K. Edwards*
138. Drug Safety Assessment in Clinical Trials, *edited by Gene S. Gilbert*
139. Design of Experiments: A No-Name Approach, *Thomas J. Lorenzen and Virgil L. Anderson*
140. Statistics in the Pharmaceutical Industry: Second Edition, Revised and Expanded, *edited by C. Ralph Buncher and Jia-Yeong Tsay*
141. Advanced Linear Models: Theory and Applications, *Song-Gui Wang and Shein-Chung Chow*
142. Multistage Selection and Ranking Procedures: Second-Order Asymptotics, *Nitis Mukhopadhyay and Tumulesh K. S. Solanky*
143. Statistical Design and Analysis in Pharmaceutical Science: Validation, Process Controls, and Stability, *Shein-Chung Chow and Jen-pei Liu*
144. Statistical Methods for Engineers and Scientists: Third Edition, Revised and Expanded, *Robert M. Bethea, Benjamin S. Duran, and Thomas L. Boullion*
145. Growth Curves, *Anant M. Kshirsagar and William Boyce Smith*
146. Statistical Bases of Reference Values in Laboratory Medicine, *Eugene K. Harris and James C. Boyd*
147. Randomization Tests: Third Edition, Revised and Expanded, *Eugene S. Edgington*
148. Practical Sampling Techniques: Second Edition, Revised and Expanded, *Ranjan K. Som*
149. Multivariate Statistical Analysis, *Narayan C. Giri*
150. Handbook of the Normal Distribution: Second Edition, Revised and Expanded, *Jagdish K. Patel and Campbell B. Read*
151. Bayesian Biostatistics, *edited by Donald A. Berry and Dalene K. Stangl*
152. Response Surfaces: Designs and Analyses, Second Edition, Revised and Expanded, *André I. Khuri and John A. Cornell*
153. Statistics of Quality, *edited by Subir Ghosh, William R. Schucany, and William B. Smith*
154. Linear and Nonlinear Models for the Analysis of Repeated Measurements, *Edward F. Vonesh and Vernon M. Chinchilli*
155. Handbook of Applied Economic Statistics, *Aman Ullah and David E. A. Giles*
156. Improving Efficiency by Shrinkage: The James-Stein and Ridge Regression Estimators, *Marvin H. J. Gruber*
157. Nonparametric Regression and Spline Smoothing: Second Edition, *Randall L. Eubank*
158. Asymptotics, Nonparametrics, and Time Series, *edited by Subir Ghosh*

Additional Volumes in Preparation

Multivariate Analysis, Design of Experiments, and Survey Sampling, *edited by Subir Ghosh*

Statistical Process Monitoring and Control, *S. Park and G. Vining*

NONPARAMETRIC REGRESSION AND SPLINE SMOOTHING

Second Edition

RANDALL L. EUBANK

Department of Statistics
Texas A&M University
College Station, Texas

CRC Press is an imprint of the
Taylor & Francis Group, an **informa** business

The first edition of this book was published as *Spline Smoothing and Nonparametric Regression* (1988).

First published in 1999 by Marcel Dekker

Published in 2020 by CRC Press
Taylor & Francis Group
6000 Broken Sound Parkway NW, Suite 300
Boca Raton, FL 3487-2742

First issued in paperback 2020

ISBN-13: 978-0-367-57921-0 (pbk)
ISBN-13: 978-0-8247-9337-1 (hbk)

Visit the Taylor & Francis Web site at
http://www.taylorandfrancis.com

and the CRC Press Web site at
http://www.crcpress.com

For Lisa, Boomer, and Tiny

Preface

When the first edition of this book, *Spline Smoothing and Nonparametric Regression*, appeared in 1988 it joined Prakasa Rao (1983) and Müller (1988) as the books concerning nonparametric regression which were widely available at that time. In the ensuing decade the number and variety of texts has increased substantially and many wonderful books are now available on data smoothing and related topics. These include texts on local polynomial smoothing by Fan and Gjibels (1996), treatises on kernel smoothing by Härdle (1990) and Wand and Jones (1995), books on spline smoothing by Wahba (1990) and Green and Silverman (1994) and more general overviews by Scott (1992) and Simonoff (1995).

The development of so many excellent smoothing related books in the last decade greatly influenced my perspective when preparing this second edition. There no longer appears to be a need for in-depth theoretical treatments of any particular smoothing methodology since these can be found in the books mentioned above, for example. Instead, I felt that an introductory treatment that opens the door for further study and investigation could be of some value and that is the viewpoint I have used in preparing the present book. It is intended primarily for a newcomer to the field of data smoothing whose knowledge of Statistics is at least that of, say, a second or third year Statistic's graduate student.

For maximum benefit, a reader of this book should have a good working knowledge of calculus, mathematical statistics, matrix theory and some elementary large sample theory (e.g., the law of large numbers, the Lindeberg-Feller central limit theorem and Slutsky's Theorem). The tools that are employed for proofs are not sophisticated and tend more toward grind-it-out linear thinking than mathematical elegance. This means that although the process of filling in omitted steps in an argument can be tedious, it can be accomplished given a determined attitude and a willingness to put pen to paper. My experience has been that students benefit more from completely understanding a simple, expandable special case of a problem than from vague cognition of some general framework. Thus, detailed analyses are typically provided only for canonical problems that allow for general developments through analogy and/or exercises.

I have taught the material in this book (with the exception of that in Chapter 6) a number of times in a course on smoothing at Texas A&M University. By selective developments wherein some topics (such as asymptotics for polynomial regression, the nonuniform design case for kernel smoothers, etc.) are left as class exercises, I have been able to cover most of Chapters 1-5 in a one semester course.

A question of possible interest concerns how this book differs from the first edition. The field of nonparametric smoothing has come a long way since 1988 and the changes made in the second edition reflect this fact. While the two books are organized similarly and share many words and even entire paragraphs in some places, the overlap in content is less than 50%. One of the most significant changes is the deletion of almost all of the material from Chapter 6 of the first edition on asymptotic theory for smoothing splines and smoothing spline variants. Detailed discussion of smoothing spline variants now seems redundant in light of the books by Wahba and Green and Silverman. The asymptotic theory for smoothing splines in the present book is developed in detail only for the linear smoothing spline with a uniform design. This estimator has very limited practical value but serves as a nice pedagogical tool since its Demmler-Reinch representation can be obtained explicitly and its asymptotic properties can be derived from a kernel perspective complete with an analysis of boundary effects. The reduced emphasis on smoothing splines is part of the reason for the reordering of the book's title.

Two other substantial changes occur in Chapter 3. First, the tedious "exact" analysis of sine/cosine series estimator asymptotics using complex exponentials from the first edition has been deleted. Instead, I focus on cosine series estimators (which are to be preferred in an asymptotic sense) and use an upper bounding approach (taught to me by Paul Speckman) to give a simplified approximate analysis of the asymptotic properties of the estimators. (This same technique is used for polynomial regression and other series estimators in the exercises.) The second change is the introduction of a bias reduction lemma (also due to Paul Speckman). This provides a general approach to removing edge effects that leads, for example, to boundary corrected series estimators and gives one way of deriving estimators for partially linear models.

Examples of other material that is included in the present text that is not in the first edition includes: regressograms, order selection for hierarchical models, asymptotic and finite sample investigations of the properties of confidence intervals and bands, estimation in partially linear models, testing lack-of-fit for parametric models, polynomial-trigonometric regression, "new" results on bandwidth selection and locally linear regression. The idea for using regressograms in the introduction came from a discussion I had with Dennis Cox a number of years ago. Many of the data analysis examples in this book are also new and now, through the wonders of LaTeX and GNUPLOT, are accompanied by associated graphical presentations in most cases. The world of technical word processing has come a long way in the last decade and the resulting flexibility has taken me much closer in 1998 to the book I first envisioned in the early 1980's.

The preparation of this book in LaTeX was assisted by Yvonne Clark, Andy Liaw and Shane Reese. Initial set-up of my computers with LaTeX 2_ε and GNUPLOT, formatting and even typing of early versions of some of the chapters was done by Andy Liaw. Shane Reese provided additional computer set-ups and some crucial problem-solving support. This book would likely not exist without the help that Andy and Shane provided.

A number of friends have provided valuable input and advice on earlier drafts of this book. These include Karen Klipple, Sang-Joon Lee, Chin-Shang Li, Hans Müller, Naisyin Wang and Suojin Wang. Chin-Shang Li (with three-plus readings) wins the prize for finding the most errors while Suojin Wang is the winner for finding the worst mistake. The precise prizes to be awarded these lucky winners will be determined at a later date.

I am grateful to the faculty and staff of the Department of Statistics at Texas A&M University for their support of this project. In particular, Ruth Hanson and Joe Newton have provided valuable aid at several stages in the development of this book. The support of my research by NSF is also gratefully acknowledged.

I am indebted to many people who have shared their ideas about smoothing with me either directly or indirectly. These include Dennis Cox, Hans Müller, Doug Nychka, John Rice and Murray Rosenblatt, whose published work on kernel and spline smoothing has had a profound influence on me, in particular, and the field of nonparametric function estimation in general. The work of Grace Wahba deserves a special mention. She has provided the driving force behind the spline smoothing concept that has taken it from relative obscurity to the level of development and popularity it enjoys today. In this regard, all of us that work in smoothing are in her debt.

On a more personal level, I have had the good fortune of working with a number of talented statisticians in the last twenty years. My initial interest in splines and the free knot problem was a by-product of joint research conducted with Pat and Phill Smith. Most of what I know about asymptotics is due to long-term collaborations with Vince LaRiccia and Paul Speckman and my life has been enriched both personally and professionally by their friendship and patient tutelage. Interactions and joint work with Jeff Hart and Suojin Wang have expanded my horizons while adding to my statistical toolbag.

I am deeply gratefully to Manny Parzen for his advice and friendship over the years and for his early guidance that helped establish the foundation on which this book was built. This book as well as the last ten years of my life would not exist without the help of Dr. Robert Grossman, whose surgical and negotiating skills rescued me (at least temporarily) from both cancer and my HMO.

Finally, I wish to thank my three best friends, Lisa, Boomer and Tiny, for their love and patience over the years it has taken to prepare this book. Words cannot properly convey my love and gratitude or thank them enough.

RANDALL L. EUBANK

Contents

i

Preface iii

1 Introduction 1

1.1 Regression Analysis 1

1.2 Nonparametric Regression 10

1.2.1 Linear Estimators 11

1.2.2 Consistency 14

1.2.3 Interval Estimation 17

1.3 Scope 21

1.4 Exercises 23

2 What Is a Good Estimator? 27

2.1 Performance Criteria 27

2.2 Estimating $P(\lambda)$ and $R(\lambda)$ 37

2.3 Order Selection in Hierarchical Models 50

2.4 Appendix 62

2.5 Exercises 63

3 Series Estimators 71

3.1 Introduction 71

3.2 Some Function Space Theory 73

3.3 Generalized Fourier Series Estimators 79

3.4 Trigonometric Series Estimators 85

3.4.1 Form of the Estimator 86
3.4.2 An Example 90
3.4.3 Consistency and Efficiency of the Estimator 97
3.4.4 Asymptotic Distribution Theory 103
3.4.5 Testing for No Effect 111
3.5 Polynomial Regression 119
3.5.1 Form of the Estimator 122
3.5.2 An Example 125
3.5.3 Large Sample Properties of the Estimator 127
3.6 Polynomial-Trigonometric Regression 130
3.7 Partially Linear Models 136
3.8 Appendix 143
3.9 Exercises 145

4 Kernel Estimators **155**
4.1 Introduction 155
4.2 Kernel Estimators 156
4.3 Consistency 164
4.4 Selecting the Kernel Functions 175
4.5 Selecting the Bandwidth 178
4.6 Higher Order and Derivative Estimators 185
4.7 Locally Linear Estimators 189
4.8 Asymptotic Distribution Theory 194
4.9 Partially Linear Models 203
4.10 Random t's 210
4.11 Extensions 214
4.12 Exercises 216

5 Smoothing Splines **227**
5.1 Introduction 227
5.2 Form of the Estimator 230
5.3 Selection of λ 239
5.4 Computation 244
5.5 Large Sample Properties 246

5.6 Smoothing Splines as Bayes Estimators 260
5.7 Extensions 273
5.8 Appendix 281
5.9 Exercises 284

6 Least-Squares Splines **291**
6.1 Introduction 291
6.2 Form of the Estimator 291
6.3 Selecting λ 294
6.4 Computational Considerations 299
6.5 Asymptotic Analysis 301
6.6 Extensions 305
6.7 Exercises 308

Bibliography **311**

Index **335**

Chapter 1

Introduction

1.1 Regression Analysis

When someone says they have performed regression analysis this usually conjures up visions of estimated regression coefficients, residual plots, etc., which have been obtained by fitting a linear model to a set of data. This is certainly one aspect of regression analysis. However, there are many other topics which also merit inclusion under this broad heading. This text focuses on a particular collection of such topics generally referred to as nonparametric regression.

To begin our study it will be helpful to focus our attention on a simple model that provides a convenient framework for the discussion of different approaches to regression analysis. Specifically, let us suppose that observations are taken on a continuous random variable Y at n predetermined values of a continuous independent variable t. Let (t_i, y_i), $i = 1, \ldots, n$, be the values of t and Y which result from this sampling scheme and assume that t_i and y_i are related by the regression model

$$y_i = \mu(t_i) + \varepsilon_i,\ i = 1, \ldots, n, \tag{1.1}$$

where the ε_i are zero mean, uncorrelated, random variables with a common variance σ^2 and the $\mu(t_i)$ are values of some unknown function μ at the

design points $t_1, \ldots, t_n$. We will assume that $0 \le t_1 \le \cdots \le t_n \le 1$ in all that follows. There is no loss of generality in making this assumption (Exercise 1) and it provides for some simplicity and specificity in subsequent notation.

The function μ in (1.1) is called the ***regression function*** or ***regression curve***. In this setting, the phrase "regression analysis" refers to methods for statistical inference about the regression function.

The determination of a suitable inferential methodology for model (1.1) will hinge on the assumptions it is possible to make about μ. A ***parametric regression model*** presumes that the form of μ is known except for finitely many unknown parameters. More specifically, it is assumed that there is a vector of parameters $\boldsymbol{\beta} = (\beta_1, \ldots, \beta_p)^T \in B$, some subset of $\mathbb{R}^p$, and a known function $\mu(\cdot; \boldsymbol{\beta})$ such that $\mu(\cdot) = \mu(\cdot; \boldsymbol{\beta})$. It is therefore clear that for a parametric regression model inference about μ is tantamount to inference about $\boldsymbol{\beta}$.

Parametric models can depend on the parameters in a linear and/or nonlinear fashion. For example, regression functions such as $\mu(t) = \beta_1 t^{\beta_2}$ and $\mu(t) = \beta_1 + \beta_2 \exp\{-\beta_3 t^{\beta_4}\}$ provide cases where parameters enter the model linearly and nonlinearly. The first regression curve $\mu(t) = \beta_1 t^{\beta_2}$ is linear in β_1 and nonlinear in β_2 while the second is linear in β_1 and β_2 and nonlinear in the other two parameters. In contrast, for a purely linear model there are known functions $x_1, \ldots, x_p$ such that

$$\mu(t) = \sum_{j=1}^{p} \beta_j x_j(t). \tag{1.2}$$

Inference about the parameters in model (1.1) when (1.2) holds is typically referred to as ***linear regression analysis***.

Regression analysis techniques for parametric models represent one approach to conducting inference about μ. Using an appropriate estimation methodology, such as least-squares, it is possible to utilize the data to estimate parameters and thereby estimate μ. The result is a fitted curve that has been selected from the family of curves allowed under the model and conforms to the data in some fashion.

Example 1.1 (Simple Linear Regression). An illustration of a regression analysis for a set of data is given in Figure 1.1. Here we have generated (i.e., simulated) a sample of $n = 50$ observations from model (1.1) using uniformly spaced design points $t_i = (2i-1)/100, i = 1, \ldots, 50$, normally distributed random errors (or ε's) with $\sigma = .2$ and $\mu(t) = t + .5\exp\{-50(t - .5)^2\}$. The data along with the true regression function are shown in the figure.

A simple linear regression model was fitted to the data using least squares. This gives an estimated mean function of the form $\widehat{\mu}(t) = b_0 + b_1(t - \bar{t})$ with

$$b_1 = \sum_{i=1}^{n} y_i(t_i - \bar{t}) / \sum_{i=1}^{n} (t_i - \bar{t})^2, \tag{1.3}$$

$$b_0 = \bar{y} = n^{-1} \sum_{i=1}^{n} y_i \tag{1.4}$$

and $\bar{t} = n^{-1} \sum_{i=1}^{n} t_i$. For this particular data set $b_0 = .577$ and $b_1 = 1.17$. The resulting fitted line is superimposed on the data in Figure 1.1. Since the true regression function is not linear the simple linear regression estimator does a rather poor job of estimating the true mean function in this case.

Parametric regression modeling is undoubtedly the most prevalent approach to regression analysis. However, there are other methods of fitting curves to data. One collection of procedures that can be used for this purpose are nonparametric regression techniques. These methods give estimates of μ that allow great flexibility in the possible form of the regression curve and, in particular, make no assumptions about a parametric form. A *nonparametric regression model* generally only assumes that the regression curve belongs to some infinite dimensional collection of functions. For example, μ may be assumed to be differentiable or differentiable with a square integrable second derivative, etc. Assumptions of this type are concerned with qualitative properties of μ and are in marked contrast to the assumptions employed in parametric modeling that entail a much greater level of specificity about μ.

An important difference between parametric and nonparametric regression methodologies is their respective degree of reliance on the information

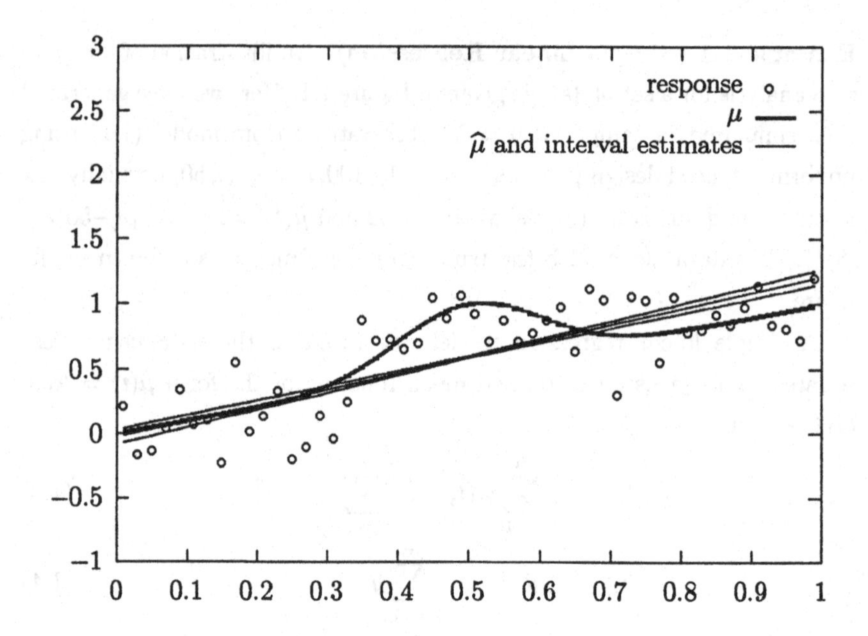

Figure 1.1: Point and Interval Estimates for Simulated Data.

about μ obtained from the experimenter and from the data. To specify a nonparametric regression model the experimenter will choose an appropriate function space that is believed to contain the unknown regression curve. This choice is usually motivated by smoothness (i.e., continuity and differentiability) properties the regression function can be assumed to possess. The data is then utilized to determine an element of this function space that is representative of the unknown regression curve. In contrast, under a parametric model, the experimenter chooses one possible family of curves, from the collection of all curves, and inputs this choice into the inferential process. The information the data can supply concerning model development is then restricted to what can be extracted from the data under this assumed parametric form. Consequently, nonparametric regression techniques rely more heavily on the data for information about μ than their parametric counterparts.

When appropriate, parametric models have some definite advantages. The corresponding inferential methods usually have nice efficiency properties. Also, the parameters may have some physical meaning which makes them interpretable and of interest in their own right. Unfortunately, possibly because of these factors, parametric models are often used when there is little available information concerning the functional form of μ. The experimenter assumes that μ has some *ad hoc* parametric representation with little or no knowledge concerning the accuracy of his or her assumptions. If the assumed parametric model is in error, the aforementioned advantages of the parametric approach will not obtain. Thus, there are few gains to be realized from using a poorly specified parametric form for μ. In fact, the use of an inappropriate parametric model can be dangerous in the sense of producing misleading or incorrect inference about the regression curve. An example of this can be found in Gasser et al (1984) who used nonparametric regression methodology to demonstrate that certain *ad hoc* parametric growth models overlooked important features of human growth.

Example 1.2 (Simple Linear Regression: Continued). To give a simple, yet concrete, example of how parametric regression methods can succeed and fail, assume again that we have data from model (1.1) with $t_i = (2i-1)/2n$ and a continuously differentiable regression function μ. Further suppose that we fit this data using the simple linear regression estimator $\widehat{\mu}$ from (1.3)-(1.4).

First let us suppose that the true regression function is actually linear in which case $\mathrm{E}\,\widehat{\mu}(t) = \mu(t)$ for all t. The variance of $\widehat{\mu}(t)$ is

$$\begin{aligned} \mathrm{Var}\,(\widehat{\mu}(t)) &= \sigma^2[n^{-1} + (t-\bar{t})^2/\textstyle\sum_{i=1}^n (t_i-\bar{t})^2] \\ &= \sigma^2[n^{-1} + 12n\frac{(t-.5)^2}{n^2-1}] \end{aligned} \tag{1.5}$$

since $\bar{t} = 1/2$ and $\sum_{i=1}^n (t_i - \bar{t})^2 = (n^2-1)/12n$ in this case. Consequently, the average (across the design) mean squared error or risk for $\widehat{\mu}$ is

$$R(\widehat{\mu}) = n^{-1}\sum_{i=1}^n \mathrm{E}\,(\widehat{\mu}(t_i) - \mu(t_i))^2 = n^{-1}\sum_{i=1}^n \mathrm{Var}\,(\widehat{\mu}(t_i)) = 2\sigma^2/n \tag{1.6}$$

which decays to zero at the parametric convergence rate n^{-1}.

The usual confidence intervals that are used to accompany a simple linear regression estimator of $\mu(t)$ are

$$\widehat{\mu}(t) \pm z_{\alpha/2}\sqrt{\mathrm{Var}\,(\widehat{\mu}(t))} \tag{1.7}$$

with $\mathrm{Var}\,(\widehat{\mu}(t))$ given in (1.5) and $z_{\alpha/2}$ the $100(1-\alpha/2)$th percentile of the standard normal distribution. In practice one would typically use a Student's t critical value rather than $z_{\alpha/2}$ in (1.7) and σ^2 would need to be estimated in the variance of $\widehat{\mu}(t)$. The basic conclusions that follow continue to hold if such changes are made, so we will deal only with the simpler intervals (1.7).

Now if $\mu(t) = \beta_0 + \beta_1(t - .5)$ and the ε_i are *iid*, we have $\widehat{\mu}(t) - \mu(t) = \sum_{i=1}^{n} \varepsilon_i w_{in}(t)$ with

$$w_{in}(t) = n^{-1} + 12n(t-.5)(t_i - .5)/(n^2 - 1). \tag{1.8}$$

The Lindeberg-Feller Central Limit Theorem (see, e.g., Section 3.4.4) has the implication that $\sum_{i=1}^{n} \varepsilon_i w_{in}/[\sigma^2 \sum_{i=1}^{n} w_{in}^2]^{1/2}$ will have a standard normal limiting distribution if

$$\max_{1\le i\le n} |w_{in}|/[\sum_{j=1}^{n} w_{jn}^2]^{1/2} \to 0$$

as $n \to \infty$. This is clearly true if we choose the w_{in} to be the $w_{in}(t)$ in (1.8). Thus,

$$P(\,\mu(t) \in \widehat{\mu}(t) \pm z_{\alpha/2}\sqrt{\mathrm{Var}\,(\widehat{\mu}(t))}\,) \to 1 - \alpha \tag{1.9}$$

as $n \to \infty$.

Equations (1.6) and (1.9) have the implication that $\widehat{\mu}$ possesses some good qualities (at least for large n) as an estimator of μ when the true regression function is a line. In this case it provides us with a highly efficient estimator and the associated confidence intervals (1.7) have the desired (asymptotic) coverage probabilities.

All of the previous analysis has been predicated on μ actually being linear. This raises the question of what might happen if we relax this assumption. To explore this issue further now suppose that

$$\mu(t) = \beta_0 + \beta_1(t - .5) + f(t),$$

where $f \neq 0$ is a continuously differentiable function satisfying $\int_0^1 f(t)dt = \int_0^1 (t-.5)f(t)dt = 0$. One finds that although the variance of $\widehat{\mu}(t)$ is still given by (1.5), we now have

$$\begin{aligned}\mathrm{E}\,\widehat{\mu}(t) &= \beta_0 + \beta_1(t-.5) + n^{-1}\sum_{i=1}^{n} f(t_i) + \frac{12n\sum_{i=1}^{n} f(t_i)(t_i-.5)}{n^2-1}(t-.5)\\ &= \beta_0 + \beta_1(t-.5) + O(n^{-1}),\end{aligned}$$

uniformly in $t \in [0,1]$, since $n^{-1}\sum_{i=1}^{n} f(t_i)$ and $n^{-1}\sum_{i=1}^{n} f(t_i)(t_i-.5)$ are quadrature approximations for $\int_0^1 f(t)dt$ and $\int_0^1 (t-.5)f(t)dt$. Using $\mathrm{E}\,(\widehat{\mu}(t)-\mu(t))^2 = \mathrm{Var}\,(\widehat{\mu}(t)) + (\mathrm{E}\,\widehat{\mu}(t)-\mu(t))^2$ we then obtain

$$\begin{aligned}R(\widehat{\mu}) = n^{-1}\sum_{i=1}^{n}\mathrm{E}\,(\widehat{\mu}(t_i)-\mu(t_i))^2 &= \frac{2\sigma^2}{n} + n^{-1}\sum_{i=1}^{n}(\mathrm{E}\,\widehat{\mu}(t_i)-\mu(t_i))^2\\ &= \int_0^1 f^2(t)dt + O(n^{-1}).\end{aligned}$$

Consequently, the average mean squared error for $\widehat{\mu}$ does not even converge to zero in this case and $\widehat{\mu}$ is, in that sense, not a consistent estimator of μ.

In terms of the confidence intervals (1.7), $(\widehat{\mu}(t) - \mathrm{E}\,\widehat{\mu}(t))/\sqrt{\mathrm{Var}\,(\widehat{\mu}(t))}$ still has a limiting standard normal distribution, but there is now an additional term that arises from the bias of the estimator: namely, $(\mathrm{E}\,\widehat{\mu}(t) - \mu(t))/\sqrt{\mathrm{Var}\,(\widehat{\mu}(t))} \sim \sqrt{n}f(t)/\sigma(1+12(t-.5)^2)^{1/2} = \sqrt{n}C_t$. If Z denotes a random variable having a standard normal distribution, we then have

$$P(\,\mu(t) \in \widehat{\mu}(t) \pm z_{\alpha/2}\sqrt{\mathrm{Var}\,(\widehat{\mu}(t))}\,) \sim P(|Z+\sqrt{n}C_t| \leq z_{\alpha/2}) \to 0$$

as $n \to \infty$ unless $C_t = 0$. Thus, the coverage levels for the parametric confidence intervals (1.7) can actually converge to zero rather than $1-\alpha$ if the form of the regression curve has been misspecified and the true mean function is not linear.

There are always questions concerning the finite sample applicability of large sample or asymptotic work such as that carried out above. However, referring back to Figure 1.1, we see that the 95% pointwise confidence intervals (1.7) that are shown in the figure only tend to capture values of the true regression curve in regions of the t variable where $\widehat{\mu}$ crosses μ. This is exactly what our large sample analysis predicted. To further explore this

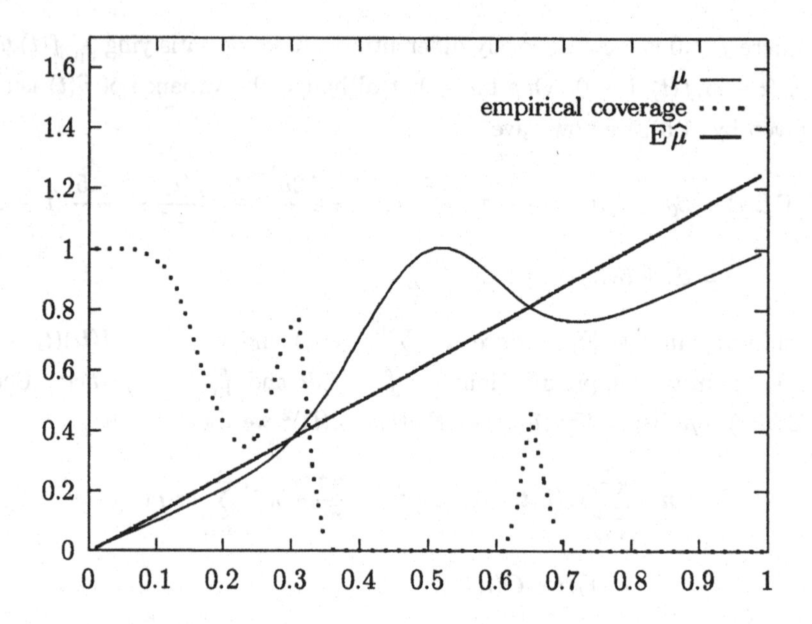

Figure 1.2: Empirical Coverages for Linear Regression Confidence Intervals.

issue we replicated the simulation experiment used to produce the data in Figure 1.1 a total of 1000 times while keeping track of the proportion of times the values of $\mu(t_i), i = 1, \ldots, 50$, were contained in their corresponding 95% pointwise confidence intervals obtained from (1.7). This gives a total of 50 proportions (or empirical coverage probabilites) each of which is based on 1000 independent observations. The resulting proportions are shown in Figure 1.2 along with the true regression curve and $\mathrm{E}\,\widehat{\mu}$. All these empirical coverages should be around .95 if the parametric confidence intervals are working effectively. Instead, the coverages are close to the nominal 95% level only when $\mathrm{E}\,\widehat{\mu}(t)$ is close to $\mu(t)$ and can otherwise be essentially zero as predicted by our asymptotics.

The previous example is informative in that it illustrates that parametric methods can fail when the true form of the regression function is misspecified. Thus, the use of parametric methods can be quite dangerous in situations where there is little that is known about the regression function. In such cases, the bulk of the information about μ lies in the data rather than the person conducting the study or experiment. Accordingly, it seems more reasonable to use inferential techniques which rely heavily on the data. It is for this reason that nonparametric regression techniques are ideally suited to problems of inference when the available knowledge about μ is limited in nature.

Nonparametric regression techniques overcome the inherent difficulty with parametric techniques: namely, that the functional form of μ must be known. However, their use is not without its price. Generally, nonparametric estimators are less efficient than the parametric variety when the parametric model is valid.

For most parametric estimators the risk, or expected squared error of estimation, will decay to zero at a rate of n^{-1}. As we shall see in the next section, the corresponding rate for nonparametric estimators is usually $n^{-\delta}$ for some number $\delta \in (0,1)$ that depends on the smoothness of μ. For example, if μ is twice differentiable $n^{-4/5}$ is an often quoted rate. Thus, nonparametric regression techniques suffer a loss of efficiency when compared to parametric methods. However, this comparison is not entirely fair for if we knew the form of μ we would certainly utilize a parametric approach to estimation in most cases. Nonparametric estimators become candidates for estimation of μ only when there is some question about an appropriate parametric form for μ. In that case if an incorrect parametric model is used the n^{-1} rate will no longer obtain and nonparametric regression techniques will fare much better than parametric methods.

The result of a nonparametric regression analysis is a curve fitted to a set of data. Since this curve is produced without assuming a parametric form for μ there will be some loss in the interpretability of estimators obtained in this fashion in that there will no longer be quantities such as estimated regression coefficients to be interpreted. However, the fitted curve itself is an estimate of the parameter μ and any functional of μ is also a parameter which can be estimated using an estimate of the regression curve. For example, if μ represents a growth curve then the point at which the velocity curve μ' attains its maximum may be of interest. This type of parameter can be estimated and interpreted from either a nonparametric or parametric viewpoint. Thus, there are regression curve based parameters (i.e., functionals of μ) about which inference can be made and interpretations can be drawn using either parametric or nonparametric techniques. In most cases where nonparametric regression is really appropriate it is parameters of this nature which are actually of interest anyway.

To summarize, parametric and nonparametric regression techniques represent two different approaches to the problem of regression analysis. Parametric methods require very specific, quantitative information from the experimenter about the form of μ that places restrictions on what the data can tell us about the regression function. Such techniques are most appropriate when theory, past experience and/or other sources are available that provide detailed knowledge about the process under study. In contrast, nonparametric regression techniques rely on the experimenter to supply only qualitative information about μ and let the data speak for itself concerning the actual form of the regression curve. These methods are best suited for inference in situations where there is little or no prior information available about the regression curve.

In conclusion, it should be noted that even though parametric and nonparametric regression models represent distinctly different approaches to regression analysis, this does not mean that the use of one approach precludes the use of the other. Indeed, nonparametric regression techniques can be used to assess the validity of a proposed parametric model. Conversely, it may be that the form of a fitted regression curve obtained by nonparametric techniques will suggest an appropriate parametric model for use in future studies. (See, e.g., Example 3 of Altman 1992.) Thus, nonparametric regression procedures may represent the final stage of data analysis or merely an exploratory or confirmatory step in the modeling process.

1.2 Nonparametric Regression

In this section we discuss the nonparametric regression problem from a general viewpoint to give an indication of the important issues which must be addressed for any particular solution. To begin, let us assume that model (1.1) is in effect and that $\mu \in W$, where W is some infinite dimensional collection of functions. The objective is inference about μ in the form of both point and interval estimation.

There are many natural choices for the function space W. Depending on one's prior beliefs about the smoothness of μ, it might be reasonable to take W to be the set of all continuous functions on $[0,1]$ or, more generally, the set of all functions with m continuous derivatives on $[0,1]$ for some non-negative integer m. More exotic choices for W that allow for discontinuities in the regression curve or its derivatives are also sometimes useful. For the present we will leave W unspecified and proceed only under the assumption that it is some rich collection of functions that can be assumed to contain the true regression curve.

1.2.1 Linear Estimators

To appreciate some of the motivations behind the various types of nonparametric regression estimators, first let us consider the problem of estimating the mean vector $\boldsymbol{\mu} = (\mu(t_1), \ldots, \mu(t_n))^T$. The corresponding response vector $\mathbf{y} = (y_1, \ldots, y_n)^T$ gives one possible estimator. This estimator is unbiased in that $\mathrm{E}\, y_i = \mu(t_i), i = 1, \ldots, n$, and has variance-covariance matrix $\sigma^2\mathbf{I}$. If observations are repeated at some values of t they would, of course, be grouped and their averages used to obtain more precise estimators. This does not alter the conclusions which follow so we will assume that the t_i are distinct in subsequent discussions.

The vector of responses will not, in general, be a suitable estimator of $\boldsymbol{\mu}$. One reason for this is that y_i has an unacceptably large squared error risk as an estimator of $\mu(t_i)$. In particular, the average squared error risk for the response vector $\mathbf{y}$ as an estimator of μ is $n^{-1}\sum_{i=1}^{n} \mathrm{E}\,(\mu(t_i) - y_i)^2 = \sigma^2$, and, hence, the quality of $\mathbf{y}$ as an estimator does not improve as the sample size increases. In Chapters 3–6 we discuss a number of estimators of $\boldsymbol{\mu}$ for which the risk will decay like $n^{-\delta}$ for some $\delta > 0$.

Another problem with the choice of $\mathbf{y}$ as an estimator is that it only provides an estimator of $\boldsymbol{\mu}$ and does not solve the problem of estimating the function μ at non-design points. If the objective is to estimate the entire function μ, it will no longer be possible to obtain an unbiased estimator since this would involve estimating more parameters (actually an infinite number of parameters) than there are data values (Exercise 3).

If μ is believed to be smooth, then the observations at points t_i near t should contain information about the value of μ at t. Thus, it should be possible to use something like a local average of the data near t to construct an estimator of $\mu(t)$. In particular, if we use this approach at a design point $t = t_i$, we should be able to obtain an estimator of $\mu(t_i)$ with smaller variance and, if the bias that is introduced is not too large, even a smaller risk than y_i.

In this text we will focus primarily on what are known as ***linear estimators*** of μ. This type of estimator includes local averages as well as many other estimators of interest. A linear estimator of $\mu(t)$ will have the form

$$\mu_\lambda(t) = \sum_{i=1}^{n} K(t, t_i; \lambda) y_i \,, \tag{1.10}$$

where $K(\cdot, t_i; \lambda)$, $i = 1, \ldots, n$, is a collection of weight functions that depend on the design and a tuning parameter or, more generally, a set of tuning parameters, λ. The reason (1.10) is called a linear estimator is because, given λ, it is a linear function of the responses $y_1, \ldots, y_n$.

If the weights in (1.10) are all nonnegative and $\sum_{i=1}^{n} K(t, t_i; \lambda) = 1$, estimator (1.10) is actually a weighted average of the y_i. It is, in fact, a weighted least-squares estimator in the sense that it is the minimizer of the weighted squared error criterion

$$\sum_{i=1}^{n} K(t, t_i; \lambda)(y_i - \theta)^2 \tag{1.11}$$

with respect to θ.

One very simple example of a linear estimator is the ***regressogram*** estimator from exploratory data analysis that owes its origin to Tukey (1961) (cf Kotz, Johnson and Read 1988). To explicitly define this estimator, let λ be a positive integer and partition the interval $[0, 1]$ into λ subintervals of the form $P_j = [\frac{j-1}{\lambda}, \frac{j}{\lambda})$, $j = 1, \ldots, \lambda - 1$ and $P_\lambda = [\frac{\lambda-1}{\lambda}, 1]$. Then, the

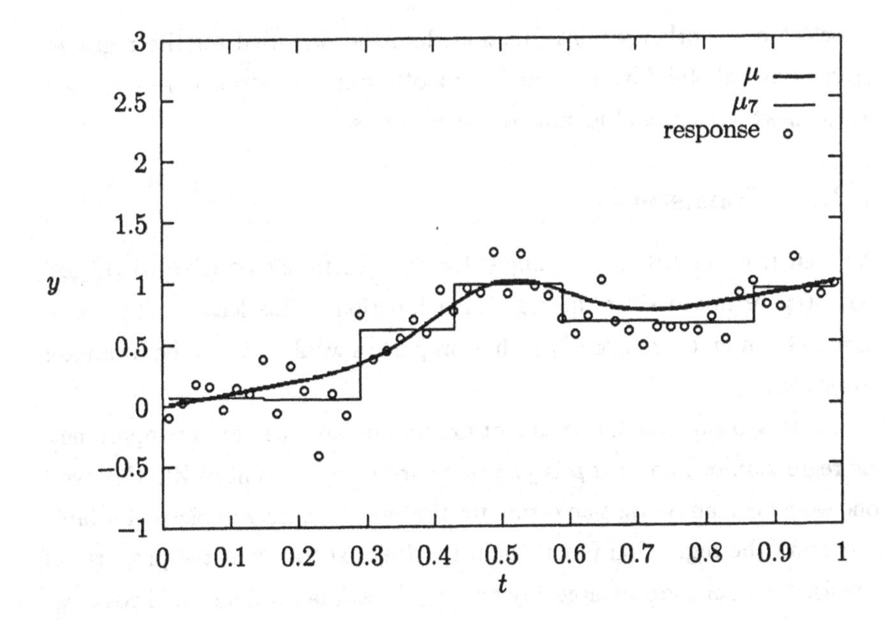

Figure 1.3: Regressogram Fit to Simulated Data.

regressogram estimator is obtained from (1.10) by using

$$K(t, t_i; \lambda) = \frac{\sum_{r=1}^{\lambda} I_{P_r}(t) I_{P_r}(t_i)}{\sum_{j=1}^{n} \sum_{r=1}^{\lambda} I_{P_r}(t) I_{P_r}(t_j)} \tag{1.12}$$

with $I_{P_r}(\cdot)$ the indicator function for the interval P_r. A little algebra reveals that in this case $\mu_\lambda(t)$ is just the average of those responses whose corresponding t ordinates fall in the interval of the partition that contains the point of estimation.

Example 1.3 (A Regressogram Example). Figure 1.3 gives a plot of a regressogram estimator for the data set in Figure 1.1 based on the choice of $\lambda = 7$. The true regression curve is plotted along with the regressogram estimator from which one can see that the estimator certainly recovers some of the key features (i.e., the locations of peaks and valleys) of μ.

The example illustrates the lack of smoothness of the regressogram estimator. It would often seem more appropriate to use a smoother, at least continuous, estimator. To obtain a smoother estimator it is only necessary to select a smoother weight function than the one used for the regressogram. Several standard choices for smooth weight functions are discussed in the next section and in subsequent chapters.

1.2.2 Consistency

As seen from (1.10) and Example 1.3, it is certainly possible to suggest potential estimators for the regression function. This leads us to a new issue. Given an estimator of μ, what properties would we like the estimator to possess?

As previously noted, one cannot expect unbiasedness for a nonparametric regression estimator if μ is just some arbitrary element of W. However, one would certainly want an estimator that will be nearly unbiased for large n over all the regression functions in W. Related to this is the property of consistency. As n grows large any reasonable estimator of μ should converge to μ in some probabilistic sense.

Of particular interest for subsequent work is the performance of an estimator $\widehat{\mu}$ as measured by the (average) risk

$$R(\widehat{\mu}) = n^{-1} \sum_{i=1}^{n} \mathrm{E}\left(\widehat{\mu}(t_i) - \mu(t_i)\right)^2 . \qquad (1.13)$$

If $R(\widehat{\mu})$ converges to zero as n becomes large this gives a form of squared-error consistency for the estimator.

Example 1.4 (The Risk for a Regressogram). It is often fairly simple to produce a consistent nonparametric estimator for μ. To illustrate this let W be the set of all continuously differentiable functions on [0, 1] and take $t_i = (2i-1)/2n, i = 1, \ldots, n$, in (1.1). If we now choose $\widehat{\mu}$ to be the regressogram estimator μ_λ corresponding to (1.12) we have that for $t \in P_j$

$$\mathrm{Var}\left(\mu_\lambda(t)\right) = \sigma^2 / n_j$$

and

$$\mathrm{E}\,\mu_\lambda(t) = \sum_{t_i \in P_j} \mu(t_i)/n_j$$

with $n_j = \sum_{i=1}^{n} I_{P_j}(t_i)$ the number of design points falling in the jth partition.

The Mean Value Theorem gives that $\mu(t_i) = \mu(t)+\mu'(\xi_{ij})(t_i-t)$ for $\xi_{ij} \in P_j$. Thus, for $t \in P_j$, $|\mathrm{E}\,\mu_\lambda(t)-\mu(t)| \le \lambda^{-1}\sup_{0\le s\le 1}|\mu'(s)|$ because $|t-t_i| \le \lambda^{-1}$ for $t, t_i \in P_j$. Consequently, the average risk for the regressogram estimator satisfies

$$\begin{aligned} R(\lambda) &= n^{-1}\sum_{i=1}^{n}\mathrm{E}\,(\mu_\lambda(t_i)-\mu(t_i))^2 \\ &= n^{-1}\sum_{j=1}^{\lambda}\sum_{t_i\in P_j}\{\mathrm{Var}\,(\mu_\lambda(t_i)) + (\mathrm{E}\,\mu_\lambda(t_i)-\mu(t_i))^2\} \\ &\le \frac{\lambda\sigma^2}{n} + \lambda^{-2}(\sup_{0\le s\le 1}|\mu'(s)|)^2, \end{aligned} \tag{1.14}$$

which converges to zero provided only that $\lambda, n \to \infty$ with $\lambda/n \to 0$. This latter condition has the practical implication that the number of partitions should be chosen sufficiently large to make the bias of the estimator small but not so large as to compromise the variability reduction that occurs from the averaging of responses within a partition.

By minimizing the upper bound in (1.14) as a function of λ, we see that a "best" choice for the number of partitions would correspond to taking λ proportional to $n^{1/3}$. This would produce an "optimal" $n^{-2/3}$ convergence rate for the regressogram estimator which is much slower than the parametric convergence rate of n^{-1}. It is, in fact, not possible to obtain such a parametric rate using a regressogram in this case. If, for example, we take λ to be fixed so that the variance term in (1.14) is $O(n^{-1})$, it turns out that the squared bias part of the risk $n^{-1}\sum_{i=1}^{n}(\mathrm{E}\,\mu_\lambda(t_i)-\mu(t_i))^2$ converges to $\int_0^1 \mu^2(t)dt - \lambda\sum_{j=1}^{\lambda}(\int_{P_j}\mu(t)dt)^2$ which is nonzero in general (Exercise 5). Conversely, if λ grows with n then the variance part of the risk decays slower than n^{-1}.

The most immediate implication of the previous example is that it is possible to produce a consistent estimator of the regression curve even when no parametric assumptions are made concerning its form. It also illustrates that while such estimators may have some (nonparametric) efficiency properties, they cannot be expected to converge to μ at a parametric rate.

Another issue that was explored in the example was the rate of convergence for the risk of the regressogram estimator. Such rates of convergence are of interest since they provide information about estimator efficiency that allows us to critique the performance of one estimator relative to another. In this regard, we will say that an estimator $\widehat{\mu}$ is *asymptotically optimal* if, uniformly in $\mu \in W$, $R(\widehat{\mu})$ decays like $n^{-\delta}$ for some $\delta > 0$ as n increases and no estimator can achieve a better rate of convergence. In the case of a parametric model one finds that $\delta = 1$ under standard regularity conditions. However, this is not true for nonparametric models and generally $\delta < 1$ in this latter case. Slower convergence rates are the price that is paid for making fewer assumptions about μ.

An important case where the best convergence rate for $R(\widehat{\mu})$ is known is for W being the set of all functions μ on $[0,1]$ which, for some integer $m \geq 1$, have $m-1$ absolutely continuous derivatives and $\int_0^1 \mu^{(m)}(t)^2\,dt \leq B$ for B a finite, positive constant. If we let L denote the set of all linear estimators of μ, Speckman (1985) has shown that if the t_i are generated by a positive, bounded design density, w, through the relation

$$\int_0^{t_i} w(t)\,dt = (2i-1)/2n,$$

then

$$\min_{\widehat{\mu}\in L}\max_{\mu\in W} R(\widehat{\mu}) = C(B,m)n^{-2m/(2m+1)}(1+o(1))\,, \tag{1.15}$$

where $o(1) \to 0$ as $n \to \infty$ and

$$C(B,m) = [B(2m+1)]^{\frac{1}{2m+1}}\left\{\frac{\sigma^2 m\int_0^1 w(t)^{1/2m}\,dt}{\pi(m+1)}\right\}^{\frac{2m}{2m+1}}. \tag{1.16}$$

Results in Golubev and Nussbaum (1990) have the implication that (1.15) continues to hold even if we minimize over all estimators, rather than just linear estimators, while Efromovich (1996) has established that, with certain modifications, (1.15) extends to include the case of random t_i. Thus, for this particular W, the best uniform rate of convergence will be $n^{-2m/(2m+1)}$ and we will say that any given estimator is mth *order* when its associated risk can be made to decay at the optimal convergence rate $n^{-2m/(2m+1)}$ over all functions in W.

Convergence rate results for other function spaces can be found in Stone (1980, 1982), Ibragimov and Hasminski (1981) and Yatracos (1988, 1989).

Related problems concerning optimal convergence rates for estimation of various functionals of μ are discussed in Fan (1991), Donoho and Low (1992), Efromovich and Low (1996) and references therein.

1.2.3 Interval Estimation

By studying consistency and rates of convergence we learn how an estimator behaves as a point estimator of the regression function. However, it is also quite important to obtain methods for assessing the precision of an estimator from data in the form of confidence intervals and bands. For this purpose we must investigate the distributional properties of an estimator.

Intuitively, the relationship between (1.10) and averages suggests that central limit theorems and other large sample results associated with means will be applicable to linear estimators. This actually turns out to be true for many types of nonparametric estimators so that it is possible to establish large sample distribution theory for most estimators of interest. However, the inherent bias associated with nonparametric estimation of μ creates substantial problems in translating such large sample theory into practically useful methods for interval estimation as illustrated in the next example.

Example 1.5 (Regressogram Confidence Intervals). Again consider the estimation of a continuously differentiable regression function by the regressogram estimator μ_λ in the case of a uniform design. Since for $t \in P_j, \mu_\lambda(t)$ is just the average of the n_j responses whose t ordinates fall in P_j, we know that $\sqrt{n_j}(\mu_\lambda(t) - \mathrm{E}\,\mu_\lambda(t))/\sigma$ has a limiting standard normal distribution provided only that $n_j \to \infty$ or, equivalently, $\lambda/n \to 0$ as $n \to \infty$. This suggests that we use

$$\mu_\lambda(t) \pm z_{\alpha/2}\frac{\sigma}{\sqrt{n_j}} \text{ for } t \in P_j, \quad j = 1, \ldots, \lambda, \tag{1.17}$$

to obtain an interval estimator of $\mu(t)$.

Figure 1.4 gives a plot of the 95% confidence intervals that were produced using (1.17) with the regressogram fit to the data in Figure 1.3. The performance of these intervals in not entirely satisfactory in that there are regions where the intervals fail to cover the regression curve. As for the simple linear regression example of Section 1.1, the experiment that produced

Figure 1.4 was replicated 1000 times to produce 50 empirical coverage probabilities corresponding to each of the design points. The results are shown in Figure 1.5. The empirical coverages for the regressogram intervals (1.17) tend to depart substantially from the nominal 95% level in many areas although their performance is still preferable to the parallel results for the parametric intervals in Figure 1.2.

An explanation for these simulation results can be obtained by observing that in order for confidence intervals such as (1.17) to actually work we need an approximate standard normal distribution for the pivotal quantity $T = \sqrt{n_j}(\mu_\lambda(t) - \mu(t))/\sigma$. However,

$$T = \frac{\sqrt{n_j}(\mu_\lambda(t) - \mathrm{E}\,\mu_\lambda(t))}{\sigma} + \frac{\sqrt{n_j}(\mathrm{E}\,\mu_\lambda(t) - \mu(t))}{\sigma} \tag{1.18}$$

so that the effectiveness of a normal approximation relies on the bias term $\sqrt{n_j}(\mathrm{E}\,\mu_\lambda(t) - \mu(t))$ being negligible or approximately zero. In particular, from a large sample standpoint we will need the bias of the regressogram estimator to decay to zero faster than $1/\sqrt{n_j}$. We have already seen that $|\mathrm{E}\,\mu_\lambda(t) - \mu(t)| \le \lambda^{-1} \sup_{0\le s\le 1} |\mu'(s)|$, so we would expect to need the number of partitions to grow in a way that ensures that $\sqrt{n_j}/\lambda \to 0$. To make this more precise it will be useful to first refine some of our previous approximations.

Observe that with a uniform design we have $n_j = n\lambda^{-1} + O(1)$ and combine this with a quadrature argument to obtain

$$\mathrm{E}\,\mu_\lambda(t) = \lambda \int_{(j-1)/\lambda}^{j/\lambda} \mu(s)ds + O(\frac{\lambda^2}{n} + \frac{\lambda}{n^2}).$$

Thus, if $\lambda \to \infty$ with $\lambda^2/n \to 0$, the Mean Value Theorem for integrals gives

$$\sqrt{n_j}(\mathrm{E}\,\mu_\lambda(t) - \mu(t)) \sim \sqrt{\frac{n}{\lambda}}(\mu(\xi_j) - \mu(t))$$

for some $\xi_j \in P_j$. Now write $t = \xi_j - u_j\lambda^{-1}$ for $u_j \in [-1, 1]$ and use Taylor's Theorem (see, e.g, Section 3.5) along with the continuity of μ' to obtain

$$\sqrt{n_j}(\mathrm{E}\,\mu_\lambda(t) - \mu(t)) \sim \sqrt{n}\mu'(t)u_j\lambda^{-3/2}.$$

This converges to zero if $n/\lambda^3 \to 0$ which precludes the "optimal" $n^{1/3}$ rate for the number of partitions.

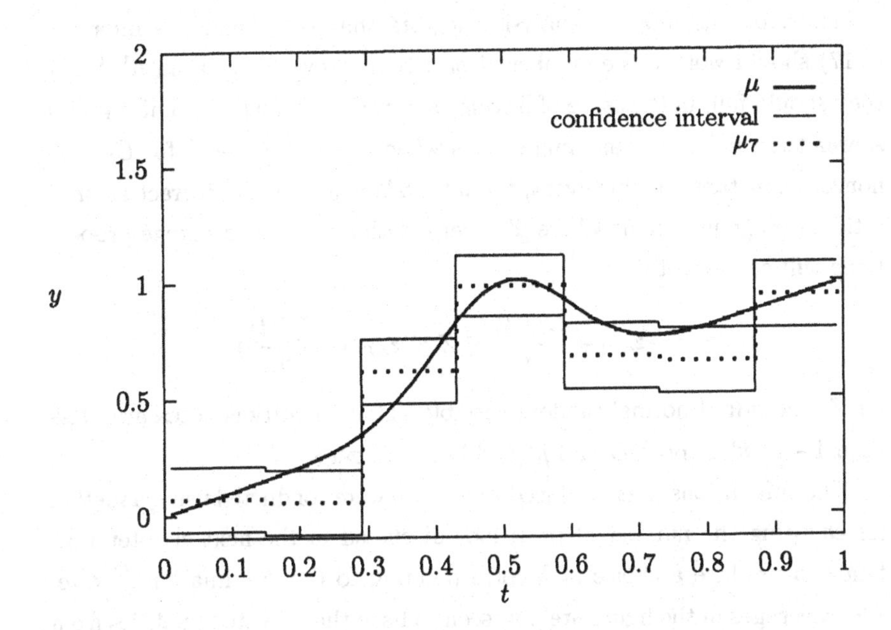

Figure 1.4: Regressogram Confidence Intervals for Simulated Data.

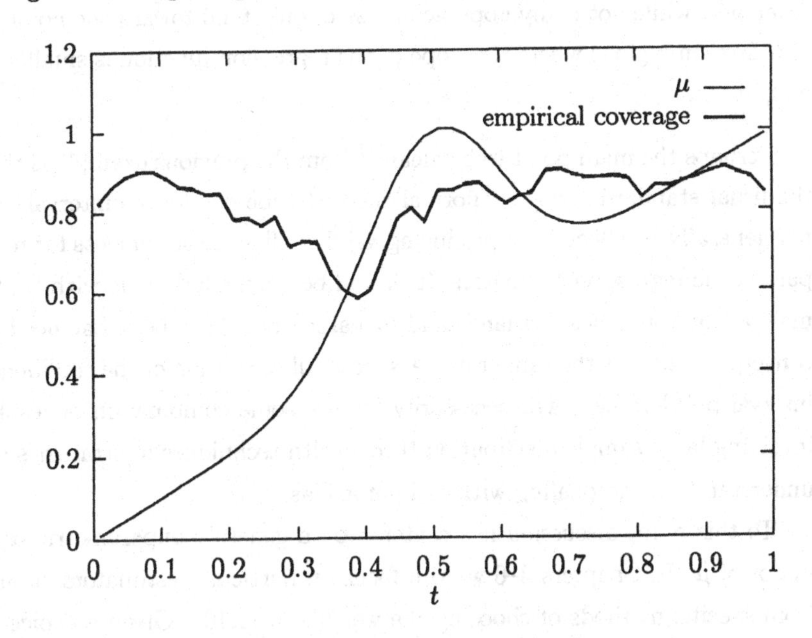

Figure 1.5: Empirical Coverages for Regressogram Confidence Intervals.

Thus, our asymptotic analysis suggests that the confidence intervals (1.17) should work if the number of partitions grows faster than $n^{1/3}$ and they should fail, in the sense of having asymptotic coverage zero, if λ grows slower than $n^{1/3}$. A remaining case is when $\sqrt{n}u_j\lambda^{-3/2} \rightarrow C$ for C some nonzero constant. In this event, the intervals will have the correct asymptotic coverage in regions where μ' is zero. Otherwise, the coverage proportions will converge to

$$P(-z_{\alpha/2} - \frac{C\mu'(t)}{\sigma} \leq Z \leq z_{\alpha/2} - \frac{C\mu'(t)}{\sigma})$$

for Z a standard normal random variable. This quantity is necessarily less than $1 - \alpha$ when both C and $\mu'(t)$ differ from zero.

The simulations that produced Figure 1.5 were conducted using a method for choosing the number of partitions discussed in the next chapter that tends to produce a choice of λ corresponding to the "optimal" $n^{1/3}$ rate. The coverages in the figure are now seen to have the expected qualities from our asymptotic analysis of this case. Specifically, we see that the empirical coverages, while not really approaching zero, only tend toward the nominal 95% level in regions where the slope of the regression function is smaller in magnitude.

Perhaps the main point to be gleaned from the previous example is that the usual standard error or "normal theory" type confidence interval will not generally be effective in producing valid confidence statements for nonparametric regression estimators. This will be particularly true with "optimal" estimators whose variance and (squared) bias have been balanced to converge to zero at the same rate. A successful resolution of the confidence interval problem here will necessarily require some combination of results involving large sample distribution theory with techniques or strategies for understanding and dealing with estimator bias.

To this point discussion has centered on a general nonparametric estimator of μ. In Chapters 3–6 we will focus on particular estimators arising from specific methods of choosing the weights in (1.10). Given a choice of the weight function, our attention turns to what properties are possessed by the resulting estimator. Apart from any other properties it may be possible to establish, the above discussions suggest that attention must also be directed toward questions of consistency, rates of convergence and asymptotic distribution theory that includes the study of both the variability and bias of the estimator.

1.3 Scope

In this final section we give a brief preview of the types of estimators to be studied in subsequent chapters. The discussion serves two purposes. First, it introduces the key estimators to be studied in the text and thereby provides other examples of linear estimators. Secondly, it provides concrete illustrations of the need for the objective methods of assessing an estimator's performance from the data that are the subject of the next chapter.

Our study of nonparametric estimators begins in Chapter 3 where it is assumed that the regression function can be represented as a linear combination of known functions $\{x_i\}_{i=1}^{\infty}$. Thus, there are coefficients $\{\beta_i\}_{i=1}^{\infty}$ such that $\mu = \sum_{i=1}^{\infty} \beta_i x_i$. This form for μ is reminiscent of the linear model (1.2) except that now there are infinitely many coefficients. If, for example, the β_i are square summable, then the coefficients must eventually decay to zero. It therefore seems reasonable to assume that there is an integer λ such that $\mu \doteq \sum_{i=1}^{\lambda} \beta_i x_i$. If this is so, μ might be viewed as following an approximate linear model and techniques for estimation from linear models could be employed for inference about μ. The resulting estimators for μ are termed series estimators.

One objective of Chapter 3 is to study the extent to which parallels of results from linear regression analysis can be developed for series estimators. There is, however, another problem which must be dealt with: the selection of λ. Since μ and, hence, the β_j are unknown there is no way to know *a priori* what will constitute a good value for λ. Instead, one must generally rely on the data to aid in such determinations.

The series estimators studied in Chapter 3 are linear estimators. The parameter λ for the weights in (1.10) is the number of functions x_j used to construct the estimator. Unfortunately, the weights for series estimators generally involve the design points in a complicated fashion. A simpler method of choosing the weights in (1.10) is to use, for example, a symmetric, unimodal probability density function K on $[-1, 1]$ and, for some real number $\lambda > 0$, estimate μ by

$$\mu_\lambda(t) = \sum_{i=1}^{n} K\left(\frac{t-t_i}{\lambda}\right) y_i \Big/ \sum_{j=1}^{n} K\left(\frac{t-t_j}{\lambda}\right). \tag{1.19}$$

This estimator is one illustration of what are called kernel estimators. These types of estimators are the subject of Chapter 4.

Estimator (1.19) is a weighted average of the responses. Since K is unimodal the heaviest weights will generally be assigned to responses whose design ordinates are near t, depending on how λ is chosen. The parameter λ is called the bandwidth and governs the amount of averaging which takes place. In order to use a kernel estimator one must first select an appropriate value for the bandwidth.

Chapter 5 focuses on inference about μ based on smoothing spline estimators. These estimators can be viewed as generalizing polynomial regression and also have ties to the kernel estimators from Chapter 4. If, for example, we assume that μ belongs to the set of all continuously differentiable functions on $[0, 1]$ with square integrable second derivatives, then an appropriate smoothing spline estimator of μ would be the function μ_λ that minimizes the penalized least-squares criterion

$$n^{-1}\sum_{i=1}^{n}(y_i - f(t_i))^2 + \lambda \int_0^1 f''(t)^2\,dt, \quad \lambda \geq 0, \tag{1.20}$$

over all functions f in this set. The resulting estimator provides a generalization of the linear regression estimator for μ. The parameter λ in (1.20) is also the parameter for the weights in (1.10). In this case λ determines the influence of the roughness penalty $\int_0^1 f''(t)^2\,dt$ and thereby governs the smoothness of the estimator. For this reason it is usually termed the smoothing parameter. Estimation of μ by smoothing splines will require the selection of an appropriate value for λ.

Chapter 6 deals with yet another estimation technique, termed least-squares spline regression, that has ties to both linear and nonlinear regression. This method also depends on a set of parameters λ that must be selected in order to conduct inference.

Thus, all the estimators to be considered in this book involve some type of tuning parameter λ. This leads to questions concerning what constitutes an optimal choice for λ and if there are techniques which can employ the data to effectively guide the selection of λ in practice. These subjects are the essence of the next chapter which provides a general discussion of how to assess the quality of an estimator and presents techniques that utilize the data to select an optimal estimator from a class of estimators.

1.4 Exercises

1. Let $\widehat{\mu}(t_1, y_1, \ldots, t_n, y_n)$ be any estimation prescription (i.e., formula) for μ. Show how to alter this prescription to make it applicable for an arbitrary design $a \leq t_1 \leq \cdots \leq t_n \leq b$ with $a \neq 0$ and/or $b \neq 1$.

2. Verify the identity

$$\begin{aligned} R(\widehat{\mu}) &= n^{-1} \sum_{i=1}^{n} \mathrm{E}\, (\widehat{\mu}(t_i) - \mu(t_i))^2 \\ &= n^{-1} \sum_{i=1}^{n} (\mathrm{E}\, \widehat{\mu}(t_i) - \mu(t_i))^2 + n^{-1} \sum_{i=1}^{n} \mathrm{Var}\, (\widehat{\mu}(t_i)) . \end{aligned}$$

3. For any fixed n and t not a design point, show that there is no linear estimator of $\mu(t)$ that is unbiased over all functions μ.

4. The data in Table 1.1 was used for Figure 1.3.

   (a) Experiment with fitting other regressogram estimators to this data using other choices for λ than the one employed in the example.

   (b) Look at plots of the bias (as a function of t) for various choices of λ. What tends to happen to estimator bias as λ is increased or decreased?

   (c) How does the variance of the estimator behave as λ is increased?

Table 1.1: Regressogram Example Data

t	y	t	y	t	y	t	y
.010	-.0937	.030	.0247	.050	.1856	.070	.1620
.090	-.0316	.110	.1442	.130	.0993	.150	.3823
.170	-.0624	.190	.3262	.210	.1271	.230	-.4158
.250	.0975	.270	-.0836	.290	.7410	.310	.3749
.330	.4446	.350	.5432	.370	.6946	.390	.5869
.410	.9384	.430	.7647	.450	.9478	.470	.9134
.490	1.2437	.510	.9070	.530	1.2289	.550	.9638
.570	.8834	.590	.6982	.610	.5729	.630	.7160
.650	1.0083	.670	.6681	.690	.5964	.710	.4759
.730	.6217	.750	.6221	.770	.6244	.790	.5918
.810	.7047	.830	.5234	.850	.9022	.870	.9930
.890	.8045	.910	.7858	.930	1.1939	.950	.9272
.970	.8832	.990	.9751				

(d) If we define the estimation loss as $L(\lambda) = \frac{1}{50}\sum_{i=1}^{50}(\mu_\lambda(t_i) - \mu(t_i))^2$, then verify that for this data set the loss is actually minimized by taking $\lambda = 8$. Compare this "best" value to the best choice for minimizing the variance or bias of the estimator. What conclusion can you draw from this concerning minimization of the loss relative to minimizing variance or bias alone?

5. For the situation considered in Example 1.4 show that if λ is any fixed, finite integer then, as $n \to \infty$, $n^{-1}\sum_{i=1}^{n}(\mathrm{E}\,\mu_\lambda(t_i) - \mu(t_i))^2$ converges to $\int_0^1 \mu^2(t)dt - \lambda\sum_{j=1}^{\lambda}(\int_{P_j}\mu(t)dt)^2$ with μ_λ the regressogram estimator. Under what conditions will the limit be zero and what are the implications of this latter case concerning the form of μ?

6. Consider the following extension of the regressogram estimator. Define the partitions $P_j = [\frac{j-1}{\lambda}, \frac{j}{\lambda}), j = 1, \ldots, \lambda - 1$, $P_\lambda = [\frac{\lambda-1}{\lambda}, 1]$ and for $t \in P_j$ let $\mu_\lambda(t) = b_{0j} + b_{1j}(t - \bar{t}_j)$ for

$$b_{1j} = \sum_{t_i \in P_j} y_i(t_i - \bar{t}_j) / \sum_{t_i \in P_j} (t_i - \bar{t}_j)^2,$$

$$b_{0j} = \bar{y}_j = n_j^{-1} \sum_{t_i \in P_j} y_i,$$

$\bar{t}_j = n_j^{-1} \sum_{t_i \in P_j} t_i$ and $n_j = \sum_{i=1}^{n} I_{P_j}(t_i)$.

(a) Develop computer code for the computation of μ_λ and use this to obtain an alternative fit to the data in Table 1.1.

(b) Show that if μ has two continuous derivatives and the number of partitions is allowed to grow at the rate $n^{1/5}$, then the squared error risk (1.13) for μ_λ decays to zero at the rate $n^{-4/5}$.

Chapter 2

What Is a Good Estimator?

2.1 Performance Criteria

In subsequent chapters we will study several families of estimators for the regression function. Each family is indexed by a parameter (often called the smoothing parameter). The selection of an estimator from a given family requires the choice of a value for this index parameter. While such a choice can be made subjectively, an objective choice will usually be preferred, at least as a starting point for subsequent estimator fine tuning. The present chapter provides an overview of several techniques which have been found useful for smoothing parameter selection. Their usefulness is not restricted to nonparametric regression problems and they can, in fact, be used to solve a variety of estimation problems, some of which will be explored in examples and exercises.

To provide a general development that is also suitable for subsequent work let us assume that a vector of observations $\mathbf{y} = (y_1, \ldots, y_n)^T$ has been obtained following the model

$$\mathbf{y} = \boldsymbol{\mu} + \boldsymbol{\varepsilon}, \tag{2.1}$$

where $\boldsymbol{\mu} = (\mu_1, \ldots, \mu_n)^T$ is an unknown mean vector and $\boldsymbol{\varepsilon} = (\varepsilon_1, \ldots, \varepsilon_n)^T$

is a vector of zero mean, uncorrelated random errors with common variance σ^2. Suppose now that we are considering a class of estimators for $\boldsymbol{\mu}$, $C(\Lambda) = \{\boldsymbol{\mu}_\lambda : \lambda \in \Lambda\}$, with Λ representing some index set. As seen below, the index parameter λ may be scalar, vector or even set valued. To simplify things we will also assume that the elements of $C(\Lambda)$ are linear estimators. This means that for each λ there is an $n \times n$ matrix $\mathbf{S}_\lambda$ such that

$$\boldsymbol{\mu}_\lambda = \mathbf{S}_\lambda \mathbf{y}.$$

Unless stated otherwise, $\mathbf{S}_\lambda$ will be assumed to be symmetric and positive semi-definite.

The problem to be considered is selection of a best estimator of $\boldsymbol{\mu}$ from among the elements of $\{\boldsymbol{\mu}_\lambda : \lambda \in \Lambda\}$. The word "best" in our problem statement has been left open for interpretation and until it is given a precise meaning there is clearly no way to formulate a solution. Thus, to effect an objective choice for an estimator we must quantify what is meant by "best." Of course, one person's idea of "best" may not coincide with another's, so there is no universally accepted measure of an estimator's performance. Nonetheless, there are certain critera which are widely accepted and used. We will focus on three of these.

Define the ***loss*** in estimating μ by

$$L(\lambda) = n^{-1} \sum_{i=1}^{n} (\mu_i - \mu_{\lambda i})^2, \tag{2.2}$$

where $\mu_{\lambda i}$ is the ith element of the n-vector $\boldsymbol{\mu}_\lambda$. Note that $L(\lambda)$ is just the squared Euclidean distance between $\boldsymbol{\mu}$ and $\boldsymbol{\mu}_\lambda$ (apart from the factor n^{-1}) and therefore represents a natural measure of the closeness of $\boldsymbol{\mu}_\lambda$ to $\boldsymbol{\mu}$. Associated with the loss is its expected value

$$R(\lambda) = \mathrm{E}\, L(\lambda) \tag{2.3}$$

which is called the ***risk***. Both $L(\lambda)$ and $R(\lambda)$ provide assessments of an estimator's performance with smaller values of the criteria being indicative of better estimation. A value of λ that minimizes the loss provides a best (from a squared error standpoint) estimator of $\boldsymbol{\mu}$ for the particular data set in question while the value of λ that minimizes the risk can be viewed as best for prediction of future responses or estimation of $\boldsymbol{\mu}$ in repeated sampling.

Another criterion that is closely related to (2.3) is the *prediction risk* (sometimes called the mean squared error of prediction). The idea behind this measure is as follows. Suppose we are planning to observe n new observations $\mathbf{y}^* = \boldsymbol{\mu} + \boldsymbol{\varepsilon}^*$, with $\boldsymbol{\mu}$ the same vector of means as in (2.1) and $\boldsymbol{\varepsilon}^*$ a vector of random errors with zero means and common variance σ^2 that are uncorrelated with each other and with $\boldsymbol{\varepsilon}$. To evaluate the performance of $\boldsymbol{\mu}_\lambda$ as a predictor of future observations we could then use the expected squared error of $\boldsymbol{\mu}_\lambda$ as a predictor of $\mathbf{y}^*$: i.e.,

$$P(\lambda) = n^{-1} \sum_{i=1}^{n} \mathrm{E}(y_i^* - \mu_{\lambda i})^2. \tag{2.4}$$

One finds that

$$P(\lambda) = \sigma^2 + R(\lambda) \tag{2.5}$$

(Exercise 1), which has the consequence that an estimator of $\boldsymbol{\mu}$ that minimizes the risk $R(\lambda)$ will also minimize $P(\lambda)$ and conversely.

Since there is a one-to-one correspondence between the estimators in $C(\Lambda)$ and the elements in the index set Λ, an optimal or *best* estimator can be characterized as one whose index value λ minimizes the risk (equivalently, the prediction risk) or the loss over all elements of Λ. Unfortunately, neither $L(\lambda)$, $R(\lambda)$ nor $P(\lambda)$ can actually be computed without knowing $\boldsymbol{\mu}$. Thus, in practice, the criterion of interest must be estimated from the data and its estimate then minimized with respect to λ to obtain an estimate of the best choice for λ. Of the three criteria, $R(\lambda)$ and $P(\lambda)$ are the most mathematically tractable and, accordingly, in the next section we will turn our attention to the problem of estimating the risk and the prediction risk. We should note, however, that the relationship between the loss and the risk allows some degree of flexibility in our thinking and we will also, on occasion, view an estimator of the minimizer of the risk as an estimator of the minimizer of the loss.

The loss, risk and prediction risk provide three possible measures of performance for an estimator of $\boldsymbol{\mu}$. There are, of course, many others. For example, instead of squared errors we might use absolute values. However, squared errors are generally easier to deal with from a mathematical viewpoint and, given no compelling reason to use absolute values, will often be preferred for this reason.

If only one of the means, μ_j, is of interest, then another possible measure is, e.g., $\mathrm{E}(\mu_j - \mu_{\lambda j})^2$. In contrast to this measure, $R(\lambda)$ and $P(\lambda)$ provide global measures of the performance of $\boldsymbol{\mu}_\lambda$; that is, they give an indication of how $\boldsymbol{\mu}_\lambda$ behaves as an estimator over all the elements of $\boldsymbol{\mu}$.

In the context of nonparametric regression the use of a global performance measure seems desirable. In that setting we have

$$y_i = \mu(t_i) + \varepsilon_i,\ i = 1, \ldots, n,$$

with μ an unknown regression function and $t_i, i = 1, \ldots, n$, design points in the interval $[0,1]$. This corresponds to model (2.1) with $\mu_i = \mu(t_i), i = 1, \ldots, n$. Thus, $R(\lambda)$ or $P(\lambda)$ provide descriptions of the efficacy of an estimator of the regression function over the entire interval of observation. There are other closely related measures of performance with similar properties. If $\mu_\lambda(\cdot)$ is an estimator of the regression function $\mu(\cdot)$, then we might consider measures such as the ***integrated loss***

$$IL(\lambda) = \int_0^1 (\mu(t) - \mu_\lambda(t))^2\, dt$$

or ***integrated risk***

$$IR(\lambda) = \int_0^1 \mathrm{E}\,(\mu(t) - \mu_\lambda(t))^2\, dt.$$

When the t_i are uniformly spaced points, $L(\lambda)$ and $R(\lambda)$ are just quadrature formulae for $IL(\lambda)$ and $IR(\lambda)$, respectively. More generally, if the t_i are unequally spaced $L(\lambda)$ and $R(\lambda)$ give quadrature approximations to modified versions of $IL(\lambda)$ and $IR(\lambda)$ that come from integrating against a weight function which reflects the spacing of the design.

At this point some examples of estimation problems which fit into the framework of model (2.1) will be helpful in clarifying and solidifying the concepts and notation. The first two examples are not particularly pertinent to our study of nonparametric regression but may be more familiar to some readers. The third example is directly related to the topic of series estimators to be studied in Chapter 3.

Example 2.1 (Variable Selection). Suppose that $\mathbf{y} = (y_1, \ldots, y_n)^T$ is obtained as the response to n values of p input variables $X_1, \ldots, X_p$. Thus, y_i is the output obtained from an input vector

$$\mathbf{x}_i = (x_{i1}, \ldots, x_{ip})^T, \; i = 1, \ldots, n.$$

Our objective is to develop a scheme for prediction of future responses to various input configurations.

Assume that the response variable Y is linearly related to the input variable so that $\mu_i = \mathbf{x}_i^T \boldsymbol{\beta}$ for some coefficient vector $\boldsymbol{\beta} = (\beta_1, \ldots, \beta_p)^T$. To estimate $\boldsymbol{\beta}$ we can use the least-squares estimator $\mathbf{b}$ obtained by minimizing the residual sum-of-squares $\sum_{i=1}^n (y_i - \mathbf{x}_i^T \mathbf{c})^2$ with respect to the vector $\mathbf{c}$. If $\mathbf{X}^T = [\mathbf{x}_1, \ldots, \mathbf{x}_n]$ has rank p, $\mathbf{b}$ is given explicitly by

$$\mathbf{b} = (\mathbf{X}^T\mathbf{X})^{-1}\mathbf{X}^T\mathbf{y}.$$

Our estimate of $\boldsymbol{\mu}$ is then

$$\widehat{\boldsymbol{\mu}} = \mathbf{X}\mathbf{b} = \mathbf{X}(\mathbf{X}^T\mathbf{X})^{-1}\mathbf{X}^T\mathbf{y}. \tag{2.6}$$

By using $\widehat{\boldsymbol{\mu}}$ one obtains an unbiased estimator of $\boldsymbol{\mu}$. However, as seen in the next section, the use of an unbiased estimator need not lead to an estimator with minimal prediction risk. Thus, define our set Λ to consist of all possible subsets of the indices of the variables $X_1, \ldots, X_p$. In general, there will be 2^p such subsets. For example, when $p = 2$, Λ consists of the sets $\{1\}$, $\{2\}$, $\{1,2\}$ and the empty set.

Corresponding to each $\lambda \in \Lambda$ (except the empty set) is a design matrix

$$\mathbf{X}_\lambda = \{x_{ij}\}_{i=1,n, j\in\lambda}. \tag{2.7}$$

For instance, if $p = 2$ and $\lambda = \{2\}$, $\mathbf{X}_\lambda = (x_{12}, \ldots, x_{n2})^T$ and, if $\lambda = \{1, 2\}$, $\mathbf{X}_\lambda$ and $\mathbf{X}$, the design matrix for the full set of predictors, coincide.

To estimate $\boldsymbol{\mu}$ we can now choose from the class of estimators $C(\Lambda) = \{\boldsymbol{\mu}_\lambda : \lambda \in \Lambda\}$ with

$$\boldsymbol{\mu}_\lambda = \mathbf{X}_\lambda(\mathbf{X}_\lambda^T\mathbf{X}_\lambda)^{-1}\mathbf{X}_\lambda^T\mathbf{y}, \; \lambda \in \Lambda, \tag{2.8}$$

under the convention that $\boldsymbol{\mu}_\lambda = \mathbf{0}$, an n-vector of all zero elements, when λ is the empty set. For a given λ, $\boldsymbol{\mu}_\lambda$ is recognized as the least-squares estimator of $\boldsymbol{\mu}$ corresponding to the predictor variable subset whose indices are in λ. Note that $\mathbf{S}_\lambda$ in this setting is, from (2.8),

$$\mathbf{S}_\lambda = \mathbf{X}_\lambda(\mathbf{X}_\lambda^T\mathbf{X}_\lambda)^{-1}\mathbf{X}_\lambda^T \tag{2.9}$$

which is easily seen to be symmetric and nonnegative definite.

As an example of a situation where variable selection methods are of interest we consider a study of housing rent data for 455 households in Mariposa County, Arizona, that was reported in Gunst, and Mason (1980). The data consists of a response variable, monthly rent for an unfurnished dwelling unit, and 17 additional predictor variables which are believed to be related to housing rent. One aspect of the study was concerned with developing a predictive equation for housing rents. Due to the costs and difficulty of maintaining information on the predictor variables it was also important to determine if some subset of the predictor variables could be substituted for the entire set of predictors with comparable (or better) results from a prediction standpoint.

To make the presentation somewhat more manageable we will concentrate here on predicting housing rent, Y, using only four of the predictor variables in the housing rent study: namely,

X_2: a measure of school quality and lack of transportation for the census tract where the dwelling unit is located,

X_3: a dichotomous (i.e., two-valued) variable which takes on the value 1 or 0 according as to whether or not the length of residence is 5 to 10 years,

X_4: the logarithm of the total floor area of the dwelling unit, and

X_5: a measure of average surface area and structural quality, 0 being worst and 3 being the best.

There is also a fifth predictor, X_1, which is defined to be unity for all observations. Some summary statistics for Y and $X_2, \ldots, X_5$ are presented in Table 2.1.

Table 2.1: Summary Measures for the Housing Rent Data

Variable	Mean	Standard Deviation
Y	129.8718	44.3970
X_2	.1171	1.0906
X_3	.1077	.3103
X_4	6.2794	.3539
X_5	2.2973	.6364

Source: Gunst and Mason (1980)

Let us now assume that housing rent, Y, follows a linear model in $\{X_1, \ldots, X_5\}$ or, more precisely, that

$$\begin{aligned} Y &= \beta_1 + \beta_2 X_2 + \beta_3 X_3 + \beta_4 X_4 + \beta_5 X_5 + \varepsilon \\ &= \sum_{j=1}^{5} \beta_j X_j + \varepsilon, \end{aligned}$$

with ε denoting a random error term. The question to be addressed is "Can an optimal prediction equation for housing rent be found involving $\{X_1, \ldots, X_5\}$?" To answer this question we must first decide on what the word "optimal" will mean in the context of this problem.

If a particular type of dwelling unit is to be studied with known or specified values $x_2^0, \ldots, x_5^0$ for $X_2, \ldots, X_5$, then a particular mean value $\mu_0 = \beta_1 + \sum_{j=2}^{5} \beta_j x_j^0$ is of interest. In that case, one might define the least-squares estimator $\mu_{\lambda 0}$, based on some subset of $\{X_1, \ldots, X_5\}$, as being optimal if it minimized $\mathrm{E}\,(\mu_0 - \mu_{\lambda 0})^2$ over all subsets of the predictor variable, λ, that are of interest. It seems unlikely in this setting that a particular type of housing unit will be a focal point of interest. Instead, one would anticipate that a least-squares predictive equation which performed well over all types of dwelling units in the study would be preferred. Thus, it is reasonable to define an optimal predictive equation for this problem as one which corresponds to the minimizer of something like the prediction risk (2.4).

Given that the selection of a variable subset to minimize (2.4) is the objective, one must now ask how this can be accomplished. If the true Y means were available for each of the sample points, it would be a simple matter to compute $P(\lambda)$ and find the optimal set of predictors. However, since only sample information is available, the true Y means are unknown and hence the risk is not computable. Thus, the choice of a variable subset must be guided by the data, possibly through sample estimates of the predictive risk in (2.4), and it becomes necessary to postpone further consideration of the optimal subset question until the next section where estimators of $P(\lambda)$ are developed.

Example 2.2 (Ridge Regression). Consider now a related problem to that of Example 2.1 where we again have predictor variables $\{X_1, \ldots, X_p\}$ which are linearly related to Y and are to be used in estimating $\boldsymbol{\mu}$. In some cases the X variables are so closely related that the $\mathbf{X}$ matrix is nearly rank deficient (i.e., has rank less than p). This has some unpleasant consequences for estimator variances and can cause difficulties when attempting to solve the normal equations $\mathbf{X}^T\mathbf{X}\mathbf{c} = \mathbf{X}^T\mathbf{y}$ from which (2.6) derives. This and other factors led Hoerl and Kennard (1970a, b) to propose what is called a *ridge regression estimator* for $\boldsymbol{\beta}$ which is defined by

$$\mathbf{b}_\lambda = (\mathbf{X}^T\mathbf{X} + \lambda\mathbf{I})^{-1}\mathbf{X}^T\mathbf{y},\ \lambda \geq 0.$$

The ridge estimator of $\boldsymbol{\mu}$ is then obtained by the relation

$$\boldsymbol{\mu}_\lambda = \mathbf{X}\mathbf{b}_\lambda = \mathbf{X}(\mathbf{X}^T\mathbf{X} + \lambda\mathbf{I})^{-1}\mathbf{X}^T\mathbf{y}. \tag{2.10}$$

Ridge estimators have a variety of optimality properties, some of which are explored in the exercises.

As seen from (2.10), ridge estimators involve a scalar parameter λ that is often called the *ridge parameter*. Thus, to choose an estimator a value of λ must be selected. Note that the choice $\lambda = 0$ gives $\boldsymbol{\mu}_0 = \hat{\boldsymbol{\mu}}$, the least-squares estimator from (2.6).

We now see that ridge estimators fit into the framework set out above. More precisely, ridge regression corresponds to the case $\Lambda = \{\lambda : 0 \leq \lambda < \infty\}$ and $C(\Lambda) = \{\boldsymbol{\mu}_\lambda : \lambda \in \Lambda\}$ with $\boldsymbol{\mu}_\lambda$ defined by (2.10) and $\mathbf{S}_\lambda = \mathbf{X}(\mathbf{X}^T\mathbf{X} + \lambda\mathbf{I})^{-1}\mathbf{X}^T$.

There are several generalizations of ridge estimators. For example, we might consider an estimator of the form

$$\boldsymbol{\mu}_\lambda = \mathbf{X}(\mathbf{X}^T\mathbf{X} + \mathbf{V})^{-1}\mathbf{X}^T\mathbf{y},$$

where, e.g., $\mathbf{V} = \text{diag}(\lambda_1, \ldots, \lambda_p)$ for $\lambda_j \geq 0$, $j = 1, \ldots, p$. In this case Λ can be viewed as the set of all p-vectors with nonnegative coordinates. More generally still, $\mathbf{V}$ could be some positive semi-definite matrix. Estimators of this type are called *generalized ridge estimators.*

Example 2.3 (Cosine Series Regression). Let us now turn to an example that derives from a nonparametric regression context. For this purpose assume that we observe n response values $y_1, \ldots, y_n$ at equally spaced design points $t_i = (2i - 1)/2n$, $i = 1, \ldots, n$. The y_i's and t_i's are related by

$$y_i = \beta_1 + \sum_{j=2}^{\infty} \beta_j \sqrt{2} \cos((j-1)\pi t_i) + \varepsilon_i,\ i = 1, \ldots, n, \tag{2.11}$$

where the ε_i are zero mean, uncorrelated random variables with common variance σ^2 and $\beta_1, \beta_2, \ldots$ are unknown parameters. The problem here is to estimate the regression function $\mu(t) = \beta_1 + \sum_{j=2}^{\infty} \beta_j \sqrt{2} \cos((j-1)\pi t)$.

Now (cf Section 3.8)

$$n^{-1} \sum_{i=1}^{n} \cos(j\pi t_i) \cos(k\pi t_i) = \frac{1}{2}\delta_{jk},\ j, k = 1, \ldots, n-1, \tag{2.12}$$

and

$$n^{-1} \sum_{i=1}^{n} \cos(j\pi t_i) = 0,\ j = 1, \ldots, n-1, \tag{2.13}$$

with δ_{jk} the Kronecker delta function which is one if $j = k$ and zero otherwise. Thus, we find that the least-squares estimators of $\beta_1, \ldots, \beta_n$ are

$$b_1 = n^{-1} \sum_{i=1}^{n} y_i \tag{2.14}$$

and

$$b_j = \sqrt{2}n^{-1}\sum_{i=1}^{n} y_i \cos((j-1)\pi t_i),\ j = 2,\ldots,n. \tag{2.15}$$

Given the form of our regression function, it seems reasonable to consider estimators of the means $\mu_i = \mu(t_i)$ such as

$$\mu_\lambda(t_i) = b_1 + \sum_{j=2}^{\lambda} b_j \sqrt{2}\cos((j-1)\pi t_i), \tag{2.16}$$

where λ is some integer between 1 and n with $\lambda = 1$ taken to mean that $\mu_1(\cdot) = b_1$. Thus, we can take $\Lambda = \{1,\ldots,n\}$ and

$$\mathbf{S}_\lambda = n^{-1}\mathbf{X}_\lambda \mathbf{X}_\lambda^T \tag{2.17}$$

with $\mathbf{X}_1 = \mathbf{1}$, an n-vector of all ones and, in general,

$$\mathbf{X}_\lambda = [\mathbf{1} \,|\, \{\sqrt{2}\cos((j-1)\pi t_i)\}_{i=1,n,j=2,\lambda}]. \tag{2.18}$$

This estimation problem is quite similar in nature to the variable selection problem of Example 2.1. To see this, think of the cosine functions as providing a collection of "predictor" variables $\{X_j\}_{j=1}^{\infty}$ that are functions of the primary predictor t with $X_1 = 1$ and $X_j(t) = \sqrt{2}\cos((j-1)\pi t)$, $j = 2,\ldots$ One then considers selecting from the ordered (or hierarchical) variable subsets $\{X_1\}, \{X_1, X_2\}, \ldots, \{X_1, \ldots, X_n\}$ rather than all possible variable subsets. There are, however, important distinctions between the present example and Example 2.1: namely, there are an infinite number of predictors in this case and, unless the β_j all vanish for $j \geq n$, there is no unbiased estimator in $C(\Lambda)$ for the regression curve $\mu(t)$ at some general point t. There is, of course, an unbiased estimator of $\boldsymbol{\mu} = (\mu(t_1), \ldots, \mu(t_n))^T$ provided by the response vector $\mathbf{y}$ that corresponds to the estimator $\boldsymbol{\mu}_\lambda$ with $\lambda = n$. But, as noted in Chapter 1, this estimator is generally not acceptable from the standpoint of accuracy and summarization of the data.

Model (2.11) is actually a nonparametric regression model like (1.1) and the estimator μ_λ in (2.16) is what is termed a cosine series regression estimator for μ. Both this model and estimator will be discussed extensively in Chapter 3. From our earlier discussion of performance criteria for nonparametric regression, we are likely to want a choice for λ that makes the cosine series estimator perform well as an estimator of μ over the entire design. Thus, estimation of the minimizer of $L(\lambda)$ would seem to provide a natural approach to selecting the number of terms in (2.16).

An example of a case where model (2.11) and estimator (2.16) are applicable is provided by the data in Figure 2.1. This data was simulated from the model

$$y_i = e^{2.5t_i} + \varepsilon_i,\ i = 1,\ldots,50, \tag{2.19}$$

with $t_i = (2i-1)/100$, $i = 1,\ldots,50$, and the ε_i are uncorrelated, zero mean, normal random errors with $\sigma = 1.5$. The regression function in (2.19) can be shown to correspond to (2.11) with $\beta_1 = (e^{2.5} - 1)/2.5$ and

$$\beta_j = 2.5\sqrt{2}\frac{((-1)^{(j-1)}e^{2.5} - 1)}{(2.5)^2 + ((j-1)\pi)^2},\ j = 2,\ldots$$

The plot shows the data along with the true regression function and a fit to the data using (2.16) with the particular choice of $\lambda = 4$. The problem we are now faced with is that of determining a good value for the number of cosine functions to be used in fitting this data set.

2.2 Estimating $P(\lambda)$ and $R(\lambda)$

In the previous section we considered three measures of the performance of an estimator of $\boldsymbol{\mu}$: the loss, the risk, and the prediction risk. In this section we concentrate our attention on obtaining estimators of the latter two measures.

As before, let $C(\Lambda) = \{\boldsymbol{\mu}_\lambda : \lambda \in \Lambda\}$ be a class of linear estimators for $\boldsymbol{\mu}$ with $\boldsymbol{\mu}_\lambda = (\mu_{\lambda 1},\ldots,\mu_{\lambda n})^T = \mathbf{S}_\lambda \mathbf{y}$ and consider estimation of the prediction risk defined in (2.4). The naive estimator of $\mathrm{E}(y_i^* - \mu_{\lambda i})^2$ in this case is the square of the residual $y_i - \mu_{\lambda i}$. Thus, one might estimate $P(\lambda)$ by the average of the squared residuals $n^{-1}RSS(\lambda)$, where

$$RSS(\lambda) = \sum_{i=1}^{n}(y_i - \mu_{\lambda i})^2$$

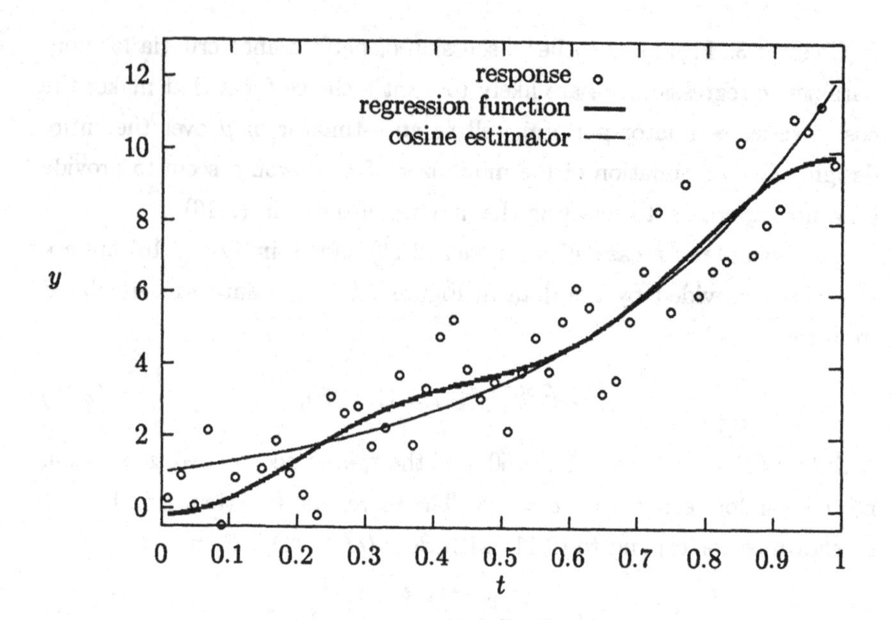

Figure 2.1: Cosine Regression Fit for Simulated Data.

is the residual sum-of-squares associated with the fitted values for the estimator.

The question that must now be asked is "How does $n^{-1}RSS(\lambda)$ perform as an estimator of $P(\lambda)$?" A partial answer can be obtained by computing the expected value of $n^{-1}RSS(\lambda)$ and comparing that to $P(\lambda)$.

To evaluate the expectation of $n^{-1}RSS(\lambda)$ note that

$$RSS(\lambda) = (\mathbf{y} - \boldsymbol{\mu}_\lambda)^T(\mathbf{y} - \boldsymbol{\mu}_\lambda) = \mathbf{y}^T(\mathbf{I} - \mathbf{S}_\lambda)^2\mathbf{y}.$$

Thus, from Lemma 2.2 of Section 2.4

$$\begin{aligned} \mathrm{E}\,RSS(\lambda) &= \boldsymbol{\mu}^T(\mathbf{I} - \mathbf{S}_\lambda)^2\boldsymbol{\mu} + \sigma^2 \mathrm{tr}\left[(\mathbf{I} - \mathbf{S}_\lambda)^2\right] \\ &= \boldsymbol{\mu}^T(\mathbf{I} - \mathbf{S}_\lambda)^2\boldsymbol{\mu} + n\sigma^2 + \sigma^2\mathrm{tr}\left[\mathbf{S}_\lambda^2\right] - 2\sigma^2\mathrm{tr}\left[\mathbf{S}_\lambda\right]. \end{aligned} \quad (2.20)$$

In contrast,

$$\begin{aligned} P(\lambda) &= \sigma^2 + R(\lambda) \\ &= \sigma^2 + n^{-1}\sum_{i=1}^{n} \mathrm{E}\,(\mu_i - \mu_{\lambda i})^2 \\ &= \sigma^2 + n^{-1}\mathrm{E}\,(\boldsymbol{\mu} - \boldsymbol{\mu}_\lambda)^T(\boldsymbol{\mu} - \boldsymbol{\mu}_\lambda) \\ &= \sigma^2 + n^{-1}\boldsymbol{\mu}^T(\mathbf{I} - \mathbf{S}_\lambda)^2\boldsymbol{\mu} + n^{-1}\sigma^2 \mathrm{tr}\left[\mathbf{S}_\lambda^2\right]. \end{aligned} \tag{2.21}$$

Comparing (2.20) and (2.21) we see that $n^{-1}RSS(\lambda)$ is a biased estimator of $P(\lambda)$ with bias equal to $-2n^{-1}\sigma^2\mathrm{tr}\,[\mathbf{S}_\lambda]$. Assuming that σ^2 is known, an unbiased estimator of $P(\lambda)$ is therefore given by

$$\widehat{P}(\lambda) = n^{-1}RSS(\lambda) + 2n^{-1}\sigma^2\mathrm{tr}\,[\mathbf{S}_\lambda]. \tag{2.22}$$

Similarly, an unbiased estimator of the risk is

$$\widehat{R}(\lambda) = \widehat{P}(\lambda) - \sigma^2. \tag{2.23}$$

The estimator $\widehat{R}(\lambda)$ is closely related to a criterion for selecting λ proposed by Mallows (1973). His criterion, often termed C_L, is

$$M(\lambda) = \frac{1}{\sigma^2}RSS(\lambda) - n + 2\mathrm{tr}\,[\mathbf{S}_\lambda].$$

Note that $M(\lambda) = n\widehat{R}(\lambda)/\sigma^2$ and, hence, $\widehat{R}(\lambda)$ and $M(\lambda)$ will be minimized at the same value of λ.

Let us pause for a moment to examine the risk in more detail. Note from (2.5) and (2.21) that

$$\begin{aligned} R(\lambda) &= n^{-1}\boldsymbol{\mu}^T(\mathbf{I} - \mathbf{S}_\lambda)^2\boldsymbol{\mu} + n^{-1}\sigma^2\mathrm{tr}\left[\mathbf{S}_\lambda^2\right] \\ &= n^{-1}\sum_{i=1}^{n}(\mu_i - \mathrm{E}\,\mu_{\lambda i})^2 + n^{-1}\sum_{i=1}^{n}\mathrm{Var}\,(\mu_{\lambda i}), \end{aligned}$$

where $\mathrm{Var}\,(\mu_{\lambda i}) = \mathrm{E}\,(\mu_{\lambda i} - \mathrm{E}\,\mu_{\lambda i})^2$. This shows that the risk breaks into two components: namely,

$$B^2(\lambda) = n^{-1}\sum_{i=1}^{n}(\mu_i - \mathrm{E}\,\mu_{\lambda i})^2$$

and

$$V(\lambda) = n^{-1} \sum_{i=1}^{n} \mathrm{Var}\,(\mu_{\lambda i}).$$

The first component, $B^2(\lambda)$, is a measure of the estimator's bias, while the second is an indication of the estimator's variability. If the elements of $C(\Lambda)$ are all unbiased for $\boldsymbol{\mu}$, minimization of $R(\lambda)$ (or $P(\lambda)$) is tantamount to selecting an estimator which minimizes $V(\lambda)$. More generally, when $C(\Lambda)$ contains biased estimators, minimization of $R(\lambda)$ will entail a balancing of the (squared) bias component and the variance component. Note that even when $C(\Lambda)$ contains an unbiased estimator it may not represent the optimal estimator. Thus, making $B^2(\lambda)$ vanish, even when that is feasible, is not always the best strategy and it is possible for a biased estimator to have a smaller risk than an unbiased one (cf Exercise 4).

If σ^2 is known, $\widehat{P}(\lambda)$ and $\widehat{R}(\lambda)$ can be used directly to evaluate the performance of $\boldsymbol{\mu}_\lambda$. However, this will generally not be the case and σ^2 will usually require estimation in (2.22) and (2.23).

In some cases the dependence of (2.22) and (2.23) on σ^2 is not a serious problem. For example, in the variable selection problem of Example 2.1 there are many estimators of σ^2. If $\#(\lambda)$ denotes the number of elements in λ, then $RSS(\lambda)/(n - \#(\lambda))$ provides an estimator of σ^2 for each $\lambda \in \Lambda$. A common approach, in this setting, is to use the estimator of σ^2 corresponding to $\lambda = \{1, \ldots, p\}$, i.e., the estimator that is obtained when all variables are used in estimating $\boldsymbol{\mu}$.

There is, of course, one case in which no estimator of σ^2 is needed for estimation of $P(\lambda)$, even when σ^2 is unknown. This occurs when $\mathrm{tr}[\mathbf{S}_\lambda] = 0$ for all $\lambda \in \Lambda$. Estimators of this type are called ***nil-trace estimators*** and are explored further in Exercises 17 and 18. Most estimators do not have the nil-trace property, however, and there are situations where there are no obvious estimators of σ^2. Consequently, we now consider two estimators of the prediction risk that do not require estimation of σ^2.

One alternative estimator of $P(\lambda)$ that avoids estimation of σ^2 can be motivated by further examination of the residual sum-of-squares. In using $n^{-1}RSS(\lambda)$ to estimate $P(\lambda)$ we implicitly used $y_i - \mu_{\lambda i}$ to estimate the prediction error associated with a future response with mean μ_i. Since $\mu_{\lambda i}$ was constructed using the data $y_1, \ldots, y_n$ we would anticipate that $\mu_{\lambda i}$ would do better in "predicting" y_i than some future response y_i^*. As a result, we would expect $n^{-1}RSS(\lambda)$ to be smaller on the average than $P(\lambda)$. This is, in fact, the case. As seen from (2.20) and (2.21) $n^{-1}RSS(\lambda)$ tends to underestimate $P(\lambda)$ by a factor of $2\sigma^2 \text{tr}[\mathbf{S}_\lambda]/n$.

An intuitive approach to correcting $n^{-1}RSS(\lambda)$ for under-estimation of $P(\lambda)$ would involve replacement of $(y_i - \mu_{\lambda i})^2$ by some quantity which gives a better estimator of $\text{E}\,(y_i^* - \mu_{\lambda i})^2$. To give some insight into what follows, consider what could be done if we had multiple observations for each mean. As a simple example, suppose we had observed

$$y_{ki} = \mu_i + \varepsilon_{ki},\ k = 1, 2,\ i = 1, \ldots, n.$$

In words, suppose that we observe two response values for each mean instead of just one. Then, an estimate of $\boldsymbol{\mu}$ could be obtained using only the first n responses and $\text{E}\,(y_i^* - \mu_{\lambda i})^2$ could be estimated using $(y_{2i} - \mu_{\lambda i})^2$. Since y_{2i} was not used in computing $\mu_{\lambda j}$ we are no longer assessing the prediction properties of the estimator by the data used for its construction. Consequently, $n^{-1}\sum_{i=1}^{n}(y_{2i} - \mu_{\lambda i})^2$, with $\mu_{\lambda i}$ based only on the y_{1j}, $j = 1, \ldots, n$, would be expected to give a better estimator of $P(\lambda)$ than $n^{-1}RSS(\lambda)$. This is, in fact, the case and it can be shown (Exercise 14) that this particular estimator is unbiased for $P(\lambda)$.

Of course, we will not generally have multiple observations for each μ_i. Also, when we do, it will usually be preferable to include them in the estimation process rather than holding them back for validation of the estimator. However, the idea of using a subsample of the data for estimating the prediction error suggests another approach that is applicable in certain situations.

In some cases it is possible to compute a parallel of, or approximation to, the estimator $\mu_{\lambda i}$ that does not involve the ith observation. In situations such as this the data can be effectively broken into n subsamples of size $n-1$ and the observation held out of each subsample used for estimation of the prediction error. Before providing a general formulation it will be helpful to consider an example of a setting that permits this type of development.

For the variable selection problem it is possible to delete the point $(\mathbf{x}_i^T, y_i)$ from consideration and, for any given variable subset, recompute the coefficient estimates. More precisely, for a variable subset with indices in λ, let $\mathbf{X}_{\lambda(i)}$ denote the matrix

$$\{x_{kr}\}_{k=1,\ldots,i-1,i+1,\ldots,n,r\in\lambda}$$

and let $\mathbf{y}_{(i)} = (y_1, \ldots, y_{i-1}, y_{i+1}, \ldots, y_n)^T$. Then

$$\mathbf{b}_{\lambda(i)} = \left(\mathbf{X}_{\lambda(i)}^T \mathbf{X}_{\lambda(i)}\right)^{-1} \mathbf{X}_{\lambda(i)}^T \mathbf{y}_{(i)}$$

is the least-squares estimator of the coefficients for the variables X_r, $r \in \lambda$, computed without the use of $(\mathbf{x}_i^T, y_i)$. An estimator of μ_i is then provided by

$$\mu_{\lambda(i)} = \mathbf{x}_{\lambda i}^T \mathbf{b}_{\lambda(i)}$$

with $\mathbf{x}_{\lambda i}$ the vector of the values for the X variables with indices in λ that corresponds to the ith response. The estimator $\mu_{\lambda(i)}$ is directly related to $\mu_{\lambda i}$ in the sense that both are least-squares estimators of μ_i involving the predictors indexed by λ. The only difference is that one involves $(\mathbf{x}_i^T, y_i)$ while the other does not. In fact, the connection is stronger than this since it can be shown that

$$\mu_{\lambda(i)} = \mu_{\lambda i} - s_i(y_i - \mu_{\lambda i})/(1 - s_i), \tag{2.24}$$

where s_i is the ith diagonal element of $\mathbf{S}_\lambda$. Similar conclusions hold for ridge regression (Exercise 5).

Returning to the general case, assume that $\mu_{\lambda(i)}$ represents a parallel of $\mu_{\lambda i}$ that is computed without the ith observation; estimator (2.24) is one example. Then, the estimator $\mu_{\lambda(i)}$ is constructed from a subsample of size $n - 1$ taken from the original data. If the ith observation is treated like an additional observation or holdout sample, our previous discussion would suggest using $(y_i - \mu_{\lambda(i)})^2$ to estimate $\mathrm{E}(y_i^* - \mu_{\lambda i})^2$ and estimating $P(\lambda)$ by

$$\mathrm{CV}(\lambda) = n^{-1} \sum_{i=1}^{n} \left(y_i - \mu_{\lambda(i)}\right)^2. \tag{2.25}$$

The function $\mathrm{CV}(\lambda)$ is called the *cross-validation criterion* and the process of selecting λ through minimization of (2.25) is called *cross validation* or merely CV. The quantity $n\mathrm{CV}(\lambda)$ is sometimes called the Prediction Sum of Squares or PRESS (Allen 1974).

Cross-validation is a method of reusing the data in a fashion which simulates what would be done if repeated observations were available. However, since there is actually only one observation for each μ_i it is no surprise that $\mathrm{CV}(\lambda)$ is generally biased for $P(\lambda)$ (see Exercise 15). (It is however unbiased for a prediction risk that is related to $P(\lambda)$; see Exercise 17.) An overview of various applications for CV is provided by Titterington (1985). For a comparison of cross validation with other sample reuse techniques see Efron (1982).

Another useful method for selecting λ is *generalized cross validation* which we abbreviate by GCV. Assuming that $\mathrm{tr}[\mathbf{S}_\lambda] < n$, the GCV criterion is defined by

$$\mathrm{GCV}(\lambda) = n^{-1}RSS(\lambda)/\left(n^{-1}\mathrm{tr}[\mathbf{I}-\mathbf{S}_\lambda]\right)^2. \tag{2.26}$$

The GCV criterion was first proposed by Craven and Wahba (1979) for use in the context of nonparametric regression. However, Golub, Heath and Wahba (1979) show that GCV can be used to solve a wide variety of problems involving estimation of the minimizers of $L(\lambda), R(\lambda)$ or $P(\lambda)$.

In most cases of interest GCV is closely related to CV. An illustration of this is provided by the variable selection problem where we find, using (2.24), that

$$\mathrm{CV}(\lambda) = n^{-1}\sum_{i=1}^{n}\left(y_i - \mu_{\lambda(i)}\right)^2 = n^{-1}\sum_{i=1}^{n}\left[\frac{y_i - \mu_{\lambda i}}{1 - s_i}\right]^2$$

and

$$\mathrm{GCV}(\lambda) = n^{-1}\sum_{i=1}^{n}\left(y_i - \mu_{\lambda(i)}\right)^2\left\{\frac{1 - s_i}{n^{-1}\mathrm{tr}[\mathbf{I}-\mathbf{S}_\lambda]}\right\}^2.$$

Thus, in this case GCV is essentially a weighted version of CV.

While the previous example illustrates that GCV and CV may be related, it also shows that CV is not a special case of GCV. Thus, the title generalized cross validation is somewhat of a misnomer. Nevertheless, we can still interpret GCV(λ) as being an attempt to account for the downward bias of $n^{-1}RSS(\lambda)$ for $P(\lambda)$ through multiplication by the factor $(n^{-1}\mathrm{tr}[\mathbf{I}-\mathbf{S}_\lambda])^{-2}$ which is necessarily greater than or equal to one. However, the best motivation for criterion (2.26) is probably provided by the so-called GCV Theorem first established by Craven and Wahba (1979). (See also Golub, Heath and Wahba 1979.)

Theorem 2.1 (The GCV Theorem). Let $\tau_j(\lambda) = n^{-1}\mathrm{tr}\left[\mathbf{S}_\lambda^j\right], j = 1, 2,$ and assume that $\tau_1(\lambda) < 1$. Then,

$$|\mathrm{EGCV}(\lambda) - P(\lambda)|/R(\lambda) \le g(\lambda), \tag{2.27}$$

where

$$g(\lambda) = \left[2\tau_1(\lambda) + \tau_1(\lambda)^2/\tau_2(\lambda)\right]/(1-\tau_1(\lambda))^2. \tag{2.28}$$

Before proving Theorem 2.1 let us pause to interpret its content. It states that when $g(\lambda)$ is small the distance between $P(\lambda)$ and EGCV(λ) will also be small relative to the intrinsic accuracy measure $R(\lambda)$. This can be roughly interpreted to mean that, in this case, GCV(λ) is nearly an unbiased estimator of $P(\lambda)$.

There are, in fact, many cases where $g(\lambda)$ will be small, particularly for larger sample sizes. For example, in the variable selection problem

$$\tau_1(\lambda) = n^{-1}\mathrm{tr}[\mathbf{S}_\lambda] = \#(\lambda)/n \le p/n$$

and, since $\mathbf{S}_\lambda$ is idempotent, $\tau_2(\lambda) = \tau_1(\lambda)$. Thus,

$$g(\lambda) = 3\tau_1(\lambda)/(1-\tau_1(\lambda))^2 \le 3\frac{p}{n}\Big/\left(1-\frac{p}{n}\right)^2 = 3\frac{p}{n} + O\left(\left(\frac{p}{n}\right)^2\right),$$

and, hence, $g(\lambda)$ will be small for $p = o(n)$. The ridge regression problem admits similar conclusions (Exercise 7).

Proof of Theorem 2.1. To establish the theorem, first observe that

$$\begin{aligned}\mathrm{EGCV}(\lambda) &= n^{-1}\mathrm{E}\mathbf{y}^T(\mathbf{I}-\mathbf{S}_\lambda)^2\mathbf{y}/(1-\tau_1(\lambda))^2 \\ &= n^{-1}\left\{\boldsymbol{\mu}^T(\mathbf{I}-\mathbf{S}_\lambda)^2\boldsymbol{\mu} + \sigma^2\mathrm{tr}\left[(\mathbf{I}-\mathbf{S}_\lambda)^2\right]\right\}/(1-\tau_1(\lambda))^2.\end{aligned}$$

Thus,

$$\begin{aligned}\mathrm{EGCV}(\lambda) - P(\lambda) &= \frac{R(\lambda)}{(1-\tau_1(\lambda))^2} + \sigma^2 \frac{1-2\tau_1(\lambda)}{(1-\tau_1(\lambda))^2} - (R(\lambda)+\sigma^2) \\ &= \frac{\tau_1(\lambda)\,(2-\tau_1(\lambda))}{(1-\tau_1(\lambda))^2} R(\lambda) - \sigma^2 \frac{\tau_1(\lambda)^2}{(1-\tau_1(\lambda))^2}.\end{aligned}$$

This implies that

$$\begin{aligned}|\mathrm{EGCV}(\lambda) \quad - \quad P(\lambda)|/R(\lambda) &= \left| \frac{\tau_1(\lambda)\,(2-\tau_1(\lambda))}{(1-\tau_1(\lambda))^2} - \frac{\sigma^2 \tau_1^2(\lambda)}{R(\lambda)\,(1-\tau_1(\lambda))^2} \right| \\ &\le \frac{\tau_1(\lambda)\,(2-\tau_1(\lambda))}{(1-\tau_1(\lambda))^2} + \frac{\tau_1(\lambda)^2/\tau_2(\lambda)}{(1-\tau_1(\lambda))^2} \le g(\lambda).\end{aligned}$$

In obtaining the last two inequalities we used the facts that $R(\lambda) \ge \sigma^2 \tau_2(\lambda)$ and $\tau_1(\lambda)(2-\tau_1(\lambda)) \le 2\tau_1(\lambda)$. •

So we now have available three criteria for selecting a good value of λ from the data: $\widehat{P}(\lambda)$, $\mathrm{CV}(\lambda)$ and $\mathrm{GCV}(\lambda)$. The strategy is to choose one (or more) of these criteria and use its minimizer $\widehat{\lambda}$ as an estimate of the minimizer of $L(\lambda)$ or $R(\lambda)$. The choice of which criterion to use will depend on the situation and will be governed by whether or not a value for σ^2 is known (or easily estimated) and the available computational resources. If σ^2 is unknown and no reasonable estimate is available, GCV or CV can be used since they do not require estimation of the error variance. As a rule the use of CV will involve more computational labor than GCV since its use entails recomputation of estimators from data subsets.

In some cases it may be feasible to compute two or even all three of our risk estimates. This is a sound practice and should be used whenever possible. By computing two or more of the measures we obtain some basis for comparison and, as a result, are less likely to be misled by an aberration associated with a particular measure. Also, when all the risk estimators are minimized by the same value of λ this should strengthen our belief that a good choice has been made for λ.

In all of our previous discussions we have treated λ as a nonrandom or deterministic variable. Thus, $R(\lambda)$ and $P(\lambda)$ are nonrandom functions of the nonrandom variable λ while $L(\lambda)$, $RSS(\lambda)$, $\widehat{R}(\lambda)$, $\widehat{P}(\lambda)$, GCV(λ) and CV(λ) are all random functions of a deterministic variable λ. However, once we begin to consider the value $\widehat{\lambda}$ which minimizes one of the criteria $\widehat{P}(\lambda)$, CV(λ) or GCV(λ) an additional level of stochasticity is added because $\widehat{\lambda}$ is data dependent and therefore random. There are cases where we may want to study $L(\widehat{\lambda})$, $P(\widehat{\lambda})$ or $R(\widehat{\lambda})$ and, hence, some confusion may arise as to whether or not the argument for the loss or risk is random or deterministic. To avoid such difficulties we will employ the convention that the symbol λ corresponds to a nonrandom variable while special notation such as $\widehat{\lambda}$ or $\widetilde{\lambda}$ signifies the random case. A good rule of thumb to follow is that *the value of the index parameter under consideration should be assumed nonrandom unless indicated otherwise.*

Example 2.4 (Variable Selection: Continued). To illustrate the use of the various criteria for estimating $P(\lambda)$ and $R(\lambda)$ described in this section we return again to the housing rent study discussed in Section 2.1. Recall that the objective was to find an optimal least-squares predictive equation involving some subset of the five predictor variables $X_1, \ldots, X_5$.

We have chosen to include the constant in all predictive equations. Under this constraint there are $2^4 = 16$ possible predictor combinations involving the variables with index subsets {1}, {1,2}, {1,3}, {1,4}, {1,5}, {1,2,3}, {1,2,4}, {1,2,5}, {1,3,4}, {1,3,5}, {1,4,5}, {1,2,3,4}, {1,2,3,5}, {1,2,4,5}, {1,3,4,5} and {1,2,3,4,5}. The collection of these subsets makes up the index set Λ. We now wish to find a variable subset which estimates the minimizer of $P(\lambda)$ over Λ.

As estimates of the minimizer of $P(\lambda)$ both the minimizers of $\widehat{P}(\lambda)$ and GCV(λ) will be considered. In the case of $\widehat{P}(\lambda)$ an estimator of σ^2 is required. For this purpose the estimator $\widehat{\sigma}^2 = (n-p)^{-1}\mathbf{y}^T(\mathbf{I} - \mathbf{X}(\mathbf{X}^T\mathbf{X})^{-1}\mathbf{X}^T)\mathbf{y}$ of σ^2 corresponding to the least-squares estimator of $\boldsymbol{\mu}$ based on all $p = 5$ predictor variables will be used. Its value is found to be $\widehat{\sigma}^2 = 711$ for this data.

Again let $\#(\lambda)$ denote the number of elements in λ. Then, due to the fact that $\mathrm{tr}\mathbf{S}_\lambda = \#(\lambda)$ for the variable selection problems, we find that our

Table 2.2: RSS, GCV and $\widehat{P}$ for the Housing Cost Study

Variable Subset	$n^{-1}RSS(\lambda)$	GCV(λ)	$\widehat{P}(\lambda)$
X_1	1966.8	1975.5	1969.9
X_1, X_2	1955.8	1973.1	1962.0
X_1, X_3	1920.8	1937.8	1927.1
X_1, X_4	1304.4	1315.9	1310.7
X_1, X_5	1278.1	1289.4	1284.3
X_1, X_2, X_3	1911.0	1936.5	1920.4
X_1, X_2, X_4	1298.2	1315.5	1307.6
X_1, X_2, X_5	1277.3	1294.3	1286.7
X_1, X_3, X_4	1259.3	1276.1	1268.7
X_1, X_3, X_5	1248.4	1265.0	1257.8
X_1, X_4, X_5	733.5	743.3	742.9
X_1, X_2, X_3, X_4	1253.9	1276.2	1266.4
X_1, X_2, X_3, X_5	1247.8	1270.0	1260.3
X_1, X_2, X_4, X_5	733.5	746.6	746.0
X_1, X_3, X_4, X_5	703.3	715.8	715.8
X_1, X_2, X_3, X_4, X_5	703.2	718.9	718.8

predictive risk estimators have the simple representations

$$\widehat{P}(\lambda) = n^{-1}RSS(\lambda) + 2\widehat{\sigma}^2\#(\lambda)/n$$

and

$$\text{GCV}(\lambda) = nRSS(\lambda)/(n - \#(\lambda))^2 .$$

As a result, these quantities can both be easily computed using the standard output of almost any linear regression software package. Table 2.2 gives a summary of the values of $\widehat{P}(\lambda)$ and GCV(λ) over all $\lambda \in \Lambda$ that was obtained by making use of this latter observation.

Examination of the values in Table 2.2 reveals that both $\widehat{P}(\lambda)$ and GCV(λ) point toward $\{X_1, X_3, X_4, X_5\}$ as the best predictor subset, i.e., both criteria chose $\widehat{\lambda} = \{1, 3, 4, 5\}$ as an estimate of the minimizer of $P(\lambda)$.

The second best choice involves all five predictors while the third best choice consists of the variables corresponding to $\widehat{\lambda}$ with the exclusion of X_3. From a practical viewpoint, there is not a substantial difference between our risk estimates for any of these three variable subsets. Thus, one would probably use X_1, X_3, X_4 and X_5 to construct a predictive equation or, if information on X_3 (length of residence) is difficult or costly to obtain, perhaps just X_1, X_4 and X_5 would be used.

Example 2.5 (Cosine Regression: Continued). In Example 2.3 we discussed fitting the data in Figure 2.1 using an estimator of the form (2.16). For this purpose we need to select a value for λ, the number of terms to be included in the estimator.

Initially, let us avoid the problem of estimating σ^2 here and choose λ using GCV. A plot of GCV(λ) over $\lambda = 1, \ldots, 31$ is shown in Figure 2.2 from which we obtain the estimated order $\widehat{\lambda} = 4$ that was used to produce the estimator of the regression function shown in Figure 2.1. Since this is a simulated data set we can actually compute $L(\lambda)$ and $R(\lambda)$ in this instance and, accordingly, the values of the loss criterion are also shown in Figure 2.2. From this we see that the loss is minimized at $\lambda = 5$ which turns out to also be the order that minimizes the risk.

For $\lambda > 31$, the GCV criterion for this data becomes increasingly large and no longer successfully tracks the loss function. We could have actually anticipated this result because g in (2.28) is an increasing function of λ in this instance and the GCV Theorem only "guarantees" that GCV(λ) will be effective in estimating $P(\lambda)$ in regions where $g(\lambda)$ is small. Note that the GCV Theorem would also lead us to expect that the difference between GCV(λ) and $L(\lambda)$ should be approximately σ^2 when $g(\lambda)$ is small which appears to be roughly the case in Figure 2.2.

It is actually not that difficult to produce good estimators of σ^2 for data such as that in this example. If one can assume that the true mean function is smooth, then we can use an estimator such as

$$\widehat{\sigma}^2 = \sum_{i=2}^{n-1} \widetilde{\varepsilon}_i^2/(n-2) \tag{2.29}$$

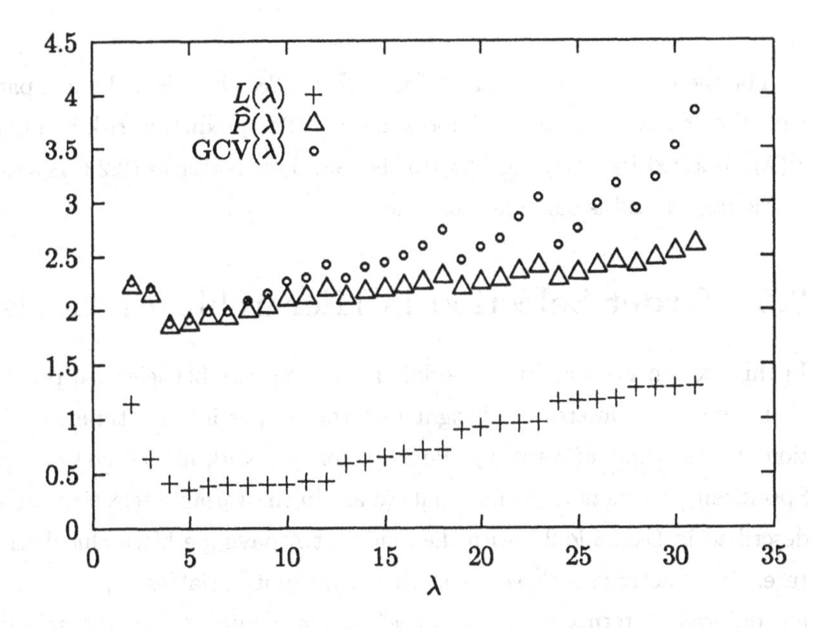

Figure 2.2: Loss, GCV and $\widehat{P}$ for Simulated Data.

with

$$\widetilde{\varepsilon}_i = (y_i - A_i y_{i-1} - B_i y_{i+1})/(1 + A_i^2 + B_i^2)^{1/2}, \; i = 2, \ldots, n-1, \qquad (2.30)$$

for $A_i = (t_{i+1} - t_i)/(t_{i+1} - t_{i-1})$ and $B_i = (t_i - t_{i-1})/(t_{i+1} - t_{i-1})$. The $\widetilde{\varepsilon}_i, i = 2, \ldots, n-1$, are called pseudo-residuals and they along with the estimator (2.29)-(2.30) derive from work in Gasser, Sroka and Jennen-Steinmetz (1986). We will refer to $\widehat{\sigma}^2$ as the GSJS estimator of σ^2 in the sequel.

The form of the pseudo-residuals and GSJS estimator can be motivated by observing that $y_i - A_i y_{i-1} - B_i y_{i+1}$ is just the residual from fitting y_i using the line that connects the points (t_{i-1}, y_{i-1}) and (t_{i+1}, y_{i+1}). Thus, if the true regression function is differentiable, the Mean Value Theorem can be invoked to see that the $\widetilde{\varepsilon}_i$ are composed primarily of contributions from the random errors in that $y_i - A_i y_{i-1} - B_i y_{i+1} = \varepsilon_i - A_i \varepsilon_{i-1} - B_i \varepsilon_{i+1} + O(n^{-1})$ if $\max|t_i - t_{i-1}| = O(n^{-1})$ and $\max|t_i - t_{i-1}|/\min|t_i - t_{i-1}|$ is bounded. From this point a little additional work establishes that the GSJS estimator is $\sqrt{n}$-consistent in that $\widehat{\sigma}^2 - \sigma^2 = O_p(n^{-1/2})$ (Exercise

20).

For the data in Figure 2.1 we found $\widehat{\sigma} = 1.28$ which is to be compared with the true value of $\sigma = 1.5$ for this case. The prediction risk estimator $\widehat{P}(\lambda)$ obtained by replacing σ^2 with the GSJS estimator in (2.22) is shown in Figure 2.1 and is also minimized at $\lambda = 4$.

2.3 Order Selection in Hierarchical Models

In this section we examine a special case of the variable selection problem that gives some interesting insight into the properties of estimator selection criteria while also setting the stage for our work in the next chapter. Specifically, let us now assume that we are in the variable selection setting described in Example 2.1 with the index set Λ having a hierarchical structure. By "hierarchical" we mean that the input variables $X_1, X_2, \ldots, X_p$ are ordered in terms of "importance" and are entered into the selection process in a sequential manner. So, one first fits the data using only the X_1 variable, then both X_1 and X_2 are used, etc. In general, the λth fitted model involves the variables $X_1, X_2, \ldots, X_\lambda$. As a result, we can think of Λ as consisting of the integers from 1 to p that correspond to each of the p possible variable subsets that are under consideration.

To simplify the presentation, the responses will be assumed to follow a canonical linear model of the form

$$y_i = \sum_{j=1}^{p} \beta_j x_{ijn} + \varepsilon_i,\ i = 1, \ldots, n, \tag{2.31}$$

where

$$\sum_{i=1}^{n} x_{ijn} x_{ikn} = n\delta_{jk},\ j, k = 1, \ldots, p, \tag{2.32}$$

and $\varepsilon_1, \ldots, \varepsilon_n$ are independent, normal random variables with mean zero and common variance σ^2. This model is essentially just the standard linear model with normal errors wherein the values of the predictor or independent variables have been orthonormalized. To further simplify matters we will also assume that the value of σ^2 is known throughout the remainder of this section.

Initially p, the total number of predictor variables under consideration, will be treated as a fixed, finite integer. The true model for the data is then presumed to contain only the first p_0 variables $X_1, \ldots, X_{p_0}$ for some integer $p_0 \leq p$. This entails that $\beta_{p_0+1} = \cdots = \beta_p = 0$ in (2.31) and that some of the coefficients $\beta_1, \ldots, \beta_{p_0}$ must be nonzero. In particular, we require that $\beta_{p_0} \neq 0$.

As in Example 2.1 let

$$\mathbf{X}_\lambda = \{x_{ijn}\}_{i=1,n,j=1,\lambda}$$

and observe that $\mathbf{X}_\lambda^T \mathbf{X}_\lambda = n\mathbf{I}$ for $\lambda = 1, \ldots, n$. Consequently, the least-squares estimator of $\boldsymbol{\mu}$ is

$$\boldsymbol{\mu}_\lambda = n^{-1}\mathbf{X}_\lambda \mathbf{X}_\lambda^T \mathbf{y}.$$

We now wish to select an estimator from the set of candidate estimators, $\{\boldsymbol{\mu}_\lambda : \lambda \in \Lambda\}$ for $\Lambda = \{1, \ldots, p\}$, using the estimated prediction risk

$$\widehat{P}(\lambda) = n^{-1} RSS(\lambda) + 2n^{-1}\sigma^2 \lambda. \tag{2.33}$$

The minimizer $\widehat{\lambda}$ of $\widehat{P}(\lambda)$ over $\lambda \in \Lambda$ will then be used as an estimator of the true model dimension p_0.

For this variable selection scenario the large sample distribution of $\widehat{\lambda}$ can be derived explicitly and is detailed in the following theorem.

Theorem 2.2. Let $u_0 = v_0 = 1$ and for $k \geq 1$ define

$$u_k = \sum_k^* \left(\prod_{i=1}^k \frac{1}{j_i!} \left(\frac{P(\chi_i^2 > 2i)}{i} \right)^{j_i} \right) \tag{2.34}$$

and

$$v_k = \sum_k^* \left(\prod_{i=1}^k \frac{1}{j_i!} \left(\frac{P(\chi_i^2 \leq 2i)}{i} \right)^{j_i} \right), \tag{2.35}$$

where χ_k^2 is a central chi-squared random variable with k degrees-of-freedom and the $\sum^*$ notation in (2.34)-(2.35) indicates summation over all k-tuples of integers $(j_1, \ldots, j_k)$ such that $j_1 + 2j_2 + \cdots + kj_k = k$. Then,

$$\lim_{n\to\infty} P(\widehat{\lambda} = \lambda) = 0, \ \lambda < p_0, \tag{2.36}$$

and

$$\lim_{n\to\infty} P(\widehat{\lambda} = \lambda) = u_{\lambda-p_0} v_{p-\lambda},\ \lambda = p_0, \ldots, p. \tag{2.37}$$

According to this theorem, our prediction risk estimator will have an asymptotically negligible chance of selecting anything less than the correct order p_0. However, there is a nonzero asymptotic chance of overfitting, i.e., of choosing a value for λ that exceeds p_0, since the limiting distribution of $\widehat{\lambda}$ is supported on the integers from p_0 all the way up to p. Note that overfitting in this case is something to be avoided because the risk $R(\lambda)$ for model (2.31) will necessarily be minimized by some value of $\lambda \leq p_0$ (Exercise 22).

Some insight into the quality of $\widehat{\lambda}$ as an estimator of p_0 can be obtained by examination of the probability of overfitting $1 - v_{p-p_0}$. Table 2.2 lists values of these probabilities for $p - p_0 = 1, \ldots, 5$. From this we see that the chances of overfitting are fairly substantial ranging up to around 29%. As might be expected, the chance of overfitting increases as the number of extraneous variables, $p - p_0$, increases. The last entry in Table 2.3 is $1 - \lim_{k\to\infty} v_k = .29$. This value can be roughly interpreted as meaning that in large samples we will have about a 29% chance of overfitting when there are very many irrelevant predictors that are being considered for inclusion in the fitted model. Zhang (1992) has also computed the expected value of $\widehat{\lambda}$ in this case and found that

$$\lim_{p-p_0\to\infty} \lim_{n\to\infty} \mathrm{E}\,\widehat{\lambda} = p_0 + .9422.$$

So, even though $\widehat{\lambda}$ has a reasonable chance of exceeding p_0, we would expect to have only about one (i.e., .9422) unnecessary predictor included in the final estimator of $\boldsymbol{\mu}$.

Proof of Theorem 2.2. To prove Theorem 2.2 we will study the limiting properties of $P(\widehat{\lambda} = \lambda)$ in two cases: namely, when $\lambda < p_0$ and when $\lambda \geq p_0$. Let us first dispense with the $\lambda < p_0$ situation.

Set

$$b_j = n^{-1} \sum_{i=1}^{n} x_{ijn} y_i$$

Table 2.3: Probability of Overfitting

$p - p_0$	$1 - v_{p-p_0}$
1	.157
2	.213
3	.239
4	.254
5	.265
∞	.290

and observe that since $n\widehat{P}(\lambda) = \sum_{i=1}^{n} y_i^2 - n\sum_{j=1}^{\lambda} b_j^2 + 2\sigma^2\lambda$, we have

$$\begin{aligned} P(\widehat{\lambda} = \lambda) &= P(\widehat{P}(\lambda) \leq \widehat{P}(k), k = 1, \ldots, p) \\ &= P(\sum_{j=k+1}^{\lambda} nb_j^2 \geq 2\sigma^2(\lambda - k), k = 1, \ldots, \lambda - 1, \\ &\qquad \sum_{j=\lambda+1}^{k} nb_j^2 \leq 2\sigma^2(k - \lambda), k = \lambda + 1, \ldots, p). \end{aligned}$$

Now use the fact that, for any two sets A and B, $P(A \cap B) \leq P(A)$, to see that

$$P(\widehat{\lambda} = \lambda) \leq P(\sum_{j=\lambda+1}^{p_0} nb_j^2 \leq 2\sigma^2(p_0 - \lambda)) \leq P(nb_{p_0}^2 \leq 2\sigma^2 p_0).$$

Under the assumptions of model (2.31), the vector $\sqrt{n}\,(b_1, \ldots, b_p)^T$ has a normal distribution with mean $\sqrt{n}\,(\beta_1, \ldots, \beta_{p_0}, 0, \ldots, 0)^T$ and variance-covariance matrix $\sigma^2\mathbf{I}$. Thus, since $\beta_{p_0} \neq 0$,

$$P(nb_{p_0}^2 \leq 2\sigma^2 p_0) = P(-\sqrt{2p_0} - \frac{\sqrt{n}\beta_{p_0}}{\sigma} \leq Z \leq \sqrt{2p_0} - \frac{\sqrt{n}\beta_{p_0}}{\sigma}),$$

with Z having a standard normal distribution.

Suppose for specificity that $\beta_{p_0} > 0$, since the other case of $\beta_{p_0} < 0$ will follow similarly. Then,

$$P(nb_{p_0}^2 \leq 2\sigma^2 p_0) \leq P(Z \leq \sqrt{2p_0} - \frac{\sqrt{n}\beta_{p_0}}{\sigma}). \qquad (2.38)$$

A useful approximation to the tail area for a standard normal distribution is given in Feller (1968, pg. 175) which states that

$$[x^{-1} - x^{-3}]e^{-x^2/2}/\sqrt{2\pi} \leq P(Z > x) \leq x^{-1}e^{-x^2/2}/\sqrt{2\pi} \qquad (2.39)$$

for any $x > 0$. For n sufficiently large this result is applicable to (2.38) and allows us to conclude that $P(\widehat{\lambda} = \lambda) = O(n^{-1/2}e^{-n\beta_{p_0}^2/2\sigma^2})$. Note that we have actually proved much more than was stated in the theorem. We have shown that not only does $P(\widehat{\lambda} = \lambda)$ decay to zero for $\lambda < p_0$ but that it actually does so exponentially fast.

We now examine the case of $\lambda \geq p_0$. In this instance we make use of the following lemma whose proof is left as an exercise (Exercise 23).

Lemma 2.1. *If $\{A_n\}$ and $\{B_n\}$ are sequences of sets with $P(B_n) \to 1$ as $n \to \infty$, then $P(A_n) - P(A_n \cap B_n) \to 0$ as $n \to \infty$.*

To apply the lemma we take $A_n = \{\widehat{P}(\lambda) \leq \widehat{P}(k), k = p_0, \ldots, p\}$ and $B_n = \{\widehat{P}(\lambda) \leq \widehat{P}(k), k = 1, \ldots, p_0 - 1\}$. Then $P(B_n) \to 1$ (Exercise 24) and, hence, $\lim_{n\to\infty} P(\widehat{\lambda} = \lambda) = \lim_{n\to\infty} P(A_n)$. However,

$$\begin{aligned} P(A_n) &= P(\sum_{j=k+1}^{\lambda} nb_j^2 \geq 2\sigma^2(\lambda - k), k = p_0, \ldots, \lambda - 1, \\ &\qquad \sum_{j=\lambda+1}^{k} nb_j^2 \leq 2\sigma^2(k - \lambda), k = \lambda + 1, \ldots, p) \\ &= P(\sum_{j=k+1}^{\lambda} nb_j^2 \geq 2\sigma^2(\lambda - k), k = p_0, \ldots, \lambda - 1) \\ &\qquad \times P(\sum_{j=\lambda+1}^{k} nb_j^2 \leq 2\sigma^2(k - \lambda), k = \lambda + 1, \ldots, p). \end{aligned}$$

Since the nb_j^2/σ^2 for $j = p_0 + 1, \ldots, p$ are independent, central, chi-squared random variables with one degree-of-freedom, we can express $P(A_n)$ as

$$P(S_j \geq 0, j = 1, \ldots, \lambda - p_0)P(S_j \leq 0, j = 1, \ldots, p - \lambda),$$

where $S_j = \sum_{i=1}^{j}(\chi_{1i}^2 - 2)$ and χ_{1i}^2 are independent, central chi-squared random variables with one degree-of-freedom. It now follows from results in

Spitzer (1956, pg. 329-330) and Shibata (1976) that this latter probability is precisely (2.37). •

The basic conclusions about $\widehat{\lambda}$ in Theorem 2.2 apply to estimated orders selected using other methods. For example, the next theorem states that the orders selected using the GCV criterion have the same large sample distribution as those obtained using the estimated predictive risk.

Theorem 2.3. *If* $\widehat{\lambda}_{\text{GCV}}$ *is the minimizer of* GCV(λ) *over* $\lambda = 1, \ldots, p$, *then* $\lim_{n\to\infty} P(\widehat{\lambda}_{\text{GCV}} = \lambda)$ *is also given by (2.36)-(2.37).*

Proof. Exercise 25.

The question that arises now is whether it is possible to devise a criterion that does not overfit, i.e., one for which $\widehat{\lambda} \xrightarrow{P} p_0$. A possible solution to the overfitting problem is to alter $\widehat{P}(\lambda)$ to where adding additional variables is more heavily penalized. This suggests looking at criteria such as $\widetilde{P}(\lambda) = n^{-1}RSS(\lambda) + n^{-1}a\lambda\sigma^2$, for some value $a > 2$. Criteria of this type have been examined by Zhang (1992) who recommends choosing a to lie somewhere between 1.5 and 5. Unfortunately, these altered criteria will still have nonzero chances for overselection even in large samples if a remains bounded.

Since using the $\widetilde{P}$ criterion with bounded a still leads to overselection, the next step is to allow $a \to \infty$ with n. To keep the penalty from becoming too severe we could choose $a = \ln(n)$ and then select λ by minimizing

$$\widetilde{P}(\lambda) = n^{-1}RSS(\lambda) + n^{-1}\ln(n)\lambda\sigma^2 \tag{2.40}$$

over all $\lambda \in \Lambda$. This particular criterion was proposed by Schwarz (1978) and Akaike (1978) and is generally referred to as either the Schwarz or BIC criterion. The next theorem indicates that (2.40) does in fact produce a consistent order selection for model (2.31).

Theorem 2.4. *If* $\widetilde{\lambda}$ *is the minimizer of* $\widetilde{P}(\lambda)$ *in (2.40) over* $\lambda = 1, \ldots, p$, *then* $\widetilde{\lambda} \xrightarrow{P} p_0$.

Proof. For $\lambda < p_0$,

$$P(\widetilde{\lambda} = \lambda) \le P(nb_{p_0}^2 \le \ln(n)(p_0 - \lambda)\sigma^2),$$

which is seen to converge to zero as before using (2.39). For $\lambda > p_0$,

$$P(\tilde{\lambda} = \lambda) \leq P(\sum_{j=p_0+1}^{p} nb_j^2 \geq \ln(n)\sigma^2) = P(\chi^2_{p-p_0} \geq \ln(n)\sigma^2),$$

where $\chi^2_{p-p_0}$ is a central chi-squared random variable with $p - p_0$ degrees-of-freedom. This bound converges to zero if $p - p_0$ is bounded. •

Example 2.6. To examine the extent that results from Theorems 2.2-2.4 can be realized in finite samples, a small simulation experiment was conducted with data generated from the model

$$y_i = \cos(3\pi t_i) + \varepsilon_i,\ i = 1, \ldots, 50,$$

for $t_i = (2i - 1)/100$ and zero mean, normal ε's with $\sigma = .5$. The data were fitted using a cosine regression estimator

$$\mu_\lambda(t) = b_1 + \sum_{j=2}^{\lambda} b_j \sqrt{2} \cos((j-1)\pi t)$$

for the b_j defined in (2.14)-(2.15) which places us in the scenario of model (2.31) with $\beta_1 = \beta_2 = \beta_3 = 0, \beta_4 = 1, \beta_j = 0, j \geq 5$ and $p_0 = 4$. This basic experiment was replicated 1000 times to give 1000 data sets each of size $n = 50$. For every one of these data sets the minimizers of the loss, $\widehat{P}$, GCV and Schwarz criteria over $\lambda = 1, \ldots, n$ were computed.

The loss was minimized at $\lambda = 4$ for every one of our 1000 data sets. The empirical relative frequency distribution for the orders selected by $\widehat{P}$ and the Schwarz criteria are shown in Figures 2.3 and 2.4, respectively. Note that neither criterion underselected (i.e., chose $\lambda < 4$) and that the orders selected by the Schwarz criterion are quite concentrated around the correct order $\lambda = 4$. In contrast, the orders selected by $\widehat{P}$ are much more variable and the chance of overselection is fairly substantial thereby reflecting what was anticipated from Theorems 2.2 and 2.4.

The simulation results for GCV were similar to those for $\widehat{P}$ except that in about 14% of the cases GCV produced extreme overselection with $\lambda = 49$ and 50. Technically, our use of $p = n$ violates the assumptions required for

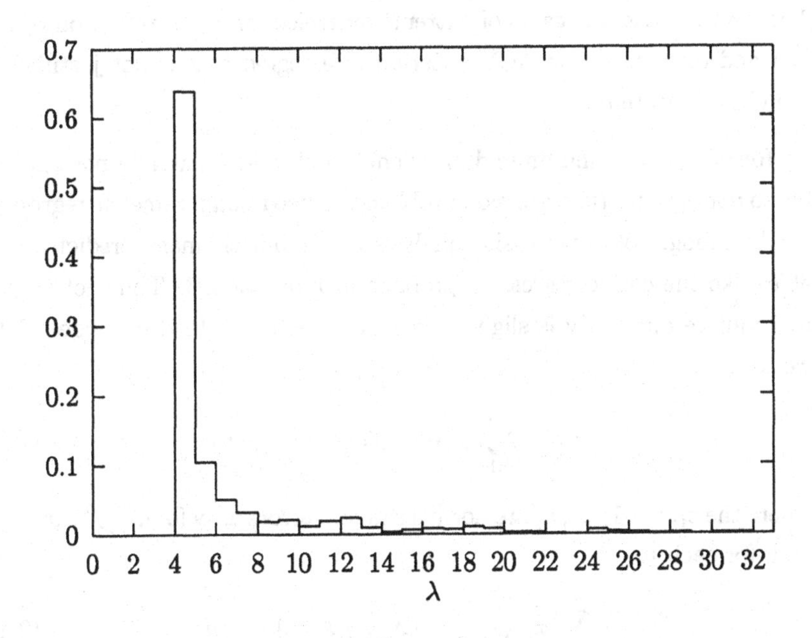

Figure 2.3: Empirical Selection Probabilities for $\widehat{P}$.

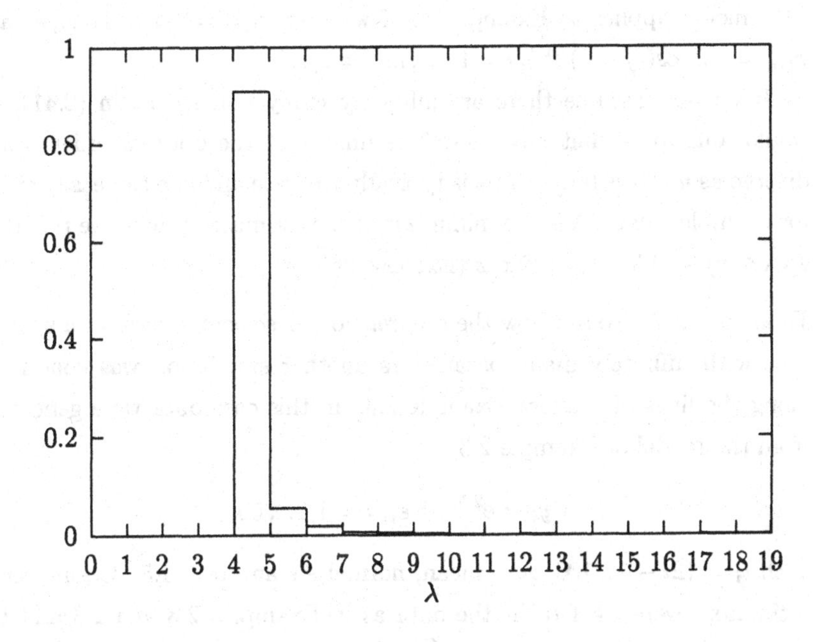

Figure 2.4: Empirical Selection Probabilities for the Schwarz Criterion.

Theorem 2.3 and the cases of extreme overselection are a reflection of this fact and our use of the GCV criterion in a region that is not justified by the GCV Theorem.

Results concerning finite dimensional models are generally not applicable to nonparametric regression problems. Indeed nonparametric regression can be thought of as regression analysis with infinitely many predictor variables like the cosine regression problem in Example 2.3. Thus, let us now reformulate our analysis slightly and assume that instead of model (2.31) we have

$$y_i = \sum_{j=1}^{\infty} \beta_j x_{ijn} + \varepsilon_i,\ i = 1, \ldots, n, \tag{2.41}$$

where the $\varepsilon_i, i = 1, \ldots, n$, are normal random errors as before, $\sum_{j=1}^{\infty} \beta_j^2 < \infty$ and, for each n

$$\sum_{i=1}^{n} x_{ijn} x_{ikn} = n\delta_{jk},\ j, k = 1, \ldots, n. \tag{2.42}$$

This model applies to Example 2.3 if we take $x_{i1n} = 1, i = 1, \ldots, n$, and $x_{ijn} = \sqrt{2}\cos((j-1)\pi t_i), i = 1, \ldots, n, j = 2, \ldots$

If we now assume there are infinitely many nonzero β_j in (2.41), we would anticipate that any "good" estimator of the optimal order would diverge as n grows large. This is typically the case and one can easily show, for example, that if $\widehat{\lambda}$ is the minimizer of the estimated predictive risk then $P(\widehat{\lambda} \le x) \to 0$ for any finite x (Exercise 26).

Example 2.7. To see how the different order selectors work in a specific case with infinitely many parameters another simulation was conducted along the lines of that for Example 2.6. In this case data were generated from the model of Example 2.3

$$y_i = e^{2.5t_i} + \varepsilon_i,\ i = 1, \ldots, 50,$$

with $t_i = (2i-1)/100$, zero mean, normal ε's and $\sigma = 1.5$. Cosine series estimators were used to fit the data as in Examples 2.3 and 2.6 and the orders that minimized the loss, $\widehat{P}$, GCV and Schwarz criteria for the estimators were computed. This experiment was then repeated 1000 times and the corresponding empirical order selection probabilities for the loss, $\widehat{P}$ and the Schwarz criterion are shown in Figures 2.5-2.7, respectively.

In contrast to what transpired for Example 2.6 we see from Figure 2.5 that in this case the minimizer of the loss has an actual distribution that includes chances for over or underselection relative to the value of $\lambda = 5$ that minimizes the risk. The same is true for the minimizers of $\widehat{P}$ and the Schwarz criteria. Although the orders selected by the loss and Schwarz criteria tend to be less variable than those from $\widehat{P}$, the distributions of the orders from $\widehat{P}$ and the loss appear more similar in terms of their centering and chance of selecting orders less than 5. The Schwarz criterion chose orders less than 5 in 72.5% of the cases compared to around 32% and 38.2% for the loss and $\widehat{P}$, respectively. Note that for this example $\ln(n) = \ln(50) = 3.9$. Thus, the Schwarz criterion places almost twice as much weight in the penalty term in (2.40) than does $\widehat{P}$. This makes it much harder for the criterion to choose a large value of λ which reduces variability of the selected orders but enhances possibilities for underselection relative to the minimizer of the risk.

To assess the performance of an order selector under model (2.41) we will need to use a different strategy than we did for the finite dimensional case. One way to measure the quality of an estimated order in this infinite dimensional case is to see how close the performance of its associated estimator of $\boldsymbol{\mu}$ is to that for the estimator using the optimal order that minimizes the risk. As a specific example, let $\widehat{\lambda}$ and λ^* be the minimizers of $\widehat{P}(\lambda)$ and $R(\lambda)$, respectively, over the index set $\Lambda_n = \{1, \ldots, [n^\delta]\}$, with $[\cdot]$ being the greatest integer function and $\delta < 1$ is some constant that governs the rate of growth for the maximum number of predictors that can be used to fit the data. One can then argue as in Shibata (1981) and Hurvich and Tsai (1995) that

$$\frac{R(\widehat{\lambda})}{R(\lambda^*)} - 1 = o_p(n^{(\delta-1)/2}). \tag{2.43}$$

This result can also be shown to apply to orders selected using the GCV criterion provided that $n^{2\delta-1} \to 0$.

The most immediate implication of (2.43) is that both the GCV and

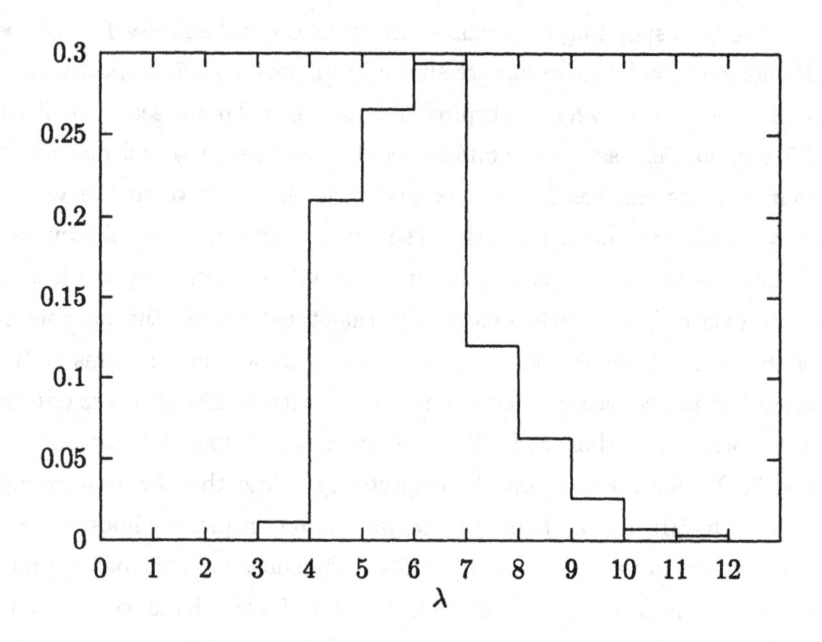

Figure 2.5: Empirical Selection Probabilities for the Loss.

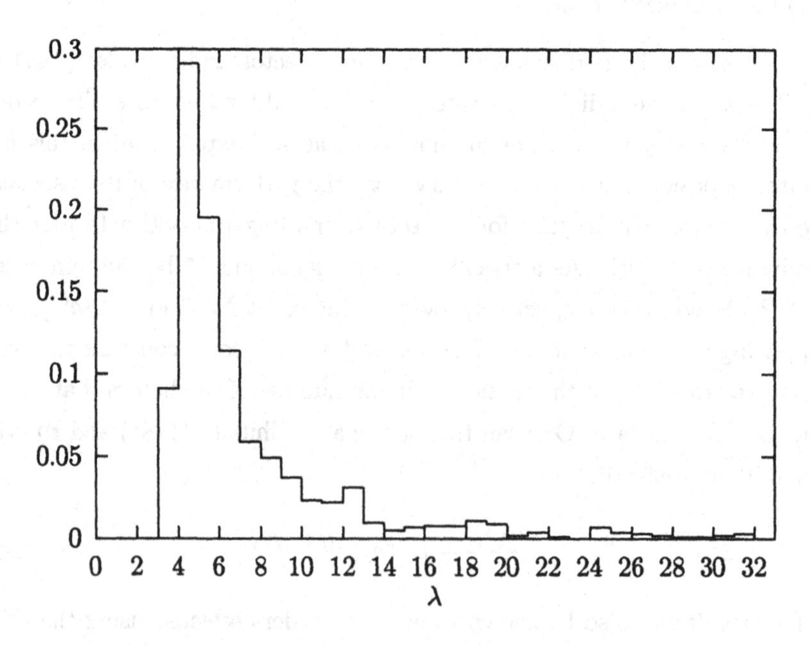

Figure 2.6: Empirical Selection Probabilities for $\widehat{P}$.

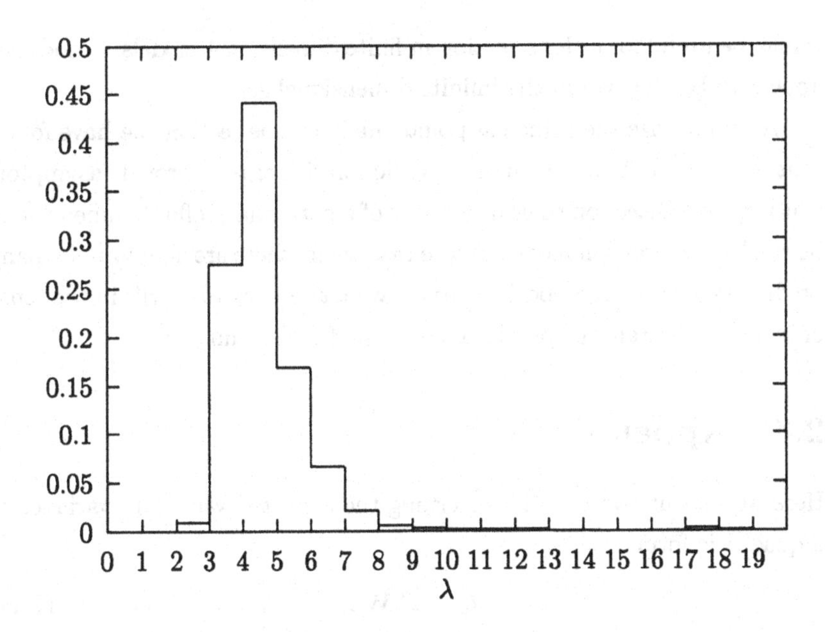

Figure 2.7: Empirical Selection Probabilities for the Schwarz Criterion.

estimated predictive risk criteria provide asymptotically optimal selection rules in the sense that they both lead to the use of estimators of μ which perform nearly as well (from a risk perspective) as the best estimator for large samples. Equation (2.43) also details a rate of convergence of the ratio $R(\widehat{\lambda})/R(\lambda^*)$ to 1. The rate becomes much faster for small values of δ and approaches $n^{-1/2}$. This makes intuitive sense because if δ is small then there are fewer candidate predictors being considered in which case $\widehat{\lambda}$ should be less variable and, accordingly, give a better approximation to λ^*. Unfortunately, if δ is chosen to be too small it can exclude important orders from consideration such as the global minimizer of $R(\lambda)$ over $\lambda \in \{1, \ldots, n\}$. Thus, some practical compromises are necessary to produce a useful and efficient choice for δ.

Finally, we ask what happens if we use an alternative criterion such as $\widetilde{P}$ in (2.40) to select the order of an estimator under model (2.41). It follows from work by Shibata (1981) that this approach does not produce orders that satisfy (2.43). Thus, although a penalty term that grows with

n may be useful for order selection in finite dimensional models, it does not appear to be effective in the infinite dimensional setting.

To summarize the principal points made in this section, we have found that both the GCV and estimated prediction risk criteria provide asymptotically optimal selection rules in the case of a particular, infinite dimensional, heirarchical regression model. In the case where there are only finitely many predictors in the model, both criteria have a tendency to overfit in the sense of including extraneous predictors into the fitted estimator.

2.4 Appendix

Here we set out two results concerning the expected value and variance of a quadratic form

$$Q = \boldsymbol{\varepsilon}^T \mathbf{W} \boldsymbol{\varepsilon}, \tag{2.44}$$

where $\mathbf{W} = \{w_{ij}\}$ is a symmetric $n \times n$ matrix and $\boldsymbol{\varepsilon} = (\varepsilon_1, \ldots, \varepsilon_n)^T$ is a vector of random variables. We first state a result concerning the mean of Q whose proof is left as an exercise (Exercise 32).

Lemma 2.2. *If* $\mathrm{E}\,\boldsymbol{\varepsilon} = \mathbf{0}$ *and* $\mathrm{E}\,\boldsymbol{\varepsilon}\boldsymbol{\varepsilon}^T = \boldsymbol{\Sigma}$, *then* $\mathrm{E}\,Q = \mathrm{tr}[\boldsymbol{\Sigma}\mathbf{W}]$.

It will also be of interest in some cases to be able to calculate the variance of Q. In this regard we have the following result.

Lemma 2.3. *Assume that* $\varepsilon_1, \ldots, \varepsilon_n$ *are iid with* $\mathrm{E}\,\varepsilon_1 = \mathrm{E}\,\varepsilon_1^3 = 0$, $\mathrm{E}\,\varepsilon_1^2 = \sigma^2$ *and* $\mathrm{E}\,\varepsilon_1^4 < \infty$. *Then,*

$$\mathrm{Var}\,(Q) = (\mathrm{E}\,\varepsilon_1^4 - 3\sigma^4) \sum_{i=1}^{n} w_{ii}^2 + 2\sigma^4 \mathrm{tr}[\mathbf{W}^2]. \tag{2.45}$$

In particular, if ε_1 *is normally distributed,* $\mathrm{Var}\,(Q) = 2\sigma^4 \mathrm{tr}[\mathbf{W}^2]$.

Proof. Equation (2.45) follows upon using the independence of the ε's and $\mathrm{E}\,\varepsilon_1 = \mathrm{E}\,\varepsilon_1^3 = 0$, to obtain

$$\begin{aligned} \mathrm{E}\,Q^2 &= \sum_i \sum_j \sum_k \sum_l w_{ij} w_{kl} \mathrm{E}\,\varepsilon_i \varepsilon_j \varepsilon_k \varepsilon_l \\ &= \sum_{i=1}^{n} w_{ii}^2 (\mathrm{E}\,\varepsilon_i^4 - 3\sigma^4) + 2\sigma^4 \sum_{i=1}^{n} \sum_{j=1}^{n} w_{ij}^2 + \sigma^4 (\sum_{i=1}^{n} w_{ii})^2. \end{aligned}$$

In the special case of normality, we find that $\mathrm{E}\,\varepsilon_1^4 = 3\sigma^4$ which produces the claimed simplification.

2.5 Exercises

1. Verify identity (2.5).

2. Establish that the ridge estimator of $\boldsymbol{\beta}$ in (2.10) is the minimizer with respect to $\mathbf{c}$ of the penalized least-squares criterion

$$\sum_{i=1}^{n} \left(y_i - \mathbf{x}_i^T \mathbf{c}\right)^2 + \lambda \mathbf{c}^T \mathbf{c}.$$

   Discuss the meaning of this criterion paying particular attention to what happens when λ is increased or decreased.

3. Consider the case where $\mathbf{y}$, conditional on $\boldsymbol{\beta}$, has an n-variate normal distribution with mean $\mathbf{X}\boldsymbol{\beta}$ and variance-covariance matrix $\sigma^2\mathbf{I}$. Let $\boldsymbol{\beta}$ be p-variate normal with mean $\mathbf{0}$ and variance-covariance matrix $\sigma_s^2\mathbf{I}$. Show that the ridge estimator $\mathbf{b}_\lambda$ with $\lambda = \sigma^2/\sigma_s^2$ is the posterior mean of $\boldsymbol{\beta}$, i.e., the mean of $\boldsymbol{\beta}$ given $\mathbf{y}$.

4. Establish the existence of biased estimators which have smaller risks than unbiased ones. [Hint: Consider the ridge regression setting and proceed as follows:

   (a) Let the singular value decomposition of $\mathbf{X}$ be $\mathbf{X} = \mathbf{U}\mathbf{D}\mathbf{V}^T$, where $\mathbf{U}$ is an $n \times n$ orthonormal matrix, $\mathbf{V}$ is a $p \times p$ orthonormal matrix and $\mathbf{D}$ is an $n \times p$ diagonal matrix whose diagonal entries are the square roots of the eigenvalues of $\mathbf{X}^T\mathbf{X}$. Using this form for $\mathbf{X}$ obtain a simple expression for $R(\lambda)$ as a sum of two functions which are, respectively, monotone increasing and decreasing in λ.

   (b) Differentiate your expression for $R(\lambda)$ and show that there is a λ_0 such that $dR(\lambda)/d\lambda < 0$ whenever $\lambda < \lambda_0$.

   (c) Conclude from (b) (Why?) that there is a λ such that $R(\lambda) < R(0)$, where $R(0)$ is the risk for the unbiased estimator $\widehat{\boldsymbol{\mu}}$.]

5. Establish a parallel of identity (2.24) for ridge regression. Derive (2.24) as a special case of your formula and use it to establish a relationship between CV and GCV in the ridge regression setting. [Hint: First consider the minimizer of

$$\sum_{k=1,k\neq i}^{n} \left(y_k - \mathbf{x}_k^T\mathbf{c}\right)^2 + \left(z - \mathbf{x}_i^T\mathbf{c}\right)^2 + \lambda\mathbf{c}^T\mathbf{c}.$$

Denoting the minimizer by $\mathbf{b}_\lambda(z)$ show that the estimators $\mathbf{b}_{\lambda(i)}$ obtained by deleting the point $(\mathbf{x}_i^T, y_i)$ is $\mathbf{b}_\lambda(\mu_{\lambda(i)})$, where $\mu_{\lambda(i)} = \mathbf{x}_i^T\mathbf{b}_{\lambda(i)}$. Now use the fact that $\mathbf{x}_i^T\mathbf{b}_\lambda(z)$ is linear in z to see that $\mu_{\lambda(i)} = \mu_{\lambda i} + s_i(\mu_{\lambda(i)} - y_i)$.]

6. For the ridge regression problem, as in Exercise 4 let $\mathbf{X} = \mathbf{U}\mathbf{D}\mathbf{V}^T$ and let $\mathbf{W}$ be the $n \times n$ unitary matrix which diagonalizes circulant matrices (e.g., Fuller 1976, Chapter 4). The typical element for $\mathbf{W}$ is $n^{-1/2}e^{2\pi ijk/n}$, $j, k = 0, \ldots, n-1$ (note: $e^{ix} = \cos x + i \sin x$ with i the complex unit satisfying $i^2 = -1$). Now transform from $\mathbf{y}$ to $\widetilde{\mathbf{y}} = \mathbf{W}\mathbf{U}^T\mathbf{y}$ and note that $\widetilde{\mathbf{y}} = \widetilde{\boldsymbol{\mu}} + \widetilde{\boldsymbol{\varepsilon}}$, where $\widetilde{\boldsymbol{\mu}} = \mathbf{W}\mathbf{D}\mathbf{V}^T\boldsymbol{\beta}$ and $\widetilde{\boldsymbol{\varepsilon}} = \mathbf{W}\mathbf{U}^T\boldsymbol{\varepsilon}$.

   (a) Derive the ridge estimator of $\widetilde{\boldsymbol{\mu}}$ and show that it has the form $\widetilde{\mathbf{S}}_\lambda\widetilde{\mathbf{y}}$ with $\widetilde{\mathbf{S}}_\lambda$ a circulant matrix. Comment on the implications of this fact.

   (b) Show that GCV is just CV in this transformed coordinate system.

   [Golub, Heath and Wahba 1979]

7. Show that $|\mathrm{EGCV}(\lambda) - P(\lambda)|/R(\lambda) \leq 3p/n + O((p/n)^2)$ for ridge regression.

8. Refer to Exercise 3. Argue that under the Bayesian formulation any estimator of λ should be invariant under an orthogonal transformation of $\mathbf{y}$ and/or $\boldsymbol{\beta}$. Show that $\mathrm{GCV}(\lambda)$ has this property but $\mathrm{CV}(\lambda)$ does not.

9. Refer to Exercise 3. Let E and E_{β} denote expectations with respect to the distributions of $\boldsymbol{\varepsilon}$ and $\boldsymbol{\beta}$, respectively. Show that $E_{\beta} E\, L(\lambda)$ and $E_{\beta} E \mathrm{GCV}(\lambda)$ are both minimized at $\lambda = \sigma^2/\sigma_s^2$.

   [Golub, Heath and Wahba 1979]

10. Suppose the estimators in $C(\Lambda)$ have the form $\boldsymbol{\mu}_\lambda = \mathbf{S}_\lambda \mathbf{y}$ but $\mathbf{S}_\lambda$ is not symmetric. Derive analogs of (2.20)–(2.23) for this case.

11. Consider the variable selection problem described in Example 2.4. In this context there are a variety of other criteria which have appeared in the literature for data driven estimator selection. Some examples include the Akaike information criterion

$$\mathrm{AIC}(\lambda) = n \ln(n^{-1} RSS(\lambda)) + 2\#(\lambda)$$

as well as

$$\mathrm{FPE}(\lambda) = (n + \#(\lambda)) n^{-1} RSS(\lambda)/(n - \#(\lambda)),$$

$$S(\lambda) = (n + 2\#(\lambda)) n^{-1} RSS(\lambda),$$

and

$$U(\lambda) = n^{-1} RSS(\lambda) \frac{n}{n - \#(\lambda)} (1 + \#(\lambda)/(n - 1 - \#(\lambda))).$$

Both AIC and FPE derive from the work of Akaike (1970, 1973, 1974), the U criterion was proposed by Hocking (1976) and Thompson (1978) and the S criterion is due to Shibata (1981). Compute the AIC, FPE, S and U criteria for the housing rent data of Example 2.4 and compare your findings with the results obtained using the predictive risk and GCV criteria.

12. Show that in the variable selection setting of Exercise 11 the FPE, AIC, S and U criteria are all closely related to GCV in the sense that for any fixed λ all four criteria can be expressed as $\mathrm{GCV}(\lambda)(1+\delta_n(\lambda))$ for individual factors $\delta_n(\lambda)$ that are small for large n and small p.

13. Twelve individuals are randomly divided into three groups of four subjects each. The first group is allowed to follow their normal sleep patterns while groups 2 and 3 are deprived of sleep for twenty-four and forty-eight hours, respectively. The experimental subjects are then tested concerning their ability to do numerical calculations. The data in Table 2.4 is the percentage of correct answers for the individuals on a test measuring arithmetic skills given at the end of the experimental period for each group.

Table 2.4: Arithmetic Skill Data

Group 1:	Group 2:	Group 3:
85	81	60
83	68	48
76	90	38
64	74	47

Let μ_1, μ_2 and μ_3 denote the population means corresponding to the three experimental groups. Compute the GCV criterion and the four criteria of Exercise 11 for the four models $\mu_1 \neq \mu_2 \neq \mu_3$, $\mu_1 = \mu_2 \neq \mu_3$, $\mu_1 \neq \mu_2 = \mu_3$ and $\mu_1 = \mu_2 = \mu_3$. According to your criteria, which model seems best? Which is second best, etc.? Discuss the similarities and differences concerning model preference indicated by the five criterion functions.

14. Let

$$y_{ki} = \mu_i + \varepsilon_{ki},\ k = 1, 2,\ i = 1, \ldots, n,$$

with the ε_{kj} being zero mean, uncorrelated random variables having common variance σ^2. Set $\mathbf{y}_k = (y_{k1}, \ldots, y_{kn})^T$, $k = 1, 2$, and define $\boldsymbol{\mu}_\lambda = \mathbf{S}_\lambda \mathbf{y}_1 = (\mu_{\lambda 1}, \ldots, \mu_{\lambda n})^T$ and $\widetilde{P}(\lambda) = n^{-1} \sum_{i=1}^{n} (y_{2i} - \mu_{\lambda i})^2$. Show that $\mathrm{E}\widetilde{P}(\lambda) = P(\lambda)$.

15. Give an example of a situation where $\mathrm{CV}(\lambda)$ is biased for $P(\lambda)$. [Hint: Try ridge regression or variable selection.]

16. Consider the case of ridge regression and, for simplicity, assume that $\mathbf{X}^T\mathbf{X} = \mathbf{I}$. Show that $R(\lambda)$ is minimized in this case by taking $\lambda = \sigma^2 p/\boldsymbol{\beta}^T\boldsymbol{\beta}$. Verify that for this choice of λ we have $ERSS(\lambda) = \text{tr}(I - \mathbf{S}_\lambda)$ and that $R(\lambda) = \sigma^2\text{tr}\mathbf{S}_\lambda/n$.

    (a) Comment on the implications of this identity concerning the use of $n^{-1}RSS(\lambda)$ for the selection of an optimal λ. In particular, does it suggest yet another data driven method for choosing the ridge parameter?

    (b) Let s_i denote the ith diagonal element of $\mathbf{S}_\lambda$. Suppose we define a confidence interval for $\mu_i = \mathbf{x}_i^T\boldsymbol{\beta}$ to be $\mathbf{x}_i^T\mathbf{b}_\lambda \pm z_{\alpha/2}[\sigma^2 s_i]^{1/2}$, where $z_{\alpha/2}$ is the $100(1-\alpha/2)$th percentage point of the standard normal distribution. In view of your identity for $R(\lambda)$ how might such an interval be expected to perform.

17. Using Exercise 10 show that $\text{CV}(\lambda)$ is an unbiased estimator of the prediction risk for a class of nil-trace estimators of $\boldsymbol{\mu}$. Work out the form of these estimators explicitly for the variable selection problem.

18. Define a class of estimators $\widetilde{C}(\Lambda)$ obtained from $C(\Lambda)$ by

$$\widetilde{\boldsymbol{\mu}}_\lambda = (\widetilde{\mu}_{\lambda 1}, \ldots, \widetilde{\mu}_{\lambda n})^T = \widetilde{\mathbf{S}}_\lambda \mathbf{y},\ \lambda \in \Lambda,$$

where

$$\widetilde{\mathbf{S}}_\lambda = (1 + \alpha(\lambda))\mathbf{S}_\lambda - \alpha(\lambda)\mathbf{I}$$

with

$$\alpha(\lambda) = \text{tr}[\mathbf{S}_\lambda]/\text{tr}[\mathbf{I} - \mathbf{S}_\lambda],$$

and $\mathbf{S}_\lambda$ corresponds to $\boldsymbol{\mu}_\lambda \in C(\Lambda)$.

    (a) Discuss the relation of the estimators in $\widetilde{C}(\Lambda)$ to those in $C(\Lambda)$. In particular, compute and discuss the form of $\alpha(\lambda)$ for the variable selection problem.

    (b) Relate $\widetilde{RSS}(\lambda) = \sum_{i=1}^n (y_i - \widetilde{\mu}_{\lambda i})^2$ to the GCV criterion for the original estimators in $C(\Lambda)$.

(c) Discuss the properties of $n^{-1}\widehat{RSS}(\lambda)$ as an estimator of the prediction risk for $\widetilde{\mu}_\lambda$. What implications does this have for the estimation of $P(\lambda)$ by $\mathrm{GCV}(\lambda)$?

[Li 1985]

19. Another view of GCV can be obtained by considering the Stein estimator of $\boldsymbol{\mu}$ corresponding to $\boldsymbol{\mu}_\lambda$ that is defined by

$$\bar{\boldsymbol{\mu}}_\lambda = \mathbf{y} - \sigma^2 \left(\mathbf{y} - \boldsymbol{\mu}_\lambda\right) / \mathbf{y}^T \mathbf{B}(\lambda)\mathbf{y},$$

where

$$\mathbf{B}(\lambda) = \{\mathrm{tr}[\mathbf{I} - \mathbf{S}_\lambda]\mathbf{I} - 2(\mathbf{I} - \mathbf{S}_\lambda)\}^{-1}(\mathbf{I} - \mathbf{S}_\lambda)^2$$

and it is assumed that the largest eigenvalue of $\mathbf{I} - \mathbf{S}_\lambda$ is less than half the trace of $\mathbf{I} - \mathbf{S}_\lambda$. The quantity $\sigma^2/\mathbf{y}^T\mathbf{B}(\lambda)\mathbf{y}$ is called a shrinkage factor. Stein's unbiased risk estimator (SURE) associated with $\bar{\boldsymbol{\mu}}_\lambda$ is

$$\mathrm{SURE}(\lambda) = \sigma^2 - \sigma^4 n^{-1} RSS(\lambda) / \left(\mathbf{y}^T\mathbf{B}(\lambda)\mathbf{y}\right)^2.$$

Stein (1981) shows that in the case of normal errors $\mathrm{SURE}(\lambda)$ is an unbiased estimate of the risk $\bar{R}(\lambda) = n^{-1}\sum_{i=1}^n \mathrm{E}\left(\mu_i - \bar{\mu}_{\lambda i}\right)^2$.

(a) A simple illustration of $\bar{\boldsymbol{\mu}}_\lambda$ can be obtained from the variable selection problem. Show that in this case $\mathbf{y}^T\mathbf{B}(\lambda)\mathbf{y} = (n - \#(\lambda))s^2(\lambda)/(n - \#(\lambda) - 2)$, where

$$s^2(\lambda) = \mathbf{y}^T(\mathbf{I} - \mathbf{X}_\lambda(\mathbf{X}_\lambda^T\mathbf{X}_\lambda)^{-1}\mathbf{X}_\lambda^T)\mathbf{y}/(n - \#(\lambda))$$

is the estimator of σ^2 for the model with variables corresponding to λ. Give an intuitive discussion of how $s^2(\lambda)$ should perform as an estimator of σ^2 under models that are either "close to" or "not close to" being correct and then discuss how the estimator $\bar{\boldsymbol{\mu}}_\lambda$ might be expected to behave in these same situations.

(b) Suppose that the largest eigenvalue of $\mathbf{I} - \mathbf{S}_\lambda$ is small relative to its trace. Then, show that

$$\mathbf{B}(\lambda) \doteq (\mathbf{I} - \mathbf{S}_\lambda)^2/\mathrm{tr}[\mathbf{I} - \mathbf{S}_\lambda]$$

and, hence, that

$$\mathrm{SURE}(\lambda) \doteq \sigma^2 - \sigma^4/\mathrm{GCV}(\lambda).$$

Discuss the implications of this relation for the selection of λ.

[Li 1985]

20. Assume that $\varepsilon_1, \ldots, \varepsilon_n$ are *iid* with $\mathrm{E}\,\varepsilon_1 = \mathrm{E}\,\varepsilon_1^3 = 0, \mathrm{E}\,\varepsilon_1^2 = \sigma^2, \mathrm{E}\,\varepsilon_1^4 < \infty$, the regression function is continuously differentiable, $\max|t_i - t_{i-1}| = O(n^{-1})$ and $\max|t_i - t_{i-1}|/\min|t_i - t_{i-1}|$ is bounded. Then show that the GSJS variance estimator (2.29)-(2.30) satisfies $\sqrt{n}(\widehat{\sigma}^2 - \sigma^2) = O_p(1)$. [Hint: Use Lemmas 2.2-2.3 and show that $\mathrm{E}(\widehat{\sigma}^2 - \sigma^2)^2 = O(n^{-1})$.]

21. Derive a version of the GSJS estimator (2.29)-(2.30) that retains its $\sqrt{n}$-consistency but allows for more than one observation at a design point. [Hint: Define the ith pseudo-residual as in (2.30) and then provide choices for the A_i and B_i that deal with the cases where $t_{i-1} = t_i = t_{i+1}$, $t_{i-1} = t_i < t_{i+1}$ or $t_{i-1} < t_i = t_{i+1}$.

22. Show that the risk associated with least-squares estimators of $\boldsymbol{\mu}$ for model (2.31) is minimized by some value of $\lambda \leq p_0$.

23. Prove Lemma 2.1.

24. For $\lambda \geq p_0$ set $B_n = \{\widehat{P}(\lambda) \leq \widehat{P}(k), k = 1, \ldots, p_0 - 1\}$. Verify that $P(B_n) \to 1$. [Hint: Show that $P(B_n^c) \to 0$ using Bonferroni's inequality.]

25. Prove Theorem 2.3. [Hint: For $\lambda < p_0$, note that $n^{-1}RSS(p_0) \xrightarrow{P} \sigma^2$. For $\lambda \geq p_0$, compute the mean and variance of $n^{-1}(n-\lambda)^2\{\mathrm{GCV}(\lambda) - \widehat{P}(\lambda)\}$ using the results in Section 2.4.]

26. Assume that there are infinitely many nonzero β_j in model (2.41) and let $\widehat{\lambda}$ be the minimizer of $\widehat{P}(\lambda)$ over $\Lambda_n = \{1, \ldots, a_n\}$ with $a_n \to \infty$ as $n \to \infty$. Show that in this case $P(\widehat{\lambda} \leq x) \to 0$ for any finite x. Interpret the meaning of this result.

27. Show how to use the selection criteria derived in Section 2.2 to choose the number of partitions for the regressogram estimator in (1.10) and (1.12) of Chapter 1. Specifically, show that for this case we may take $\Lambda = \{1, \ldots, n\}$ with the estimators in $\mathcal{C}(\Lambda)$ having the form $\mathbf{X}_\lambda(\mathbf{X}_\lambda^T\mathbf{X}_\lambda)^{-1}\mathbf{X}_\lambda^T\mathbf{y}$ for a particular matrix $\mathbf{X}_\lambda$. Use this to derive simple expressions for the $\widehat{P}$, CV and GCV criterion functions.

28. Use the selection criteria of Section 2.2 to determine a "best" choice for the number of partitions for the regressogram fit to the data in Table 1.1. In particular, check that the unbiased risk (using the GSJS variance estimator) and GCV criteria both choose $\lambda = 7$ partitions for this data while the loss and risk are minimized by 8 and 9 partitions, respectively.

29. Compute the AIC, FPE, S and U criteria from Exercise 11 for regressogram fits to the data in Table 1.1 and compare your findings with the results of Exercise 28.

30. Modify your linear regressogram program from Exercise 6 of Chapter 1 to include automated selection of the number of partitions using GCV and the unbiased risk criterion $\widehat{P}$. Use the GSJS estimator in (2.29)-(2.30) to estimate σ^2 in $\widehat{P}$.

31. Conduct a simulation along the lines of those in Examples 2.6-2.7 using your modified linear regressogram program from Exercise 30. Consider, for example, a case where the true regression function is a line (so that the optimal number of partitions is one) and, e.g., $\mu(t) = \cos(2\pi t)$ to give an "infinite dimensional" case.

32. Prove Lemma 2.2.

Chapter 3

Series Estimators

3.1 Introduction

Assume that we observe the values of a response variable Y at n predetermined values of an independent variable t. The resulting bivariate observations $(t_1, y_1), \ldots, (t_n, y_n)$ follow the nonparametric regression model

$$y_i = \mu(t_i) + \varepsilon_i,\ i = 1, \ldots, n, \tag{3.1}$$

where $\boldsymbol{\varepsilon} = (\varepsilon_1, \ldots, \varepsilon_n)^T$ is a vector of zero mean, uncorrelated, random errors having common variance σ^2, μ is an unknown regression function that we wish to estimate and $0 \le t_1 \le \cdots \le t_n \le 1$. In this chapter we encounter our first variety of nonparametric regression estimator known as a series estimator.

To motivate what follows let us think, for the moment, about estimation of μ in a linear parametric model. In that case there is a finite integer λ and functions $x_1, \ldots, x_\lambda$ such that $\mu(t) = \sum_{j=1}^{\lambda} \beta_j x_j(t)$ for some set of regression coefficients $\beta_1, \ldots, \beta_\lambda$. Estimation of μ is, therefore, equivalent to estimation of $\beta_j, j = 1, \ldots, \lambda$. This can be accomplished in a variety of ways, least squares being one possibility.

The use of a linear model for μ assumes that the basic form of μ is known. This will frequently not be the case. When knowledge of μ is more limited one may prefer to broaden the class of functions to which μ is assumed to belong. For example, one may only be willing to assume that μ is smooth, in the sense of being continuous and differentiable, or some other assumption of this nature.

When μ is assumed to belong to some general (infinite dimensional) function class (3.1) is called a *nonparametric regression model*. A model of this type may be closer to reality than a linear model. However, the improvement in realism is not without its price, for it is no longer clear how μ should be estimated. It is therefore natural to ask if there is some parallel of the linear model for our more general set of assumptions. For example, if μ is only assumed to be differentiable, can we express the regression function as a linear combination of (possibly infinitely many) known functions? If the answer to this and other similar questions is in the affirmative, there are some obvious advantages. Foremost of these is the fact that now estimation of μ becomes equivalent to estimation of regression coefficients. Thus, we can attempt to adapt what is known from linear models to aid us in determining how to estimate more general regression functions.

As we shall see in the next section, there are many classes of functions whose elements admit representations as infinite dimensional linear models, i.e., for which $\mu = \sum_{j=1}^{\infty} \beta_j x_j$ for some set of known functions $\{x_j\}_{j=1}^{\infty}$. In this instance one approach is to approximate μ by $\sum_{j=1}^{\lambda} \beta_j x_j$, for some integer λ, and then estimate $\beta_j, j = 1, \ldots, \lambda$, to obtain an estimate of μ. The result is what is known as a *series estimator*. The selection of an appropriate value of λ then falls into the category of problems addressed in Chapter 2.

In the next section a short review of some relevant function space theory is given that will provide a framework for subsequent work. Then, in Section 3.3, orthogonal series estimators of μ are developed from a general point of view. Sections 4 and 5 of this chapter are devoted to the study of two important special cases of orthogonal series estimators: trigonometric series and polynomial regression estimators. These two estimator types result from the representation of μ as a weighted sum of infinitely many trigonometric functions and polynomials, respectively. Although polynomial regression is widely used, it generally does not perform as well in practice as the other estimators of μ to be studied in Chapters 4–6. However, these other estimators have strong ties to polynomial regression and cosine series estimators. Thus, one reason for studying orthogonal series estimators is as a motivational tool that provides a point of departure into the work of subsequent chapters.

3.2 Some Function Space Theory

In the previous section a relaxation of standard parametric assumptions was made where the regression function was only assumed to lie in some general class of functions. In this section we discuss several such function classes which play an important role in future developments.

One of the most important function spaces is $L_2[0,1]$, the set of all square integrable functions on the interval $[0,1]$. For two functions $\mu_1, \mu_2 \in L_2[0,1]$ we define their *norms* by

$$||\mu_i|| = \left\{ \int_0^1 \mu_i^2(t)dt \right\}^{1/2}, i = 1, 2, \tag{3.2}$$

and *inner product* by

$$< \mu_1, \mu_2 > \ = \int_0^1 \mu_1(t)\mu_2(t)dt. \tag{3.3}$$

Note that (3.3) is well defined due to the Cauchy-Schwarz inequality

$$| < \mu_1, \mu_2 > |^2 \leq ||\mu_1||^2 ||\mu_2||^2. \tag{3.4}$$

The space $L_2[0,1]$ represents a rich collection of functions. However, it is not entirely suited to our purpose. Two functions, μ_1 and μ_2, are viewed as being the same in $L_2[0,1]$ if $||\mu_1 - \mu_2|| = 0$. Thus, μ_1 and μ_2 are deemed to be identical if they differ on sets of Lebesgue measure zero. This means that, strictly speaking, the elements of $L_2[0,1]$ are equivalence classes of functions which are equal almost everywhere with respect to Lebesgue measure. Consequently, the evaluation of an element of $L_2[0,1]$ at points in $[0,1]$ is not a well defined operation and restrictions are required on the types of functions that can be permitted in order to ensure that model (3.1) is mathematically valid. Since it is often reasonable to assume that μ in (3.1) is a smooth function, one way to accomplish this is to restrict attention to collections of functions that are smooth in the sense of satisfying various continuity and differentiability conditions.

There are several subsets of $L_2[0,1]$ which possess the types of properties we require in a function space. One of these is

$$C^m[0,1] = \{\mu : \mu^{(j)} \text{ is continuous}, j = 0, \ldots, m\}. \tag{3.5}$$

In words, $C^m[0,1]$ is the set of all functions on $[0,1]$ with m continuous derivatives. When $m = 0$ we write $C^0[0,1] = C[0,1]$ for the set of all continuous functions on $[0,1]$. Obviously $L_2[0,1] \supset C[0,1] \supset C^1[0,1] \ldots$

For some of our work it will not be necessary for μ to have m continuous derivatives. It will often suffice for $\mu^{(m)}$ to be square integrable. Thus, we define the mth order Sobolev space

$$W_2^m[0,1] = \left\{ \mu : \begin{array}{c} \mu^{(j)} \text{ is absolutely continuous ,} \\ j = 0, \ldots, m-1, \text{ and } \mu^{(m)} \in L_2[0,1] \end{array} \right\}. \tag{3.6}$$

Note that $W_2^m[0,1] \supset C^m[0,1]$ and, when $m = 0$, $W_2^0[0,1] = L_2[0,1]$.

If we restrict attention entirely to functions in $W_2^m[0,1]$ for $m \geq 1$ or $C^m[0,1]$ with $m \geq 0$ the ambiguity associated with point evaluation disappears; because when $\mu_1, \mu_2 \in C[0,1]$ and $||\mu_1 - \mu_2|| = 0$ this means that $\mu_1 \equiv \mu_2$. Both $W_2^m[0,1]$ and $C^m[0,1]$ may be equipped with their own norms and studied independently of any connection with $L_2[0,1]$. On the other hand, the fact that they are subsets of $L_2[0,1]$ means that the properties of $L_2[0,1]$ functions described below will be applicable to the elements of $W_2^m[0,1]$ and $C^m[0,1]$ as well.

It is easy to verify that $L_2[0,1]$ is a vector space, i.e., if $\mu_1, \mu_2 \in L_2[0,1]$ then $\mu_1 + \mu_2 \in L_2[0,1]$, etc. Now for finite dimensional vector spaces such as Euclidean p-space, $\mathbb{R}^p$, it is always possible to represent any element in the vector space as a linear combination of basis elements. One might expect an analogous property would hold for an infinite dimensional vector space such as $L_2[0,1]$. This is, in fact, the case. To make this result precise a few definitions are required.

Definition 3.1. *Two functions* $\mu_1, \mu_2 \in L_2[0,1]$ *are said to be orthogonal if* $< \mu_1, \mu_2 >= 0$. *We denote this by* $\mu_1 \perp \mu_2$.

Definition 3.2. *A sequence of functions* $\{x_j\}_{j=1}^{\infty}$ *is said to be orthonormal if the* x_j *are pairwise orthogonal and* $||x_j|| = 1$ *for all* j.

Definition 3.3. *A sequence of functions* $\{x_j\}_{j=1}^{\infty}$ *is said to be a complete orthonormal sequence* (CONS) *if* $\mu \perp x_j$ *for all* j *entails that* $\mu = 0$.

Recalling the definitions of orthogonality and orthonormality in $\mathbb{R}^p$, it is seen that Definitions 3.1 and 3.2 provide obvious parallels of these concepts for $L_2[0,1]$. Definition 3.3 states that $\{x_j\}_{j=1}^{\infty}$ is a CONS provided that the only function orthogonal to all the x_j is the zero function. There are many CONS's for $L_2[0,1]$. Two important examples are the following.

Example 3.1 (Trigonometric Functions). There are three ways to construct a CONS for $L_2[0,1]$ using trigonometric functions. The most common approach is to use $x_1(t) \equiv 1$ and then take

$$x_{2j}(t) = \sqrt{2}\cos(2j\pi t)$$

and

$$x_{2j+1}(t) = \sqrt{2}\sin(2j\pi t)$$

for $j = 1, 2, \ldots$ It is also possible to construct a CONS using either the sine or cosine functions separately. Of particular interest in Section 3.4 is the cosine sequence with $x_1(t) \equiv 1$ and

$$x_j(t) = \sqrt{2}\cos((j-1)\pi t),\ j = 2, \ldots \tag{3.7}$$

The sine series $x_j(t) = \sqrt{2}\sin(j\pi t), j = 1, \ldots$ is also a CONS for $L_2[0,1]$.

The orthonormality property of all the trigonometric CONS's can be established directly. For verification of their completeness see, e.g., Davis (1975, Chapter XII).

Example 3.2 (Legendre Polynomials). Another CONS for $L_2[0,1]$ can be derived from the polynomial functions $q_j(t) = t^{j-1}, j = 1, 2, \ldots$. To construct the CONS one proceeds as follows. First take

$$x_1(t) = q_1(t)/||q_1|| \equiv 1\ . \tag{3.8}$$

Now define basis functions recursively via the formula

$$x_j(t) = \left[q_j(t) - \sum_{k=1}^{j-1} < q_j, x_k > x_k(t) \right] / || q_j - \sum_{k=1}^{j-1} < q_j, x_k > x_k ||, \quad (3.9)$$

for $j = 2, \ldots$

The functions defined by (3.8)–(3.9) are known as the Legendre polynomials. It is easy to see that they are orthonormal. A proof that they are complete can be found in Davis (1975, Chapter XII).

The recursive method used to derive the Legendre polynomials is called the *Gram-Schmidt orthonormalization process.* As can be seen from (3.8)–(3.9) there is no reason to restrict its use to the polynomial functions. It can be used, more generally, to orthonormalize any sequence of functions.

The above examples illustrate the existence of a CONS for $L_2[0,1]$. The following proposition states that any CONS actually gives us a basis for $L_2[0,1]$. Its proof can be found in standard texts concerning functional analysis such as Luenberger (1969).

Proposition 3.1. *Let* $\{x_j\}_{j=1}^{\infty}$ *be a* CONS *for* $L_2[0,1]$ *and for* $\mu \in L_2[0,1]$ *define*

$$\beta_j = < \mu, x_j >, j = 1, \ldots \quad (3.10)$$

Then, $\sum_{j=1}^{\lambda} \beta_j x_j$ *is a best approximation to* μ *in the sense that*

$$|| \mu - \sum_{j=1}^{\lambda} \beta_j x_j || \leq || \mu - \sum_{j=1}^{\lambda} b_j x_j ||$$

for all $\mathbf{b} = (b_1, \ldots, b_\lambda)^T \in \mathbb{R}^\lambda$. *Moreover, as* $\lambda \to \infty$

$$|| \mu - \sum_{j=1}^{\lambda} \beta_j x_j ||^2 \to 0. \quad (3.11)$$

Proposition 3.1 states that the function sequence $\sum_{j=1}^{\lambda} \beta_j x_j$ converges to μ in $L_2[0,1]$ norm. Thus, μ and $\sum_{j=1}^{\infty} \beta_j x_j$ are identical functions when viewed as elements of $L_2[0,1]$. However, due to the properties of Lebesgue measure, this does not signify pointwise convergence in any sense; although that will also be true in most cases we consider.

As a result of Proposition 3.1 we can represent μ (in the $L_2[0,1]$ sense) as a linear combination of the basis functions $\{x_j\}_{j=1}^{\infty}$. The optimal coefficients (3.10) are called *generalized Fourier coefficients* while $\sum_{j=1}^{\infty} \beta_j x_j$ is called a *generalized Fourier series expansion* for μ. The generalized Fourier coefficients can be shown to satisfy

$$\sum_{j=1}^{\infty} \beta_j^2 = \|\mu\|^2, \tag{3.12}$$

which is known as *Parseval's relation.*

Example 3.3 (The Cosine Series). Using Proposition 3.1 the *cosine series expansion* for a function $\mu \in L_2[0,1]$ is

$$\beta_1 + \sqrt{2}\sum_{j=2}^{\infty} \beta_j \cos((j-1)\pi t) \tag{3.13}$$

with cosine Fourier coefficients

$$\beta_1 = \int_0^1 \mu(t)dt \tag{3.14}$$

and

$$\beta_j = \sqrt{2}\int_0^1 \mu(t)\cos((j-1)\pi t)dt,\ j = 2,\dots \tag{3.15}$$

The series in (3.13) actually converges pointwise to μ under the assumption that $\mu \in W_2^1[0,1]$. To see this, note that in this case we may integrate by parts to obtain

$$\begin{aligned} \beta_j &= -\frac{\sqrt{2}}{(j-1)\pi}\int_0^1 \mu'(t)\sin((j-1)\pi t)dt \\ &= -\frac{1}{(j-1)\pi}\beta_j',\ j = 2,\dots \end{aligned} \tag{3.16}$$

This expression has the consequence that the cosine Fourier coefficients for μ are absolutely summable. Specifically, we have

$$\sum_{j=2}^{\infty} |\beta_j| \le [\sum_{j=2}^{\infty} (\beta_j')^2]^{1/2} [\sum_{j=1}^{\infty} (j\pi)^{-2}]^{1/2}$$

due to the Cauchy-Schwarz inequality and the fact that $\sum_{j=2}^{\infty} (\beta_j')^2 < \infty$ because $\mu' \in L_2[0,1]$ and the sine functions are a CONS.

Since the cosine Fourier coefficients are absolutely convergent, it follows that the series (3.13) converges uniformly and, because of the continuity of the cosines, the limit function must be continuous. Let us call this limit function $\widetilde{\mu}$. Then, again due to the uniform convergence of (3.13), we can integrate the series term by term to obtain $<\widetilde{\mu}, x_j>=<\mu, x_j>, j=1,\ldots$ for $\{x_j\}_{j=1}^{\infty}$ the cosine sequence. But this means that $\widetilde{\mu}-\mu \perp x_j$ for all j and, hence, that $\widetilde{\mu} \equiv \mu$.

It will be useful in what follows to work with a generalization of $L_2[0,1]$ obtained by weighting the integrals in (3.2)–(3.3) using a continuous, positive density function w that is supported on $[0,1]$. We will call this space $L_2(w)$ and define the norms and inner products for $\mu_1, \mu_2 \in L_2(w)$ by

$$||\mu_i||_w = \left\{\int_0^1 \mu_i^2(t)w(t)dt\right\}^{1/2}, i=1,2, \tag{3.17}$$

and

$$<\mu_1, \mu_2>_w = \int_0^1 \mu_1(t)\mu_2(t)w(t)dt, \tag{3.18}$$

respectively. Note that for any function μ

$$||\mu||(\inf_{0\le t\le 1} w(t))^{1/2} \le ||\mu||_w \le ||\mu||(\sup_{0\le t\le 1} w(t))^{1/2}$$

so that $\mu \in L_2[0,1]$ if and only if $\mu \in L_2(w)$. In addition, convergence of a series to μ in $L_2[0,1]$ norm will entail convergence in $L_2(w)$ norm and conversely.

This new weighted L_2 space possesses the same essential qualities as $L_2[0,1]$. In particular, there are CONS's $\{x_j\}_{j=1}^{\infty}$ such that $<x_i, x_j>_w = \delta_{ij}, i,j=1,\ldots$ and $||\mu - \sum_{j=1}^{\lambda}\beta_j x_j||_w \to 0$ as $\lambda \to \infty$ for

$$\beta_j = <x_j, \mu>_w \tag{3.19}$$

the jth generalized $L_2(w)$ Fourier coefficient. Parseval's relation now takes the form

$$\sum_{j=1}^{\infty}\beta_j^2 = ||\mu||_w^2. \tag{3.20}$$

We will continue to refer to $\sum_{j=1}^{\infty} \beta_j x_j$ as a generalized Fourier series expansion for μ.

One specific CONS for $L_2(w)$ is provided by using the Gram-Schmidt algorithm on the polynomials as in Example 3.2 except with $||\cdot||$ and $<\cdot,\cdot>$ replaced by $||\cdot||_w$ and $<\cdot,\cdot>_w$. These polynomials are referred to as Jacobi polynomials when w is a beta density function on $[0, 1]$. A CONS using the trigonometric functions of Example 3.1 can be derived similarly.

3.3 Generalized Fourier Series Estimators

We now want to use the work in the previous section to develop a nonparametric estimator for μ. Before proceeding to this task, it will be helpful to first make some assumptions concerning the form of the design points $t_1, \ldots, t_n$. Throughout the remainder of this chapter and at other points throughout the book, we will assume that the t_i are generated by a continuous, positive design density w on $[0, 1]$ through a relationship such as

$$\int_0^{t_i} w(t)dt = (2i-1)/2n,\ i = 1, \ldots, n. \tag{3.21}$$

In words, this means that t_i is the $100\frac{(2i-1)}{2n}$th percentile of the w density. The canonical case of a uniform design corresponds to $w(t) \equiv 1$.

Assumption (3.21) is made primarily for mathematical convenience and the actual form of this density need not be known in order to compute any of the estimators that follow. Indeed, in practice it is unlikely that one actually has such a density or would necessarily even use this approach to designing an experiment. The viewpoint to be taken here is that w represents an idealized density that could have generated the actual design points that were used in the data.

Let us now assume that $\mu \in L_2(w)$ and that the generalized $L_2(w)$ Fourier series expansion $\sum_{j=1}^{\infty} \beta_j x_j$ for the β_j in (3.19) converges to μ uniformly in t as $\lambda \to \infty$. Under these assumptions on μ, model (3.1) becomes

$$y_i = \sum_{j=1}^{\infty} \beta_j x_j(t_i) + \varepsilon_i,\ i = 1, \ldots, n. \tag{3.22}$$

Thus, the data follow a linear model with infinitely many unknown regression coefficients.

Since $\mu \in L_2(w)$ we know from (3.20) that $\sum_{j=1}^{\infty} \beta_j^2 < \infty$, so the β_j must eventually decay to zero. Consequently, it seems reasonable to assume that there is an integer λ such that $\mu \doteq \sum_{j=1}^{\lambda} \beta_j x_j$ and, hence,

$$y_i \doteq \sum_{j=1}^{\lambda} \beta_j x_j(t_i) + \varepsilon_i, i = 1, \ldots, n. \tag{3.23}$$

Apart from the "$\doteq$" in (3.23), this looks like a linear model. Consequently, it seems reasonable to attempt to apply inferential techniques for linear models to this setting. We will lay out the essential ingredients to such an estimation paradigm in the remainder of this section and then explore its consequence in two special cases in Sections 3.4 and 3.5.

If one ignores the fact that we only have approximate equality in (3.23), then standard linear regression ideas would suggest that we estimate the Fourier coefficients $\beta_1, \ldots, \beta_\lambda$ in (3.23) by the minimizer of

$$RSS(\mathbf{c}; \lambda) = \sum_{i=1}^{n} \left(y_i - \sum_{j=1}^{\lambda} c_j x_j(t_i) \right)^2 \tag{3.24}$$

with respect to $\mathbf{c} = (c_1, \ldots, c_\lambda)^T$. This practice can be justified from the fact that $\beta_1, \ldots, \beta_\lambda$ minimize $\|\mu - \sum_{k=1}^{\lambda} c_k x_k\|_w^2$ and $n^{-1} RSS(\mathbf{c}; \lambda)$ provides a natural estimator of this norm difference.

Now define the design matrix by

$$\mathbf{X}_\lambda = \{x_j(t_i)\}_{i=1,n,j=1,\lambda}. \tag{3.25}$$

If $\mathbf{X}_\lambda$ has full column rank the unique minimizer of (3.24) is

$$\mathbf{b}_\lambda = (b_{\lambda 1}, \ldots, b_{\lambda\lambda})^T = \left(\mathbf{X}_\lambda^T \mathbf{X}_\lambda\right)^{-1} \mathbf{X}_\lambda^T \mathbf{y}, \tag{3.26}$$

where $\mathbf{y} = (y_1, \ldots, y_n)^T$ is the vector of response values. The resulting *generalized Fourier series estimator* of $\mu(t)$ is defined to be

$$\mu_\lambda(t) = \sum_{j=1}^{\lambda} b_{\lambda j} x_j(t). \tag{3.27}$$

The other parameter that requires estimation in our model is σ^2. For this purpose we might use the linear regression variance estimator

$$\sigma_\lambda^2 = RSS(\lambda)/(n-\lambda), \tag{3.28}$$

where $RSS(\lambda) = RSS(\mathbf{b}_\lambda; \lambda)$ is the residual sum-of-squares associated with the estimator (3.27). However, there are some problems with the use of this estimator in practice involving how the order λ should be chosen to produce the best estimator of σ^2. Unlike the finite dimensional linear models setting there is generally no choice for λ that makes this variance estimator unbiased. An application of Lemma 2.2 shows that the bias of σ_λ^2 is $(n-\lambda)^{-1}\boldsymbol{\mu}^T(\mathbf{I}-\mathbf{S}_\lambda)\boldsymbol{\mu}$ with $\mathbf{S}_\lambda = \mathbf{X}_\lambda\left(\mathbf{X}_\lambda^T\mathbf{X}_\lambda\right)^{-1}\mathbf{X}_\lambda^T$ and $\boldsymbol{\mu} = (\mu(t_1), \ldots, \mu(t_n))^T$. This bias can be quite substantial unless λ is chosen correctly. In addition, a good choice of λ for estimating σ^2 will not generally be a good choice for estimating μ using μ_λ.

There are actually a number of good variance estimators that can be used in the nonparametric regression setting which, unlike σ_λ^2, do not require the selection of an auxillary tuning parameter. Many of these use differences between responses at neighboring design points to effectively remove the influence of the deterministic component (i.e., μ) from the variance estimation problem. Perhaps the simplest of these is

$$\hat{\sigma}_{\mathrm{R}}^2 = (n-1)^{-1}\sum_{i=2}^{n}(y_i - y_{i-1})^2 \tag{3.29}$$

which was proposed in Rice (1984a), for example. If $\mu \in C^1[0,1]$, then $y_i - y_{i-1} = \varepsilon_i - \varepsilon_{i-1} + \mu(t_i) - \mu(t_{i-1}) = \varepsilon_i - \varepsilon_{i-1} + O(n^{-1})$ under assumption (3.21). Thus, $\hat{\sigma}_R^2$ is made up primarily of contributions from the random errors which are what contain all the information about σ^2. Rice (1984a) also suggested that we use the residuals from successive straight line fits to triples of points to construct a variance estimator. This results in the GSJS estimator $\hat{\sigma}^2$ of (2.29)–(2.30) from Chapter 2 that will be used extensively in this book. Other possible variance estimators include

$$\hat{\sigma}_{HKT}^2 = (n-2)^{-1}\sum_{i=2}^{n-1}(.809y_{i-1} - .5y_i - .309y_{i+1})^2 \tag{3.30}$$

proposed by Hall, Kay and Titterington (1990) who also give a nice overview of the variance estimation problem. Like the GSJS estimator, both (3.29) and (3.30) can be shown to provide $\sqrt{n}$-consistent estimators of σ^2. Further discussion of variance estimation and other possible estimators can be found in Gasser, Sroka and Jennen-Steinmetz (1986), Müller and Stadtmüller (1987a), Buckley, Eagleson and Silverman (1988), Buckley and Eagleson (1989), Hall and Marron (1990) and Thompson, Kay and Titterington (1991).

Thus, we have now developed estimators for the two principal parameters in our model: μ and σ^2. There is, however, much more that can be done using the basic regression framework that was used to produce our estimator of μ. For example, if $\beta_j = 0$, then we do not need the function x_j to approximate μ. Thus, to ascertain if x_j should be used in fitting the data we might examine a linear models type test statistic for the hypothesis that $\beta_j = 0$. Specifically, we might test the hypothesis using

$$Z_j = b_{\lambda j} / \left[\sigma^2 C_j\right]^{1/2}, \tag{3.31}$$

where C_j is the jth diagonal element of $\left(\mathbf{X}_\lambda^T \mathbf{X}_\lambda\right)^{-1}$ and σ^2 is replaced by a suitable estimator in the event of an unknown variance. The null hypothesis would be rejected at level α if $|Z_j|$ exceeded $z_{\alpha/2}$, the 100(1 - $\alpha/2$)th percentage point of the standard normal distribution. Similar considerations would suggest using an interval estimator for $\mu(t)$ of the form

$$\mu_\lambda(t) \pm z_{\alpha/2} \left[\sigma^2 \mathbf{x}_\lambda(t)^T \left(\mathbf{X}_\lambda^T \mathbf{X}_\lambda\right)^{-1} \mathbf{x}_\lambda(t)\right]^{1/2}, \tag{3.32}$$

with $\mathbf{x}_\lambda(t) = (x_1(t), \ldots, x_\lambda(t))^T$. For diagnostic analysis of a fit one could again borrow from standard regression methods and look at residual plots, influence measures, etc.

The strategy we have just outlined for inference about μ is intuitively appealing coming as it does from a much developed and well understood area of Statistics. However, at this point a word of caution is needed, for the relationship between μ and $\sum_{j=1}^{\lambda} \beta_j x_j$ is only approximate and the true mean function will not, in general, be finite dimensional. This means that the assumptions that are required for the use of standard linear regression techniques are not strictly met in our setting and, accordingly, one must be extremely careful when attempting their routine application to series estimators. There are cases, such as inference about the β_j, where this approach can be quite effective. But there are also instances, such as interval estimation of μ, where a linear models approach can fail. We will explore these issues more thoroughly in subsequent sections.

The fact that (3.23) is only an approximation raises even more fundamental questions such as whether or not μ_λ is even a consistent estimator of μ. Clearly, inference of any kind is of doubtful value unless this is the case.

To explore the consistency issue further, define the loss and risk for estimating μ with μ_λ by

$$L_n(\lambda) = n^{-1} \sum_{i=1}^{n} (\mu(t_i) - \mu_\lambda(t_i))^2 \tag{3.33}$$

and

$$R_n(\lambda) = \mathrm{E}\, L_n(\lambda) = n^{-1} \sum_{i=1}^{n} (\mu(t_i) - \mathrm{E}\,\mu_\lambda(t_i))^2 + \sigma^2\lambda/n. \tag{3.34}$$

We will work with $R_n(\lambda)$ for the present and will say that μ_λ is consistent if $R_n(\lambda) \to 0$ as $n \to \infty$. (See Exercise 5 however.)

The first term in the risk is the sum of the squared biases. Intuitively, this should decay as the number of terms or order λ for the estimator grows since by increasing λ we are including more and more terms in our approximation to μ. Thus, for $R_n(\lambda)$ to converge to zero we would anticipate the necessity that $\lambda \to \infty$ as $n \to \infty$.

The second term in the risk, $\sigma^2\lambda/n$, corresponds to the variance of the estimator μ_λ. This term grows with λ so that $R_n(\lambda)$ is bounded away from zero unless $\lambda/n \to 0$. This gives the conclusion that consistency of μ_λ should require λ to grow at a rate smaller than n.

To actually establish consistency one must now attempt to characterize the asymptotic behavior of $n^{-1}\sum_{i=1}^{n}(\mu(t_i) - \mathrm{E}\mu_\lambda(t_i))^2$. It is difficult to get very far here without making some explicit assumptions about the elements in the CONS. However, our intuition would suggest that its behavior should depend on the unestimated Fourier coefficients and the ability of the $b_{\lambda j}$ to estimate the β_j.

Shibata (1981) gives results which are relevant to the consistency of least-squares estimators for model (3.22). Li (1984) focuses on the existence of consistent estimators for some functional of the β_j, e.g., the value of μ at a fixed point t. He gives necessary and sufficient conditions for the existence of a consistent estimator and actually constructs an estimator which is consistent. Other consistent estimators that are closely related to estimator (3.27) are studied by Rutkowski (1982) and Rafajlowicz (1987, 1988).

Suppose now that we have decided to use an estimator of the form (3.27) to estimate μ. To determine which estimator to use it becomes necessary to select a value of λ. The choice of λ will typically govern how smooth or wiggly the estimator will be. Usually, larger values of λ will result in estimators that come closer to interpolation of the (t_i, y_i) than smaller values and, hence, will tend to be more wiggly. In particular, since we have assumed that $\mathbf{X}_\lambda$ has full column rank in (3.25), the value $\lambda = n$ corresponds to the interpolation case with $\mu_n(t_i) = y_i$, $i = 1, \ldots, n$.

The problem of selecting an optimal λ for series estimators is the same as the problem discussed in Section 2.3 of selecting the optimal order in a heirarchical regression model with infinitely many parameters. Thus, to estimate the minimizer of $L_n(\lambda)$ or $R_n(\lambda)$ we can use techniques such as unbiased risk estimation, cross-validation or generalized cross-validation that were studied in Chapter 2. In this settting the GCV choice for λ is the minimizer $\widehat{\lambda}$ of

$$\mathrm{GCV}(\lambda) = n^{-1} RSS(\lambda) / (1 - \lambda/n)^2. \tag{3.35}$$

An unbiased risk estimator is obtained similarly by minimizing

$$\widehat{P}(\lambda) = n^{-1} RSS(\lambda) + 2\sigma^2\lambda/n \tag{3.36}$$

when σ^2 is known. In general, the variance in (3.36) will have to be estimated using, e.g., the GSJS estimator. As noted in Section 2.3, under certain conditions the estimators $\widehat{\lambda}$ from the unbiased risk and GCV criteria can be asymptotically optimal in that

$$\frac{R_n(\widehat{\lambda})}{\inf_{\lambda \leq \lambda_n} R_n(\lambda)} - 1 = o_p(n^{(\delta-1)/2}), \tag{3.37}$$

where $\lambda_n \asymp n^\delta$ is the maximum allowable order (i.e., number of terms) in the estimator and δ is the parameter that governs how fast it can grow relative to the sample size. A more detailed treatment of this subject can be found in Hurvich and Tsai (1995).

Up to this point we have given only a generic treatment of series estimators in that we have not chosen a specific form for the CONS. The remainder of this chapter focuses on specific choices for the CONS and concentrates on investigating the properties of the resulting series estimators. In the next section we treat the particular case of the cosine CONS of Example 3.1. We will see that it is possible to establish consistency for μ_λ in this case and to justify, asymptotically, certain aspects of the plan of attack discussed above for diagnostic and inferential analysis. The details will be representative of what one might expect to encounter for some general orthonormal basis.

3.4 Trigonometric Series Estimators

In this section we discuss the properties of the particular type of series estimator, called ***trigonometric series estimators***, which results when the trigonometric functions of Example 3.1 are used for a CONS in (3.27). The resulting estimators correspond to the classical concept of Fourier series estimation that was, at one time, popular in the field of time series analysis.

The stimulus for the use of Fourier methods by time series analysts was their recognition that, when observing data over time, some aspects of real world phenomena tended to exhibit cycles or periodicities. In some cases the cycles were often while in other cases they were less frequent. For example, retail sales experience yearly cycles (e.g., at Christmas) as well as longer term upward and downward trends. Thus, they began to think about analyzing the frequency composition of a time series.

When thinking of periodic behavior perhaps the first functions which come to mind are the sines and cosines. Thus, it was natural for time series analysts to consider the use of regression models involving sines and/or cosines to describe time series behavior. The idea was to use linear combinations of functions such as 1 and $\cos(j\pi t), j = 1, \ldots, \lambda$, for some nonnegative integer λ, to model a series. The high frequency (repeats often) aspects of the series should be picked up by the higher frequency elements of the model: i.e., $\cos(j\pi t)$ for large j. The low frequency components of the series, such as trends, would then be modeled by the constant and other low frequency terms.

For the problems we will be interested in t may or may not represent time. Nonetheless insight into the properties of trigonometric series estimators can be gained through analogy with the time series setting. We may think of μ in model (3.1) as representing a trend or low frequency component while the random errors, since they will tend to oscillate rapidly about the regression curve, can be thought of as high frequency in nature. From this viewpoint a series estimator using the first few terms of a trigonometric CONS seems to represent a very natural choice for estimating the mean function.

There are three possible trigonometric CONS's to choose from in Example 3.1 which raises the question of whether one of these choices is, in some sense, superior to the other two. We will eventually see that the cosine CONS has features which makes it preferable in most cases for the purpose of nonparametric regression analysis. Thus, we will initially focus our attention on series estimators constructed from the cosine functions.

To make the analysis which follows more mathematically tractable, it will be useful to now specialize our consideration to the case of a uniform design on $[0, 1]$. Specifically, we will assume throughout the remainder of this section that

$$t_i = (2i - 1)/2n, \; i = 1, \ldots, n. \tag{3.38}$$

3.4.1 Form of the Estimator

In this section we will derive two explicit, closed form, expressions for the cosine series estimator, μ_λ, under the assumption that (3.38) holds. These will be valuable in the sequel and also provide further insight into how series estimators actually work.

Let $x_1(t) \equiv 1$, define

$$x_j(t) = \sqrt{2}\cos((j-1)\pi t),\ j = 2, \ldots,$$

and, for some integer $\lambda \geq 1$, consider fitting the function $\sum_{j=1}^{\lambda} c_j x_j(t)$ to a set of data by the method of least squares. As noted in Section 3.3, the estimators of the cosine Fourier coefficients are provided by (3.26) with

$$\mathbf{X}_\lambda = \{x_j(t_i)\}_{i=1,n,j=1,\lambda}. \tag{3.39}$$

The form of the estimator can be simplified somewhat through the use of Lemma 3.4 from Section 3.8. Under the design (3.38) we have

$$n^{-1}\sum_{i=1}^{n}\cos(j\pi t_i)\cos(k\pi t_i) = \frac{1}{2}\delta_{jk},\ j,k = 1, \ldots, n-1, \tag{3.40}$$

and

$$n^{-1}\sum_{i=1}^{n}\cos(j\pi t_i) = 0,\ j = 1, \ldots, n-1. \tag{3.41}$$

In words, (3.40)–(3.41) mean that the cosine functions exhibit a similar orthonormality property when summed across the design (3.38) to what transpires when they are integrated over $[0,1]$. This greatly simplifies the form of the estimator and much of the mathematical analysis that follows.

We now see that $\mathbf{X}_\lambda^T\mathbf{X}_\lambda = n\mathbf{I}$ and the estimated cosine Fourier coefficients (3.26) become

$$b_j = n^{-1}\sum_{i=1}^{n} y_i\sqrt{2}\cos((j-1)\pi t_i),\ j = 2, \ldots, \lambda, \tag{3.42}$$

with $b_1 = \bar{y} = n^{-1}\sum_{i=1}^{n} y_i$ the response average. The resulting estimator of μ is then given by

$$\mu_\lambda(t) = b_1 + \sum_{j=2}^{\lambda} b_j\sqrt{2}\cos((j-1)\pi t). \tag{3.43}$$

We will refer to the b_j in (3.42) as the *sample cosine Fourier coefficients.* Note that they do not actually involve λ in this case and, accordingly, we have chosen to omit the second λ subscript that was used in (3.26).

Another useful expression for μ_λ can be obtained by observing that

$$\begin{aligned} \mu_\lambda(t) &= \bar{y} + \sum_{j=2}^{\lambda} b_j \sqrt{2} \cos((j-1)\pi t) \\ &= n^{-1} \sum_{i=1}^{n} y_i K_\lambda(t, t_i), \end{aligned} \tag{3.44}$$

where

$$\begin{aligned} K_\lambda(t,s) &= 1 + 2\sum_{j=2}^{\lambda} \cos((j-1)\pi t)\cos((j-1)\pi s) \\ &= 1 + \cos\left(\frac{\lambda\pi(t-s)}{2}\right) D\left(\frac{t-s}{2}; \lambda - 1\right) \\ &\quad + \cos\left(\frac{\lambda\pi(t+s)}{2}\right) D\left(\frac{t+s}{2}; \lambda - 1\right) \end{aligned} \tag{3.45}$$

for

$$D(u;k) = \frac{\sin(k\pi u)}{\sin(\pi u)} \tag{3.46}$$

the famous Dirichlet kernel from Fourier series analysis. In obtaining this result we have used $2\cos(\frac{1}{2}(x+y))\cos(\frac{1}{2}(x-y)) = \cos(x) + \cos(y)$ along with the result

$$\sum_{j=1}^{k} \cos(j\pi x) = \cos\left(\frac{(k+1)\pi x}{2}\right) D\left(\frac{x}{2}; k\right) \tag{3.47}$$

from Gradshteyn and Ryzhik (1980, Section 1.342).

Using (3.41) the weight function K_λ is seen to satisfy

$$\sum_{i=1}^{n} K_\lambda(t, t_i) = n.$$

Thus, (3.44)–(3.45) have the implication that $\mu_\lambda(t)$ is a weighted average (if we allow for negative weights) of the y_i with weights obtained from the function (3.45).

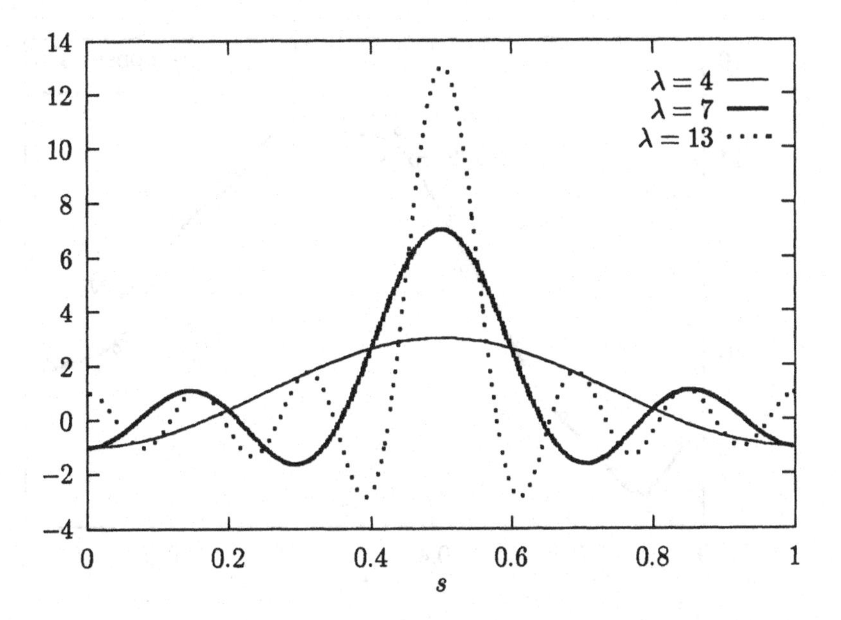

Figure 3.1: $K_\lambda(.5, s)$ for $\lambda = 4, 7$ and 13.

Graphs of $K_\lambda(.5, \cdot)$ are shown in Figure 3.1 for several values of λ which provide an indication of how responses are weighted in producing an estimator of $\mu(.5)$. Note that as λ grows larger the function becomes more peaked near the point of estimation $t = .5$. This has the consequence that for larger λ larger weights will be placed on responses whose t ordinates are near .5.

The weight function also has a rather displeasing property in that as λ grows large it oscillates more and more rapidly away from its peak. This means that the estimator is adding in or subtracting off contributions from observations far from the point of estimation and doing so in an increasingly rapid fashion as λ grows large. One might hope that the contributions from these oscillations will tend to knock each other out, on the average, in which case $\mu_\lambda(t)$ is essentially a local average of observations near t. The role of λ is then to determine how dependent the estimator is on local behavior of the data. Increasing λ would correspond to increasing local dependence and result in a more wiggly estimator.

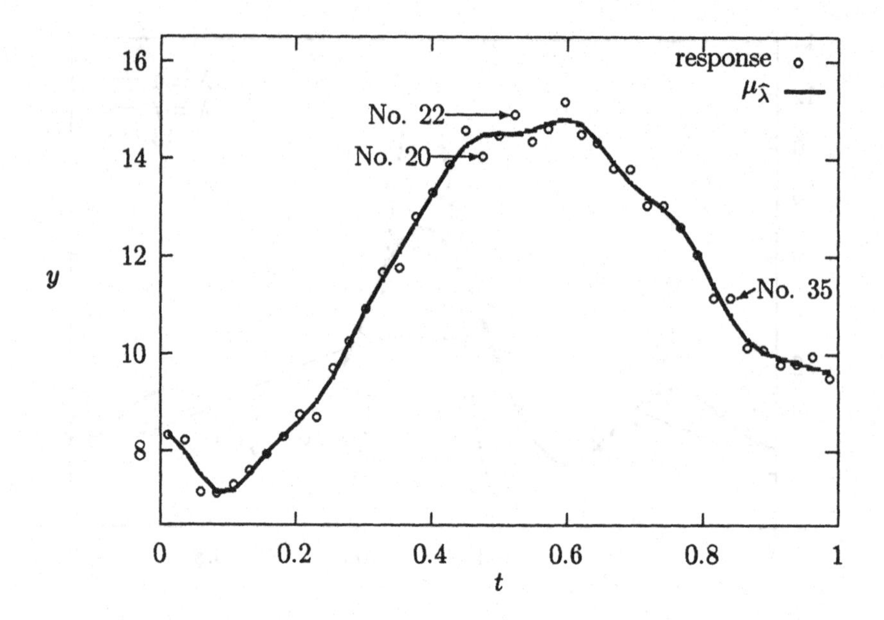

Figure 3.2: The Voltage Drop Data.

3.4.2 An Example

Before proceeding to study the large sample properties of estimator (3.43), it will be useful to first explicitly analyze a data set using the paradigm of Section 3.3 in the context of cosine series estimation. For this purpose we will use data on the voltage drop in the battery of a guided missile motor during its flight given in Montgomery and Peck (1982). A plot of the observed voltage drops as a function of time is shown in Figure 3.2. (The actual data can be found in Exercise 9.) Observations were taken at half second intervals between take-off and 20 seconds of flight. In the plot the time axis has been rescaled to conform with (3.38).

To estimate the regression function we will fit the data with an estimator of the form

$$\mu_\lambda(t) = b_1 + \sqrt{2}\sum_{j=2}^{\lambda} b_j \cos((j-1)\pi t),$$

where the b_j are defined in (3.42). The first step is to select an appropriate value of λ. Figure 3.2 shows three possible fits corresponding to the values of $\lambda = 7, 14$ and 30. The fit with $\lambda = 7$ appears to be too "smooth" in the sense that it is not able to resolve the valley that occurs near the left end of the plot. In contrast, the choice of $\lambda = 30$ seems to give too "rough" a fit in that it follows the data too slavishly. The fit with $\lambda = 14$ represents a compromise that seems to adequately resolve the left boundary behavior of the data while not focusing unduly on each individual response.

As demonstrated by Figure 3.2, there can be a substantial difference in fits with different choices of the order or smoothing parameter λ. Thus, unless one has some preconceived notion concerning a good value of λ, it would seem wise to let the data guide the order selection by using either the GCV or unbiased risk criterion $\widehat{P}$ from (3.35)–(3.36). The values of these two criteria for the voltage drop data are shown in Figure 3.4 over the range $\lambda = 3$ to $\lambda = 37$. Outside this range the GCV criterion becomes quite large and these areas have been omitted to visually enhance the behavior of the order selection criteria near their minima. In this case it turns out that both GCV and $\widehat{P}$ choose the value $\widehat{\lambda} = 14$ with the oversmoothed fit having $\lambda = 7$ being preferred by the Schwarz criterion (2.40). The fit with $\widehat{\lambda} = 14$ is the one that is shown with the data in Figure 3.2. In computing $\widehat{P}$ (and the Schwarz criterion) we used the GSJS variance estimator (2.29)–(2.30) which was found to have the value $\widehat{\sigma}^2 = .0698$ for this data.

One could now conduct further fine tuning and experiment with other choices of λ while using the GCV and unbiased risk order estimate of $\widehat{\lambda} = 14$ as just a guideline for the proper amount of smoothing for the data. In particular, other choices of λ might be tried in an attempt to remove what appear to be spurious wiggles in the peak of the fit in Figure 3.2. Unfortunately, attempts to accomplish such goals meet with only limited success here due to problems with the boundary behavior of cosine series estimators that will be discussed and remedied subsequently. Thus, let us proceed as if we are satisfied with the GCV and unbiased risk choice of $\widehat{\lambda} = 14$. The next step is to then examine the nature of the fit using diagnostics such as those from standard linear regression analysis. In particular, we will look at the leverage values, studentized residuals and measures of influence associated with the fit.

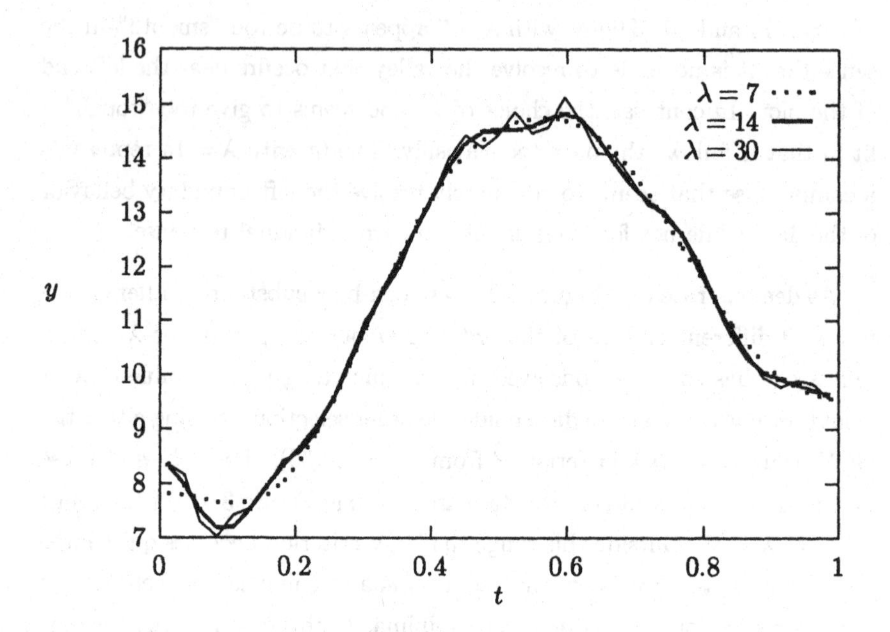

Figure 3.3: Cosine Series Fits to the Voltage Drop Data.

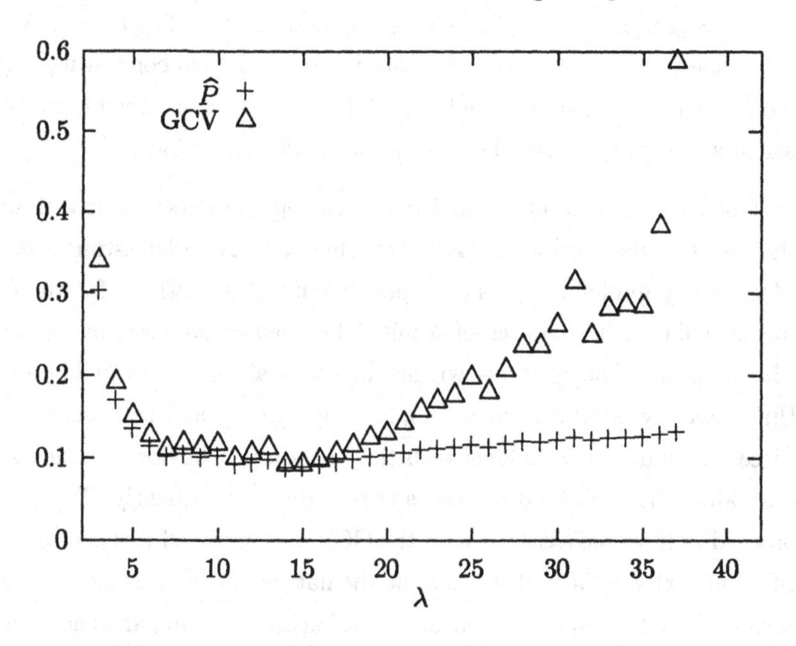

Figure 3.4: The GCV and $\widehat{P}$ Criteria for the Voltage Drop Data.

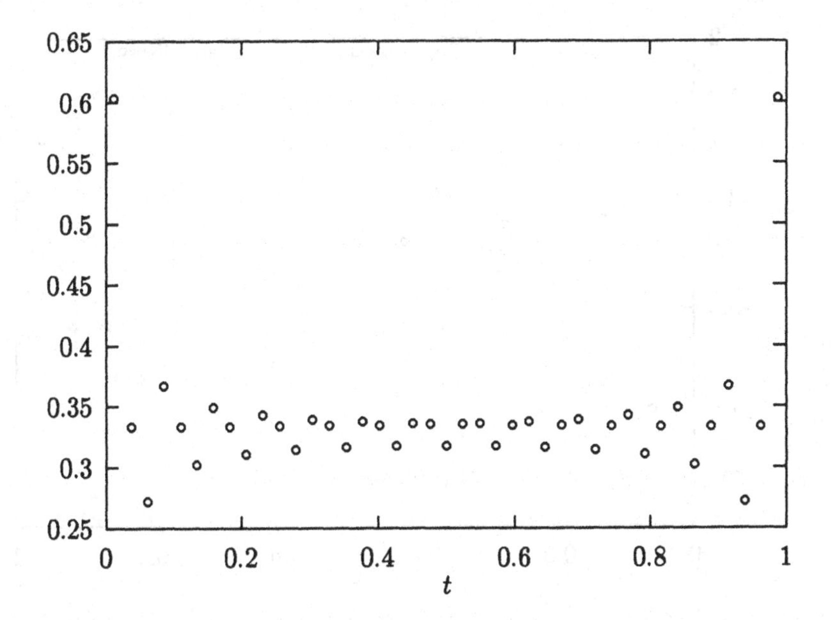

Figure 3.5: Leverage Values for the Cosine Series Fit.

The leverage values are defined as the diagonal elements s_i of the hat matrix $\mathbf{S}_{\widehat{\lambda}} = \mathbf{X}_{\widehat{\lambda}}(\mathbf{X}_{\widehat{\lambda}}^T\mathbf{X}_{\widehat{\lambda}})^{-1}\mathbf{X}_{\widehat{\lambda}}^T$ with $\mathbf{X}_{\widehat{\lambda}}$ the design matrix corresponding to $\lambda = \widehat{\lambda} = 14$ in (3.39). In this case, they have the simple form

$$s_i = n^{-1}[\widehat{\lambda} + \cos(\widehat{\lambda}\pi t_i)D(t_i; \widehat{\lambda} - 1)] \tag{3.48}$$

for D defined in (3.46). A plot of the s_i as a function of their t ordinates is given in Figure 3.5. None of the leverage values are appreciably large except those for the t_i that are near the boundaries of the design region $[0, 1]$. This indicates that the response values corresponding to the t_i in the boundary area have greater potential to substantially influence their own fits.

Figure 3.6 gives a plot of the studentized residuals

$$r_i = \frac{y_i - \mu_{\widehat{\lambda}}(t_i)}{\widehat{\sigma}\sqrt{1 - s_i}}, i = 1, \ldots n, \tag{3.49}$$

that were obtained from the fit with $\widehat{\lambda} = 14$ for the voltage drop data. Only three of these residuals have magnitudes that are close to the upper

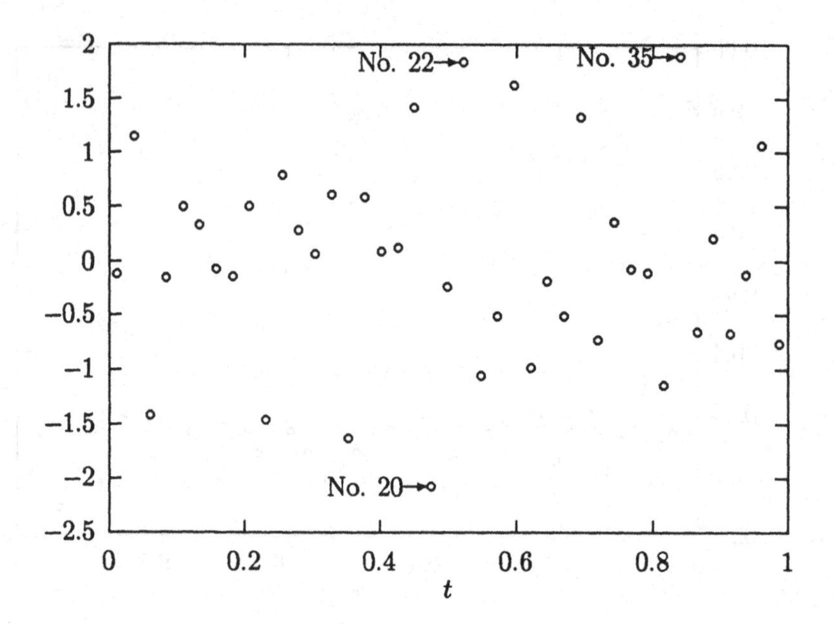

Figure 3.6: Studentized Residuals for the Cosine Series Fit

bound of 2 that gives a rough 5% level cut-off for "statistical significance" of a residual. These are the two responses marked as observations number 20 and 22 at the peak of the fitted curve and the observation marked as number 35 whose t ordinate is closer to the upper boundary.

The most important question is, of course, what observations are actually influential on the fit. To answer this one might examine values of the Cook's distance measures (Cook and Weisberg 1982)

$$d_i = \frac{s_i}{\hat{\lambda}(1 - s_i)} r_i^2 \ , \ i = 1, \ldots, n, \tag{3.50}$$

which are shown in Figure 3.7. From this plot we see that the fit is most influenced by observations 20, 22 and 35 that are also marked in Figure 3.2. From Figures 3.2 and 3.5–3.6 we can conclude that the influence of these observations is a result of their corresponding responses not being fit very well by the estimator. As a rough cut-off for what represents a "significantly" large Cook's distance value in this case we could use .52 which is the 10th percentile of an F distribution with $\hat{\lambda} = 14$ numerator and $n - \hat{\lambda} = 27$ denominator degrees-of-freedom. None of the observations

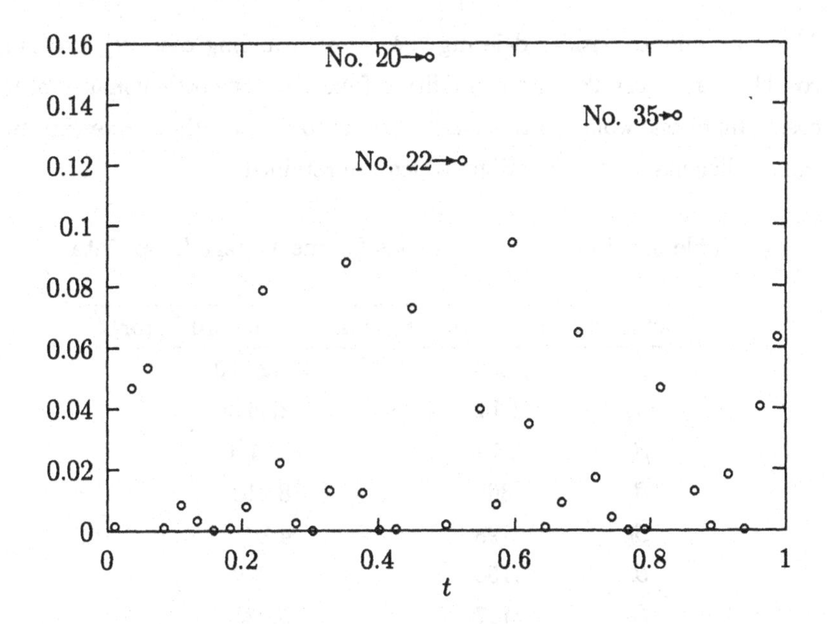

Figure 3.7: Cook's Distances for the Cosine Series Fit.

appear to be excessively influential from this viewpoint. Thus, the conclusion seems to be that any "problem" observations that were discovered in our diagnostic analysis are not particularly influential on the fit and, as a result, the voltage drop fit in Figure 3.2 appears satisfactory at this point.

The values of the estimated Fourier coefficients $\{b_j, j = 1, \ldots, 14\}$ are given in Table 3.1. Next to each coefficient we have also listed the value of the coefficient divided by an estimate of its standard error: namely, $\widehat{\sigma}/\sqrt{n} =$ .055. If μ actually followed a linear model involving the first 14 terms in the cosine series and the ε_i's were normally distributed, these values would provide tests of the hypotheses that the true Fourier coefficients which are being estimated are actually zero. Of course it is doubtful that μ follows such a linear model. Nonetheless, this interpretation is still approximately true and we can regard statistics of this type as providing indications of the importance of each of the fitted terms in estimating μ (Exercise 7).

For our purposes it will suffice to regard a term as contributing significantly to the fit if its coefficient exceeds twice its estimated standard error.

Thus, we might consider deleting the terms involving $\cos(7\pi t)$, $\cos(9\pi t)$, $\cos(11\pi t)$ and $\cos(12\pi t)$ from the fitted function. The orthogonality of the cosine functions would in this case allow us to do so without re-estimating the coefficients for the functions which are retained.

Table 3.1: Coefficient Estimates for the Voltage Drop Data

Coefficient	Estimate	Estimate/(Standard Error)
β_1	11.336	2742.789
β_2	-1.140	-27.642
β_3	-2.194	-53.196
β_4	.368	8.910
β_5	.198	4.801
β_6	.156	3.779
β_7	.127	3.083
β_8	.006	.140
β_9	.087	2.097
β_{10}	.044	1.074
β_{11}	.116	2.819
β_{12}	.022	.535
β_{13}	.024	.593
β_{14}	.112	2.725

The final step in our analysis involves the construction of interval estimators for $\mu(t)$. Figure 3.8 gives a plot of nominal 95% level intervals that were obtained using (3.32) with $\alpha = .05$ and $t = t_i, i = 1, \ldots, n$. Note that in this case the intervals take the simple form

$$\mu_{\hat{\lambda}}(t_i) \pm 1.96\hat{\sigma}\sqrt{s_i}, i = 1, \ldots, n, \tag{3.51}$$

with the s_i defined in (3.48). For display purposes the interval end points at each of the t_i have been connected in Figure 3.8 to give two continuous curves which enclose the fitted regression curve. These are not confidence bands, however, and, strictly speaking, the confidence levels for the intervals

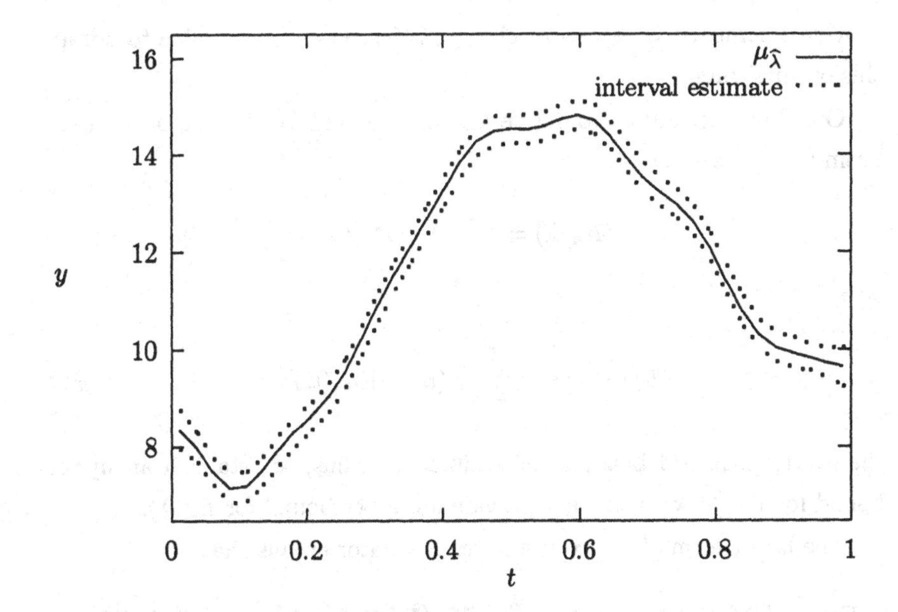

Figure 3.8: Fit and Interval Estimates for the Voltage Drop Data.

that are shown apply only to each design point separately and not uniformly across the design.

This example serves to illustrate several points. First it shows how GCV can be used to select λ and how modeling practices familiar from linear regression analysis can be used in the context of nonparametric regression series estimators. While the distribution theory for statistics used in linear models may not hold precisely in the nonparametric setting, examination of the values of statistics such as standardized coefficients, studentized residuals, etc., can still provide a valuable aid in the data fitting process.

3.4.3 Consistency and Efficiency of the Estimator

In this section we will study the large sample behavior of the risk $R_n(\lambda)$ in (3.34) in the case where μ_λ is given by (3.43) and the design is uniform as in (3.38). Let us initially assume that $\mu \in W_2^1[0,1]$ in which case the best rate of decay for R_n is $n^{-2/3}$ (cf Section 1.2.2). Thus, we begin our analysis by trying to determine whether or not cosine series estimators provide a competitive estimator choice in the sense of being able to obtain this optimal rate.

Our first step will be to derive an upper bound for the size of the risk. From (3.34) we know that

$$R_n(\lambda) = B_n^2(\lambda) + \sigma^2\lambda/n$$

with

$$B_n^2(\lambda) = n^{-1}\sum_{i=1}^{n}(\mu(t_i) - \mathrm{E}\mu_\lambda(t_i))^2 \tag{3.52}$$

the average squared bias for the estimator. Thus, to establish an upper bound for $R_n(\lambda)$ we need only provide an upper bound for $B_n^2(\lambda)$.

The linear form of the cosine series estimator entails that

$$\mathrm{E}\boldsymbol{\mu}_\lambda = (\mathrm{E}\mu_\lambda(t_1),\ldots,\mathrm{E}\mu_\lambda(t_n))^T = \mathbf{X}_\lambda\left(\mathbf{X}_\lambda^T\mathbf{X}_\lambda\right)^{-1}\mathbf{X}_\lambda^T\boldsymbol{\mu} = n^{-1}\mathbf{X}_\lambda\mathbf{X}_\lambda^T\boldsymbol{\mu}$$

with $\mathbf{X}_\lambda$ the design matrix from (3.39) and $\boldsymbol{\mu} = (\mu(t_1),\ldots,\mu(t_n))^T$. This means that $\mathrm{E}\boldsymbol{\mu}_\lambda$ is just the least-squares "fit" to the vector $\boldsymbol{\mu}$ using the cosine functions. More precisely, $\mathrm{E}\boldsymbol{\mu}_\lambda$ is the minimizer of

$$n^{-1}\sum_{i=1}^{n}(\mu(t_i) - a_i)^2$$

over all vectors $\mathbf{a} = (a_1,\ldots,a_n)^T$ of the form $\mathbf{a} = \mathbf{X}_\lambda\mathbf{c}$ for $\mathbf{c} \in \mathbb{R}^\lambda$. In particular, if we define $\widetilde{\mu}_\lambda = \sum_{j=1}^{\lambda}\beta_j x_j$ using the cosine Fourier coefficents β_j for μ in (3.14)–(3.15) and the cosine functions x_j in (3.7), then

$$B_n^2(\lambda) \le n^{-1}\sum_{i=1}^{n}(\widetilde{\mu}_\lambda(t_i) - \mu(t_i))^2. \tag{3.53}$$

The sum in (3.53) can be viewed as a quadrature approximation for $\int_0^1(\widetilde{\mu}_\lambda(t) - \mu(t))^2dt$. To assess the size of the quadrature error define the empirical design distribution function by

$$W_n(t) = \frac{\text{number of } t_i \le t}{n}.$$

It is easy to see that under (3.38)

$$\sup_{0\le t\le 1} |W_n(t)-t| = 1/2n.$$

Thus, we may integrate by parts to obtain

$$\begin{aligned}
&| \; n^{-1}\sum_{i=1}^{n}(\widetilde{\mu}_\lambda(t_i)-\mu(t_i))^2 - \int_0^1 (\widetilde{\mu}_\lambda(t)-\mu(t))^2 dt| \\
&= | \int_0^1 (\widetilde{\mu}_\lambda(t)-\mu(t))^2 d(W_n(t)-t)| \\
&= 2| \int_0^1 (\widetilde{\mu}'_\lambda(t)-\mu'(t))(\widetilde{\mu}_\lambda(t)-\mu(t))(W_n(t)-t)dt| \\
&\le 2||\widetilde{\mu}_\lambda-\mu|| \, ||\widetilde{\mu}'_\lambda-\mu'|| \sup_{0\le t\le 1}|W_n(t)-t| \qquad (3.54)
\end{aligned}$$

with $||\cdot||$ the $L_2[0,1]$ norm. Now, since $\mu \in W_2^1[0,1]$ we saw in Example 3.3 that $\beta_j = -\beta'_j/(j\pi)$ with

$$\beta'_j = \sqrt{2}\int_0^1 \mu'(t)\sin((j-1)\pi t)dt$$

the Fourier coefficients for μ' under the sine CONS for $L_2[0,1]$. Thus, using Parseval's relation (3.12) along with the cosine series expansion for μ and sine series expansion for μ' we obtain

$$||\widetilde{\mu}'_\lambda - \mu'||^2 = \sum_{j=\lambda+1}^{\infty}(\beta'_j)^2$$

and

$$||\widetilde{\mu}_\lambda - \mu||^2 = \sum_{j=\lambda+1}^{\infty}\frac{(\beta'_j)^2}{(j\pi)^2} \le (\lambda\pi)^{-2}\sum_{j=\lambda+1}^{\infty}(\beta'_j)^2 .$$

Consequently, as $n, \lambda \to \infty$

$$\begin{aligned}
B_n^2(\lambda) &= \int_0^1 (\widetilde{\mu}_\lambda(t)-\mu(t))^2 dt + o((n\lambda)^{-1}) \\
&\le \frac{1}{(\lambda\pi)^2}\sum_{j=\lambda+1}^{\infty}(\beta'_j)^2 + o((n\lambda)^{-1}). \qquad (3.55)
\end{aligned}$$

Using our upper bound for B_n^2 we can now see that if λ and n grow large, then

$$R_n(\lambda) \leq \frac{1}{(\lambda\pi)^2}\sum_{j=\lambda+1}^{\infty}(\beta_j')^2 + \frac{\sigma^2\lambda}{n} + O(n^{-1})$$
$$\leq \frac{1}{(\lambda\pi)^2}||\mu'||^2 + \frac{\sigma^2\lambda}{n} + O(n^{-1}). \quad (3.56)$$

It therefore follows that the cosine series estimator is consistent provided only that $\lambda \to \infty$ in such a way that $\lambda/n \to 0$. In words, this means that the number of terms included in the estimator should increase with n but not as fast as the sample size. This was anticipated from our earlier discussion in Section 3.3.

It also follows from (3.56) that if the order for μ_λ is allowed to grow at the rate $n^{1/3}$ we will then have $R_n(\lambda) = O(n^{-2/3})$ and, consequently, it is indeed possible for the cosine series estimator to obtain the optimal risk convergence rate for estimation of $W_2^1[0,1]$ functions. We can take this one step further and obtain a "best" choice for λ by minimizing the lead terms in the second bound in (3.56) with respect to λ. This gives the asymptotically optimal order

$$\lambda_n^* = [2n||\mu'||^2/(\sigma\pi)^2]^{1/3} \quad (3.57)$$

for which the associated risk is

$$R_n(\lambda_n^*) \leq (2^{1/3} + 2^{-2/3})(\frac{||\mu'||\sigma^2}{\pi})^{2/3}n^{-2/3} + O(n^{-1}). \quad (3.58)$$

Some important insight can be gained by closer inspection of the asymptotically optimal order and risk in (3.57) and (3.58). We see for example that the bound for $R_n(\lambda_n^*)$ is an increasing function of σ^2 while λ_n^* decreases with σ^2. This is intuitively reasonable because large values of σ^2 correspond to situations where the noise level in the data is high which makes it harder to estimate μ. In such situations the risk can be reduced most effectively by doing more averaging or smoothing which is accomplished by keeping λ small.

We also see that λ_n^* and the bound for $R_n(\lambda_n^*)$ are increasing functions of $||\mu'||$. The value of $||\mu'||$ represents the average squared slope of the regression function. It will be near zero for very flat regression curves and larger for functions with many wiggles or peaks and valleys. Thus, $||\mu'||$ can be viewed as a measure of the complexity of the regression function with larger values indicating increased estimation difficulty. In cases where $||\mu'||$ is large the bias component tends to dominate the risk so that estimation precision is improved and bias is reduced by increasing the number of terms used in the estimator.

Returning to our original bound (3.55) for the average squared bias, we see $B_n^2(\lambda)$ can actually converge to zero much faster than λ^{-2} depending on the rate of decay of the unestimated cosine Fourier coefficients $\beta_j, j = \lambda+1,\ldots$ To illustrate this point suppose, for example, that $\mu \in W_2^2[0,1]$. Then, integration by parts gives

$$\beta_j = \frac{\sqrt{2}}{((j-1)\pi)^2}[(-1)^{(j-1)}\mu'(1) - \mu'(0)] - \frac{1}{((j-1)\pi)^2}\beta_j'' \tag{3.59}$$

with

$$\beta_j'' = \sqrt{2}\int_0^1 \mu''(t)\cos((j-1)\pi t)dt$$

the cosine Fourier coefficients for μ''. Since $\mu'' \in L_2[0,1]$, $|\beta_j''| \to 0$ as $j \to \infty$ and, hence, $\beta_j = \frac{\sqrt{2}}{((j-1)\pi)^2}[((-1)^{(j-1)}\mu'(1) - \mu'(0)) + o(1)]$. Using an integral estimate to approximate $\sum_{j=\lambda}^{\infty} j^{-4}$ then gives that

$$B_n^2(\lambda) \leq \frac{2}{3\pi^4\lambda^3}[|\mu'(1)| + |\mu'(0)|]^2(1+o(1)) + o((n\lambda)^{-1}). \tag{3.60}$$

Consequently, by choosing the number of terms in the estimator to be of order $n^{1/4}$ the average squared bias and the risk can be made to decay at the rate $n^{-3/4}$ for a $W_2^2[0,1]$ regression function. This has the implication that additional smoothness of the regression function translates, as one would hope, into improved estimator efficiency.

From Section 1.2.2 we know that the best rate of convergence for the risk when estimating a $W_2^2[0,1]$ function is $n^{-4/5}$ which is faster than the cosine series rate of $n^{-3/4}$. We will see that there are, in fact, many estimators that are capable of obtaining the optimal $n^{-4/5}$ rate. Thus, it does not appear that the cosine series estimator represents a viable alternative for estimation of functions in $W_2^2[0,1]$ and we are led to wonder why the cosine series estimator fails in this case.

The root of the problem lies in identity (3.59). From this we see that if $\mu'(0) = \mu'(1) = 0$, then a parallel of our previous analysis shows that when λ is of order $n^{1/5}$ the risk for μ_λ will, in fact, decay at the optimal $n^{-4/5}$ rate. Similarly, if $\mu \in W_2^3[0,1]$ with $\mu'(0) = \mu'(1) = 0$ the optimal risk for μ_λ will decay at the optimal $n^{-6/7}$ rate. If $\mu \in W_2^4[0,1]$ with $\mu'(0) = \mu'(1) = \mu''(0) = \mu''(1) = 0$, the risk for μ_λ can be made to decay at an optimal $n^{-8/9}$ rate, etc. What determines whether such optimal convergence rates can be obtained is whether or not the function being estimated satisfies the same types of boundary behavior as the cosine functions. Thus, for example, the cosine estimator can be a *second order estimator*, i.e., obtain the $n^{-4/5}$ rate, if the regression function behaves like the cosine functions in the sense that its derivative vanishes at 0 and 1.

In practice it seems unlikely that the boundary behavior of the regression function can be expected to match that of the cosine functions. When this fails to happen (3.60) suggests that the average squared bias of the estimator will asymptotically be a function only of values of derivatives of μ at 0 and/or 1. Consequently, choices of λ that attempt to minimize the risk in such cases will tend to focus primarily on fitting at the boundaries of [0, 1]. This will cause the selected order to be larger than might otherwise be desirable for estimation at points in the interior of [0, 1].

An illustration of the practical impact of boundary effects on estimation can be seen from the fit to the voltage drop data in the previous section. In that case it seems quite unlikely that the derivative of the true regression function vanishes at 0 and this has created problems in constructing a satisfactory estimator. The GCV choice of λ in this case, which is attempting to estimate the minimizer of the risk, has produced an estimator that appears to resolve the boundary behavior of the data at the expense of introducing seemingly extraneous wiggles elsewhere in the interior of [0, 1].

Even though the boundary properties of cosine series estimators make them less than ideal for general use in nonparametric regression, they are still to be preferred over the other two trigonometric CONS's discussed in Example 3.1. The boundary problems associated with these other trigonometric bases are such that they are incapable of obtaining the optimal convergence rate for even a $W_2^1[0,1]$ function. This issue is explored further in Exercises 25-29.

Trigonometric series estimators have some decided advantages from a computational standpoint. They tend to produce well conditioned design matrices for designs that are near uniform which makes it relatively easy to compute their associated Fourier coefficients. This is to be compared with polynomial regression estimators of the next section whose design matrices are notoriously ill-conditioned. Thus, the boundary difficulties associated with trigonometric estimators seem quite disappointing initially. However, we will see in Section 3.6 that it is actually relatively easy to fix these boundary problems in a way that does not appreciably compromise the good computational features of the estimators.

To conclude this section, we should perhaps observe that the direct, practical applications of the results in this section are somewhat limited at this point. For instance, (3.57) provides us with no real help in choosing λ from data since estimation of λ_n^* requires estimation of $||\mu'||$, a formidable problem in its own right. It is therefore worthwhile to note that the reason for conducting an asymptotic analysis is generally not the production of new estimation techniques and concepts but is rather to obtain a better understanding of an estimation problem. The insights gained in this way often turn out to have practical consequences as will be demonstrated in much of the work that follows.

3.4.4 Asymptotic Distribution Theory

The previous section dealt with the properties of the cosine series estimator as a point estimator of the regression function. We would also like to have interval estimates of μ that could be constructed using μ_λ. To develop such intervals it now becomes necessary to study the large sample distribution theory associated with the estimator.

From our work on the form of the estimator in Section 3.4.1, we know that $\mu_\lambda(t) = n^{-1}\sum_{i=1}^n K_\lambda(t, t_i)y_i$ with $K_\lambda(\cdot,\cdot)$ given by (3.45). Since the weights applied to the responses sum to one, $\mu_\lambda(t)$ is a weighted average of the responses. Thus, we might expect there to be a central limit effect that would endow an appropriate normalization of $\mu_\lambda(t)$ with an approximate normal distribution for large n. If this is so, it would provide a seemingly natural starting point for the development of confidence intervals for $\mu(t)$.

Let us now assume that the random errors $\varepsilon_1, \ldots, \varepsilon_n$ are *iid* with mean zero and variance σ^2. Then, we may write

$$\mu_\lambda(t) - \mathrm{E}\mu_\lambda(t) = \sum_{i=1}^{n} w_{in}(t)\varepsilon_i \tag{3.61}$$

with

$$w_{in}(t) = n^{-1}K_\lambda(t, t_i),\ i = 1, \ldots, n. \tag{3.62}$$

The large sample distributional properties of sums such as (3.61) can be addressed using the following central limit theorem whose proof is a direct application of the classical Lindeberg-Feller Central Limit Theorem (cf Serfling 1980, Section 1.9.2).

Lemma 3.1 (Central Limit Theorem). *Let $\varepsilon_1, \ldots, \varepsilon_n$ be iid random variables with $\mathrm{E}\varepsilon_1 = 0$ and $\mathrm{Var}(\varepsilon_1) = \sigma^2 < \infty$ and define $S_n = \sum_{i=1}^n w_{in}\varepsilon_i$. Then*

$$\frac{S_n}{\sigma\sqrt{\sum_{i=1}^n w_{in}^2}} \xrightarrow{d} Z, \tag{3.63}$$

where Z has a standard normal distribution, if

$$\max_{1\le i\le n} |w_{in}| / \sqrt{\sum_{i=1}^{n} w_{in}^2} \to 0. \tag{3.64}$$

To apply Lemma 3.1 we will take $w_{in} = w_{in}(t), i = 1, \ldots, n$, in (3.62) in which case it follows directly from (3.45) that

$$\sup_{0\le t\le 1} \max_{1\le i\le n} |w_{in}(t)| \le (2\lambda - 1)/n. \tag{3.65}$$

We now need to compute $\sum_{i=1}^n w_{in}^2$ or, equivalently, the variance of $\mu_\lambda(t)$. Using (3.40)–(3.41) this is found to be

$$\begin{aligned} \mathrm{Var}(\mu_\lambda(t)) &= \sigma^2 \sum_{i=1}^{n} w_{in}^2(t) \\ &= \frac{\sigma^2}{n}[1 + 2\sum_{j=1}^{\lambda-1} \cos^2(j\pi t)]. \end{aligned} \tag{3.66}$$

The variance becomes $\sigma^2(2\lambda - 1)/n$ when $t = 0$ or 1. For $t \in (0, 1)$ we have from (3.47) that

$$\begin{aligned} 2\sum_{j=1}^{\lambda-1}\cos^2(j\pi t) &= \lambda - 1 + \sum_{j=1}^{\lambda-1}\cos(2j\pi t) \\ &= \lambda - 1 + \cos(\lambda\pi t)D(t; \lambda - 1). \end{aligned} \tag{3.67}$$

Since $|\cos(\lambda\pi t)D(t; \lambda - 1)| \leq 1/|\sin(\pi t)|$, (3.66) then gives

$$\operatorname{Var}(\mu_\lambda(t)) = \frac{\sigma^2\lambda}{n} + O(n^{-1}) \tag{3.68}$$

for any fixed $t \in (0, 1)$. By combining (3.65) and (3.68) with Lemma 3.1 we have proved the following result.

Theorem 3.1. *Assume that* $\varepsilon_1, \ldots, \varepsilon_n$ *are iid with* $\mathrm{E}\,\varepsilon_1 = 0$ *and* $\operatorname{Var}(\varepsilon_1) = \sigma^2 < \infty$. *Then, if* $\lambda/n \to 0$,

$$\frac{\mu_\lambda(t) - \mathrm{E}\,\mu_\lambda(t)}{\sqrt{\operatorname{Var}(\mu_\lambda(t))}} \xrightarrow{d} Z, \tag{3.69}$$

where Z *has a standard normal distribution.*

This theorem states that, as expected, the cosine series estimator can be recentered and rescaled in such a way that it has an approximate standard normal distribution for large n provided that the number of terms included in the estimator is not allowed to grow as fast as the sample size. This condition seems quite natural intuitively since values of λ that are near n produce estimators that tend to almost interpolate the data. In such cases there is very little or no averaging being done by the estimator which precludes the possibility of a central limit effect.

The question now arises as to how Theorem 3.1 can be used to develop interval estimators for $\mu(t)$. To address this problem first write

$$\frac{\mu_\lambda(t) - \mu(t)}{\sqrt{\operatorname{Var}(\mu_\lambda(t))}} = \frac{\mu_\lambda(t) - \mathrm{E}\mu_\lambda(t)}{\sqrt{\operatorname{Var}(\mu_\lambda(t))}} + \frac{\mathrm{E}\mu_\lambda(t) - \mu(t)}{\sqrt{\operatorname{Var}(\mu_\lambda(t))}}. \tag{3.70}$$

The first term on the right hand side of (3.70) is asymptotically standard normal due to Theorem 3.1. Thus, $(\mu_\lambda(t) - \mu(t))/\sqrt{\operatorname{Var}(\mu_\lambda(t))}$ can be used as a pivotal quantity for the construction of interval estimators provided that $(\mathrm{E}\mu_\lambda(t) - \mu(t))/\sqrt{\operatorname{Var}(\mu_\lambda(t))}$ is asymptotically negligible. The resulting intervals (using (3.68) for $t \in (0,1)$ and $\lambda \to \infty$) will have the asymptotic form

$$\mu_\lambda(t) \pm z_{\alpha/2}\sigma\sqrt{\frac{\lambda}{n}}. \tag{3.71}$$

The large sample validity of intervals such as (3.71) relies on the second term on the right hand side of (3.70) being small. This is tantamount to saying that λ must be chosen to insure that the bias of $\mu_\lambda(t)$ is negligible relative to its standard error. Unfortunately, this condition poses a problem of substantial practical significance.

To appreciate the difficulties involved, begin by assuming that $\mu \in W_2^1[0,1]$ and observe that

$$|\mathrm{E}\mu_\lambda(t) - \mu(t)| \le \sqrt{2}\sum_{j=1}^{\lambda}|\beta_j - \mathrm{E}b_j| + \sqrt{2}\sum_{j=\lambda+1}^{\infty}|\beta_j|. \tag{3.72}$$

From our work in Section 3.4.3 and the Cauchy-Schwarz inequality we know that

$$\sqrt{2}\sum_{j=\lambda+1}^{\infty}|\beta_j| \le \sqrt{2}\|\mu'\|/(\pi\sqrt{\lambda}). \tag{3.73}$$

Thus, we will have a bound on the bias of $\mu_\lambda(t)$ if we can assess the magnitude of the bias of the estimated cosine Fourier coefficients. For this purpose we have the following lemma.

Lemma 3.2. *If $\mu \in W_2^1[0,1]$, there is a constant A independent of j and n such that $|\mathrm{E}b_j - \beta_j| \le An^{-1}, j = 2,\ldots,n$. This rate improves to n^{-2} if $\mu \in W_2^2[0,1]$.*

Proof. First observe that since $\mu' \in L_2[0,1]$, the Fourier cosine series for μ converges uniformly on [0, 1] which allows us to write $\mathrm{E}b_j = \sum_{r=1}^{\infty}\beta_{r+1}d_{jr}$ with

$$d_{jr} = 2n^{-1}\sum_{i=1}^{n}\cos(j\pi t_i)\cos(r\pi t_i).$$

The d_{jr} are evaluated explicitly in Section 3.8. Using the results given there we obtain

$$\mathrm{E}b_j = \beta_j + \sum_{r=1}^{\infty}(\beta_{j+4rn} + \beta_{-j+4rn}) - \sum_{r=1}^{\infty}(\beta_{j+(4r-2)n} + \beta_{-j+(4r-2)n}).$$

Now use $\beta_j = -\beta'_j/((j-1)\pi)$ for $\beta'_j = \sqrt{2}\int_0^1 \mu'(t)\sin((j-1)\pi t)dt$ along with the square summability of the β'_j and the Cauchy-Schwarz inequality to complete the proof. A further integration by parts establishes the result for $\mu \in W_2^2[0,1]$. •

Using Lemma 3.2 and (3.72)-(3.73) gives

$$|\mathrm{E}\mu_\lambda(t) - \mu(t)| \le \sqrt{2}\|\mu'\|/(\pi\sqrt{\lambda}) + O\left(\frac{\lambda}{n}\right). \tag{3.74}$$

Since $\mathrm{Var}(\mu_\lambda(t))$ is of exact order λ/n we have therefore shown that $(\mathrm{E}\mu_\lambda(t) - \mu(t))/\sqrt{\mathrm{Var}(\mu_\lambda(t))} = o(1)$ if $\sqrt{n}/\lambda \to 0$. We state this result formally as follows.

Corollary 3.1. *If $n, \lambda \to \infty$ in such a way that $\lambda/n \to 0$ and $\lambda^2/n \to \infty$, then*

$$\frac{\mu_\lambda(t) - \mu(t)}{\sqrt{\mathrm{Var}(\mu_\lambda(t))}} \xrightarrow{d} Z, \tag{3.75}$$

where Z has a standard normal distribution and

$$P(\,\mu(t) \in \mu_\lambda(t) \pm z_{\alpha/2}\sqrt{\mathrm{Var}(\mu_\lambda(t))}\,) \to 1-\alpha. \tag{3.76}$$

On the surface this result seems useful in that it states we need only choose our order sequence to grow large slower than the sample size but faster than $\sqrt{n}$ in order for the interval estimators in (3.71) to be asymptotically valid. The problem is that this precludes asymptotically "optimal" choices for the order such as (3.57) which grows at the exact rate $n^{1/3}$ and does not satisfy the $\lambda/\sqrt{n} \to \infty$ condition that is needed for valid interval estimation.

The problem with optimal choices for λ is that they attempt to balance the variance and squared bias components of the estimator which is the wrong thing to do for interval estimation where squared bias needs to be of smaller order than the variance. A standard panacea that is often prescribed to "solve" this difficulty is to choose λ sufficiently larger than optimal to make the estimator's bias small relative to its standard error. This recommendation carries with it the associated problem of determining how much larger than optimal λ should be chosen. While there are many ways to accomplish this in an asymptotic sense, it is difficult to translate such vague recommendations into something that can be used in a practical setting where μ is unknown and λ must actually be estimated from the data.

To illustrate the kinds of problems that bias can cause for interval estimation in practice, we conducted a simulation experiment similar in spirit to that of Example 1.5 except that we used cosine series estimators instead of regressograms. As in that case we generated 1000 samples of size $n = 50$ from a nonparametric regression model with uniformly spaced design points $t_i = (2i-1)/100, i = 1, \ldots, 50$, normally distributed random errors (or ε_i's) with $\sigma = .2$ and $\mu(t) = t + \exp\{-50(t-.5)^2\}$. Cosine series estimators were fit to all the data sets with λ chosen using an unbiased risk estimator using the GSJS variance estimator. The interval estimators (3.51) were then constructed at each of the design points and the corresponding empirical coverage proportions (over the 1000 samples) were computed. Figure 3.9 gives a plot of a typical data set and cosine fit from the simulation with the associated 95% interval estimators and true regression function shown in Figure 3.10. Figure 3.11 shows the empirical coverages that were produced by the simulation using a nominal 95% level. The empirical levels are seen to depart substantially from the nominal level particularly in areas such as the boundary of $[0, 1]$ and peak of the regression function where estimator bias may be the most substantial.

The basic conclusion of our discussions concerning interval estimation is that the construction of effective, practical interval estimation methods will require the development of methods that take into account the effect of the bias of μ_λ on the intervals. This is a rather difficult task for series estimators where it is hard to get precise expressions for pointwise estimator bias, even asymptotically. In the next chapter we will discuss kernel estimators where it is much easier to suggest forms of bias estimators that can be used to produce bias corrected intervals. One of the conclusions that follows from the work with kernel estimators is that, under certain restrictions, conservative confidence intervals that account for estimator bias can be produced in that setting by using Bonferroni intervals without explicit bias

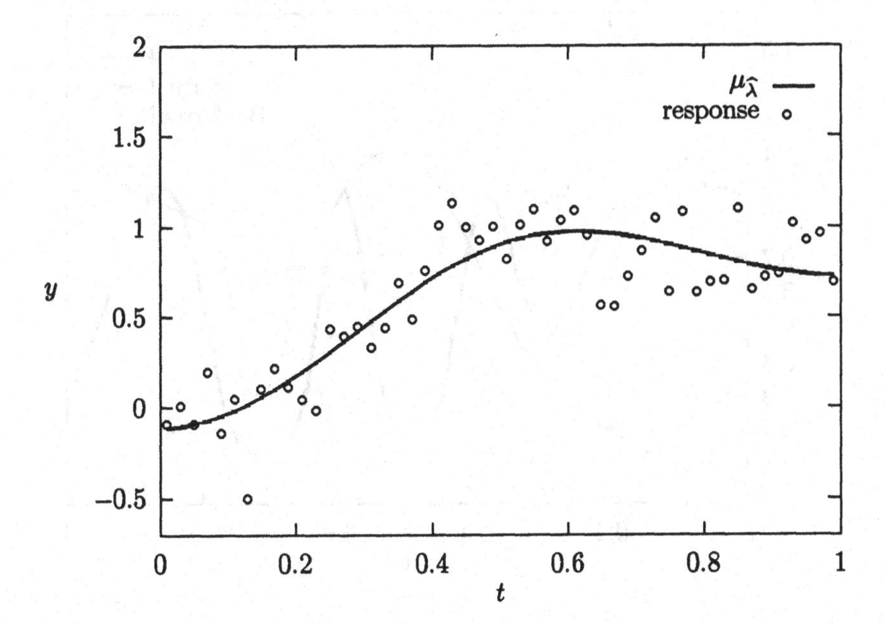

Figure 3.9: Cosine Series Fit to Simulated Data.

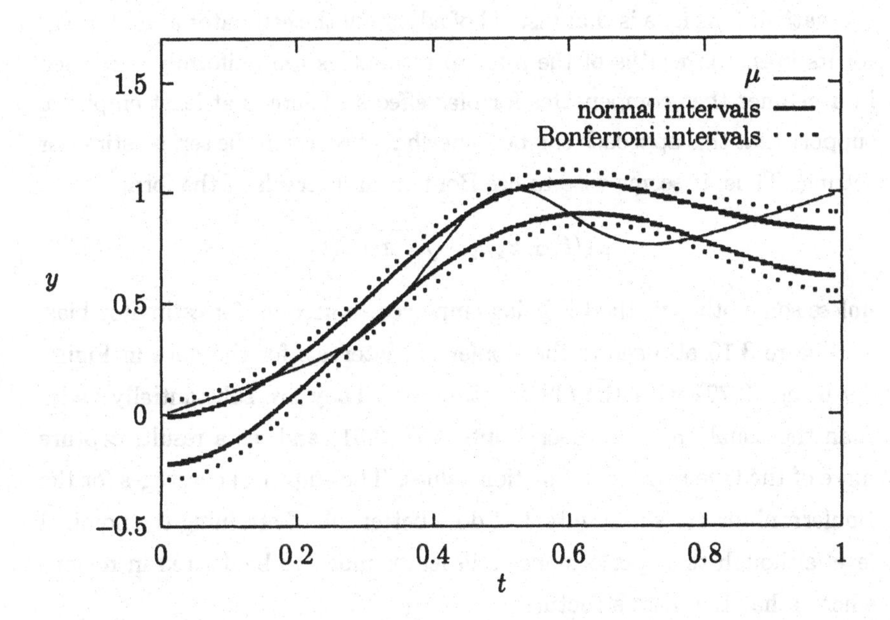

Figure 3.10: Interval Estimates for Simulated Data.

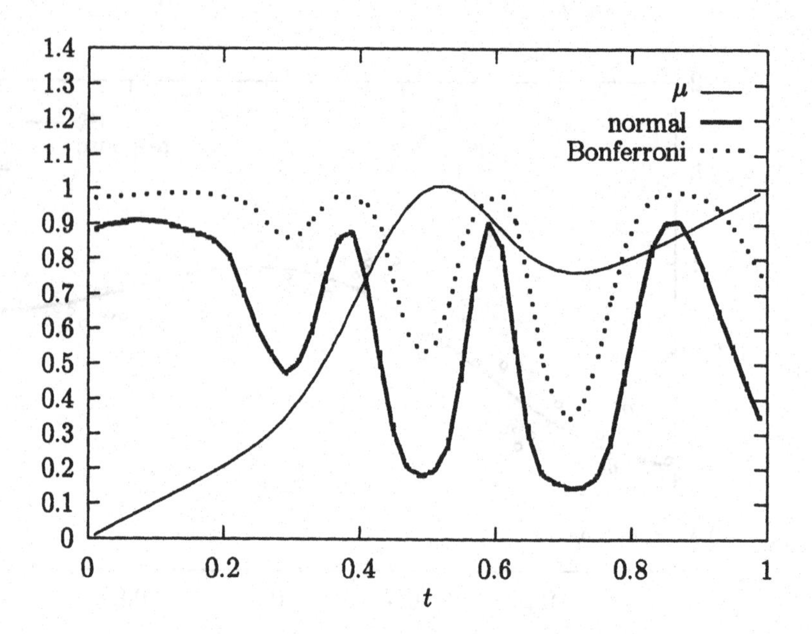

Figure 3.11: Empirical Coverages for Cosine Interval Estimates.

estimation. The idea is that instead of adjusting the estimator at each point for its bias, the widths of the interval estimators are uniformly expanded in a manner that compensates for bias effects. There is at least empirical support that this approach can be somewhat effective in the series estimator setting. Thus, it seems wise to use Bonferroni intervals of the form

$$\mu_\lambda(t) \pm z_{\alpha/(2n)}\sqrt{\mathrm{Var}(\mu_\lambda(t))} \tag{3.77}$$

unless some other method is being employed to account for estimator bias.

Figure 3.10 also shows the Bonferroni intervals for the data in Figure 3.9 using (3.77) with the GSJS estimator. They are substantially wider than the usual "normal theory" intervals (3.51) and, as a result, capture more of the true regression function values. The empirical coverages for the Bonferroni intervals in Figure 3.11 do a better job of attaining the nominal level although their performance still leaves much to be desired in regions where μ has the most structure.

Interval estimators such as (3.51) are pointwise in that they have nominal coverage level $1-\alpha$ at each point t but not uniformly for all t. Intervals that have coverage $1-\alpha$ for all $t \in [0,1]$ are called confidence bands. We will see in Chapter 4 that the Bonferroni approach can produce asymptotically valid confidence bands for kernel type nonparametric regression estimators although it is unclear to what extent this work is applicable in the series estimator setting. A promising approach to confidence band construction that does appear to apply to series estimators has been developed by Sun and Loader (1994).

Another inferential issue that was raised in Section 3.3 concerned the use of the sample Fourier coefficients to conduct tests about the true Fourier coefficients using standard, linear regression testing methodology. The news here is much more positive and this approach is actually seen to work quite well in an asymptotic sense. Specifically, using Lemmas 3.1–3.2, it can be shown (Exercise 7) that

$$\frac{\sqrt{n}(b_j - \beta_j)}{\sigma} \xrightarrow{d} Z \tag{3.78}$$

with Z a standard normal random variable. Equation (3.78) continues to hold if we replace σ by any $\sqrt{n}$-consistent estimator such as the GSJS estimator which gives an asymptotic justification for our discussions concerning the estimated cosine Fourier coefficients for the voltage drop data in Section 3.4.2.

3.4.5 Testing for No Effect

In Chapter 1, we mentioned that nonparametric regression techniques could be used in a confirmitory capacity to provide information about the lack-of-fit of parametric models. There is now a fairly substantial literature on this subject with a general treatment and overview given in Hart (1997). We will not attempt to explore all the aspects of the testing problem here but will, instead, focus on a very simple case. Specifically, we will treat only the problem of testing for no effect in regression using a cosine series estimator. While the practical aspects of this problem are somewhat limited, the basic structure of the tests developed for this setting generalizes quite easily to other lack-of-fit problems and thereby provides a stepping stone into the literature on the subject.

Let us now assume that our data follows the model

$$y_i = \mu(t_i) + \varepsilon_i,\ i = 1, \ldots, n, \tag{3.79}$$

where $t_i = (2i-1)/2n$ and, for simplicity, the $\varepsilon_i, i = 1, \ldots, n$, are *iid* normal random variables with mean zero and known variance σ^2. We wish to test the hypothesis that the predictor variable t has no effect on the response variable or, equivalently that the mean function is a constant. Thus, the null hypothesis to be tested is

$$H_0 : \mu(\cdot) = \theta, \theta \in \mathbb{R}. \tag{3.80}$$

Let μ be some generic regression function in $W_2^1[0,1]$. Then, we can always write

$$\mu = \theta + f, \tag{3.81}$$

where

$$\theta = \int_0^1 \mu(t)dt \tag{3.82}$$

and $f = \mu - \theta$ represents the departure or perturbation away from the null model. Accordingly, H_0 is equivalent to testing that $f = 0$ in (3.81).

One approach to testing H_0 would be to estimate f in (3.81) and then compare the estimator to the zero function in some fashion. The first step in such a process is to obtain an estimator of f. We will discuss how this can be done and will then see that this leads to several natural ways of conducting our comparison.

Using expression (3.43) for example we see that a cosine series fit to the data will automatically provide an estimator of f of the form

$$f_\lambda(t) = \sqrt{2}\sum_{j=2}^{\lambda} b_j \cos((j-1)\pi t), \tag{3.83}$$

where the b_j are the sample cosine Fourier coefficients (3.42). The b_j provide estimators of the cosine Fourier coefficients

$$\beta_j = \sqrt{2}\int_0^1 f(t)\cos((j-1)\pi t)dt,\ j = 2, \ldots \tag{3.84}$$

for f (and μ) which must all vanish under the null hypothesis. Note that under this formulation the null model parameter θ is just the Fourier coefficient β_1 for μ that corresponds to the constant function in the cosine CONS.

Under model (3.79) and the null hypothesis, the $\sqrt{n}b_j/\sigma, j = 2,\ldots,n$, are *iid* standard normal random variables which make them natural building blocks for the construction of test statistics about H_0. They can be combined and used in a variety of ways to produce tests with different features and properties. For example, we might consider a general class of statistics of the form

$$T = n\sum_{j=2}^{n} w_j b_j^2 \tag{3.85}$$

where $w_2,\ldots,w_n$ are pre-specified, non-negative weights. Under H_0, T is a weighted sum of *iid* chi-squared random variables with one degree-of-freedom. One can then tailor the test statistic's sensitivity to alternatives by selecting the weights in (3.85) appropriately. Two specific choices for the weights that will be of particular interest for the discussions here are $w_j = 1, j = 2,\ldots,\lambda, w_j = 0, j = \lambda+1,\ldots,n$, and $w_j = 1/((j-1)\pi)^2, j = 2,\ldots,n$. These choices result in the test statistics

$$T_1(\lambda) = \frac{n}{\sigma^2}\sum_{j=2}^{\lambda} b_j^2 = \frac{1}{\sigma^2}\sum_{i=1}^{n} f_\lambda^2(t_i) \tag{3.86}$$

and

$$T_2 = \frac{n}{\sigma^2}\sum_{j=2}^{n}\frac{b_j^2}{((j-1)\pi)^2}, \tag{3.87}$$

respectively. Note the (3.86) follows from Parseval's relation applied to the Euclidean norm for $\mathbb{R}^n$. It shows that $T_1(\lambda)$ can be viewed as measuring the distance between the cosine series estimator of f and the zero function by averaging the values of f_λ^2 across the design.

The two statistics (3.86)–(3.87) use the sample Fourier coefficients in very different ways. While $T_1(\lambda)$ uses only $\lambda - 1$ of the sample Fourier coefficients, it gives equal weight to all the coefficients that are included in the statistic. In contrast, T_2 uses all the sample Fourier coefficients, but heavily downweights or damps the contributions from coefficients corresponding to higher frequency cosine functions. One might expect these two very different ways of using the b_j to have substantial ramifications concerning the relative power of the two tests against different types of alternatives to H_0. Intuitively, it would seem that T_2 would be best suited for detecting alternatives where f is primarily aligned with low frequency cosine functions such as $\cos(\pi t)$ and $\cos(2\pi t)$. The $T_1(\lambda)$ statistic might then be better for detecting higher frequency alternatives provided that λ could be chosen correctly.

This brings up the question of how λ should be chosen for (3.86). This could be accomplished using an unbiased risk estimator, for example, and we will return to that possibility momentarily. However, since the true mean function is finite dimensional under H_0, our work in Section 2.3 suggests that we should use a consistent order selector to choose λ. In particular, we could take λ to be the minimizer of the Schwarz criterion, i.e., the maximizer $\widetilde{\lambda}$ of (cf (2.40) and Exercise 15)

$$\widetilde{M}(\lambda) = n\sum_{j=2}^{\lambda} b_j^2 - \sigma^2 \lambda \ln(n) \tag{3.88}$$

over $\lambda = 2, \ldots, n$.

Thus, we propose to test H_0 using

$$T_1 = T_1(\widetilde{\lambda}) \tag{3.89}$$

with $\widetilde{\lambda}$ obtained from (3.88). Using our work in Section 2.3 we can show (Exercise 16) the following result for this test statistic.

Theorem 3.2. *Under H_0, T_1 converges in distribution to a central chi-squared random variable with 1 degree-of-freedom. If H_0 is false, then $P(T_1 \leq C) \to 0$ for any fixed constant C.*

The first part of the theorem suggests that we can test the no effect hypothesis by comparing T_1 to critical values from a chi-squared distribution. Specifically, if $\chi^2_{1,\alpha}$ is the $100(1-\alpha)$th percentile for the chi-squared distribution with one degree-of-freedom, then the test obtained by rejecting H_0 whenever $T_1 \geq \chi^2_{1,\alpha}$ will have asymptotic level α. The second part of the theorem has the consequence that the test is consistent against all alternatives to H_0 in that

$$P(T_1 \geq \chi^2_{1,\alpha} | H_0 \text{ is false }) \to 1$$

for $\mu \in W_2^1[0,1]$ (Exercise 16).

In practice, the chi-squared approximation for T_1 provided by Theorem 3.2 tends to be somewhat inadequate in small samples. The problem stems from occasional overselection of λ by the order selector under H_0 which produces inflated values for the test statistic. To account for this one can use either a second order approximation (Kallenberg and Ledwina 1995) or simulated critical values. This latter approach is what was used to produce critical values for T_1 in the simulation experiment that is discussed below.

Another approach to testing H_0 can be obtained by going back and reconsidering the order selection ideas that led us to T_1. We know that when H_0 is true the actual order of the regression model is $\lambda = 1$. Thus, we might take a different approach and reject H_0 whenever $\widehat{\lambda}$ exceeds 1 with $\widehat{\lambda}$ the maximizer of a criterion such as

$$M(\lambda) = n \sum_{j=1}^{\lambda} b_j^2 - C\sigma^2 \lambda \tag{3.90}$$

over $\lambda = 1, \ldots, n$, for C some nonnegative constant. If, for example, we take $C = 2$, this gives the estimator of λ that one would obtain by minimizing the unbiased risk criterion (3.36) for cosine series estimation of μ.

Arguments similar to those used to establish Theorem 2.2 can be used to derive the null distribution of the statistic $\widehat{\lambda}$ (Exercise 18). The result can be formally stated as follows.

Theorem 3.3. Let $\widehat{\lambda}$ be the maximizer of (3.90) over $\lambda = 1, \ldots, n$. Then,

$$P(\widehat{\lambda} = \lambda) = u_{\lambda-1} v_{n-\lambda}, \ \lambda = 1, \ldots, n, \tag{3.91}$$

where $u_0 = v_0 = 1$ and for $k \geq 1$

$$u_k = \sum_{k}^{*} \left(\prod_{i=1}^{k} \frac{1}{j_i!} \left(\frac{P(\chi_i^2 > iC)}{i} \right)^{j_i} \right) \tag{3.92}$$

and

$$v_k = \sum_k^* \left(\prod_{i=1}^k \frac{1}{j_i!} \left(\frac{P(\chi_i^2 \le iC)}{i} \right)^{j_i} \right) \tag{3.93}$$

for χ_k^2 a central chi-squared random variable with k degrees-of-freedom and the sums in (3.92)-(3.93) are over all k-tuples of integers $(j_1, \ldots, j_k)$ such that $j_1 + 2j_2 + \cdots + kj_k = k$.

If we use $\hat{\lambda}$ for testing and reject when $\hat{\lambda} > 1$, then the chance of overselection by $\hat{\lambda}$ under H_0 is actually our chance for a Type I error. Thus, we want $P(\hat{\lambda} = 1 | H_0 \text{ is true}) = v_{n-1} = 1 - \alpha$ for α our desired Type I error probability. Using results in Spitzer (1956) it can be shown that

$$\lim_{n \to \infty} v_n = \exp\{-\sum_{j=1}^{\infty} \frac{P(\chi_j^2 > jC)}{j}\}.$$

Thus, an asymptotic α level test is obtained by choosing C in (3.90) to be the number C_α that satisfies

$$1 - \alpha = \exp\{-\sum_{j=1}^{\infty} \frac{P(\chi_j^2 > jC_\alpha)}{j}\}. \tag{3.94}$$

This can be accomplished numerically and values of C_α that have been obtained in this fashion are shown in Table 3.2 for several standard choices for α.

Table 3.2: Some values of C_α

α	.29	.2	.1	.05	.01
C_α	2	2.38	3.22	4.18	6.74

Notice from Table 3.2 that the unbiased risk choice of $C = 2$ in (3.90) produce a test with a rather large level of .29. To produce tests with more standard levels such as .05 or .1 it is necessary to penalize larger orders more heavily and thereby reduce the chance of overselection under H_0.

From the form of criterion (3.90), one can show (Exercise 19) that rejecting H_0 when $\hat{\lambda} > 1$ is equivalent to rejecting if

$$T_3 \ge C_\alpha, \tag{3.95}$$

where

$$T_3 = n \max_{2 \le \lambda \le n} [\lambda^{-1} \sum_{j=2}^{\lambda} \frac{b_j^2}{\sigma^2}]. \tag{3.96}$$

Thus, using the order selection test statistic is tantamount to assessing systematic departures of the b_j from zero through examination of the maximum partial average of their squares. One can also use (3.96) to develop a fairly immediate proof of consistency for an order selection test.

Now let us return to the statistic T_2 in (3.87). This particular statistic can be derived from a Bayesian perspective where one models $f(\cdot)$ in (3.81) as $bZ(\cdot)$ for $Z(\cdot)$ a Brownian motion process and b an unknown constant. A test for H_0 can then be derived by testing that $b = 0$. This and other related approaches to testing H_0 are studied in Cox and Koh (1989), Barry and Hartigan (1990) and Buckley (1991).

It is straightforward to show that (Exercise 20) T_2 has the same limiting null distribution as the random variable $\sum_{j=1}^{\infty} Z_j^2/(j\pi)^2$ where $\{Z_j\}_{j=1}^{\infty}$ is a sequence of *iid* standard normal random variables. But this is the same limiting distribution as for the Cramér-von Mises statistic from classical goodness-of-fit analysis. Consequently, asymptotic critical values for T_2 can be obtained from Table 3.8.4 of Shorack and Wellner (1986), for example. It is also easy to show that a test based on T_2 will be consistent against essentially any fixed alternative to H_0 (Exercise 20).

Optimal benchmarks for large and finite sample test comparisons are provided by the best or most powerful test for H_0 against specific choices for f in (3.81). For example, results in Lehmann (1986, Chapter 7) give (Exercise 14) that a most powerful test of $H_0 : f(\cdot) = 0$ versus $H_a : f(\cdot) = \gamma g(\cdot)$, for g a specified function and γ some unknown coefficient, can be obtained by rejecting H_0 for large values of

$$T_4 = (\sum_{i=1}^{n} g(t_i)y_i)^2 / \left(\sigma^2 \sum_{i=1}^{n} g^2(t_i)\right). \tag{3.97}$$

This statistic has a limiting chi-squared distribution with one degree-of-freedom under H_0 (Exercise 14). The practical utility of T_4 is quite limited since, in contrast to T_1, T_2 and T_3, its use requires a specific choice for the direction of the alternative.

We now have four tests for H_0 and would like some way of comparing their abilities to detect alternatives to the no effect model. Ideally, one would simply compare the power functions for the tests. However, this is virtually impossible in most cases and we must devise some other method for power comparison. One way to accomplish this is to empirically approximate the tests' power functions by Monte Carlo methods and we will give examples of this approach subsequently. Another way to compare the tests is through large sample analysis of their relative powers via asymptotic relative efficiency (ARE). Expressions for ARE's associated with our tests have been obtained by Eubank (1999) who shows that T_1, T_2 and T_4 are all equally efficient asymptotically for detecting alternatives that are constant multiples of $\cos(\pi t)$. For higher frequency alternatives (i.e., alternatives with cosine series expansions containing functions other than $\cos(\pi t)$) the T_1 statistic can produce a test that is asymptotically as efficient as the T_4 test and is asymptotically more efficient than T_2.

To illustrate the performance of the tests with data we conducted a small simulation experiment. The data for the experiment was generated from model (3.79) using samples of size $n = 50$ and standard normal, random errors. Two choices were made for f in (3.81): a linear alternative

$$f(t) = \gamma\sqrt{12}(t - 1/2) \tag{3.98}$$

and a cosine alternative

$$f(t) = \gamma\sqrt{2}\cos(3\pi t). \tag{3.99}$$

Note that $\gamma^2 = \int_0^1 f(t)^2 dt$ for both (3.98) and (3.99) and the error variance is one. Thus, the value of γ represents a signal-to-noise ratio whose magnitude can be manipulated to obtain alternatives that pose varying degrees of detection difficulty.

Alternative (3.98) has cosine Fourier coefficients $\beta_j = \gamma\sqrt{24}((-1)^{(j-1)} - 1)/((j-1)\pi)^2$, which decay rapidly to zero. The primary contribution to f comes from the β_1 term in its cosine series expansion. Thus, this alternative is low frequency and our discussion of ARE's leads us to anticipate that all four tests will have comparable powers in this case. Figure 3.12 contains plots of empirical power curves (as a function of γ) that resulted from simulations using 1000 samples of size 50 from model (3.81) with f the linear function (3.98) and nominal level $\alpha = .05$. As expected, all four tests have similar powers with T_2 having almost identical power to the optimal test T_4.

Figure 3.13 is analogous to Figure 3.12 except that it corresponds to results for the cosine alternative (3.99). In this case, the alternative is higher frequency and, as a result, T_1 is seen to have better power and be more competitive with T_4 than either of T_2 or T_3.

Both the empirical and large sample work point to T_1 as the statistic of choice for testing the no effect hypotheses. Testing H_0 using T_2 would only seem advisable in cases where there is *a priori* information that the likely alternatives are low frequency in nature.

The basic ideas for testing no effect that have been presented here can be generalized in a number of ways to develop tests for many other types of parametric models. The interested reader can refer to Hart (1997) for a more detailed treatment.

3.5 Polynomial Regression

Trigonometric series estimators are important if for no other reason than the fact that most other nonparametric regression estimators are closely related to them. For the same reason polynomial regression estimators represent an important cornerstone in the theory of nonparametric regression. In this section we give a brief treatment of polynomial regression from the viewpoint of series estimators.

Suppose now that we have chosen our CONS $\{x_j\}_{j=1}^{\infty}$ by using the Gram-Schmidt process in $L_2(w)$ on the power functions as in Example 3.2. If the corresponding generalized Fourier series for μ converges uniformly we can then write model (3.1) as an infinite dimensional polynomial regression model. To estimate μ we can then approximate its polynomial Fourier series expansion by the λth partial sum, $\sum_{j=1}^{\lambda} \beta_j x_j$, and estimate the β_j through least squares. The resulting estimator is obtained from formula (3.27) and will be called a *polynomial regression estimator*.

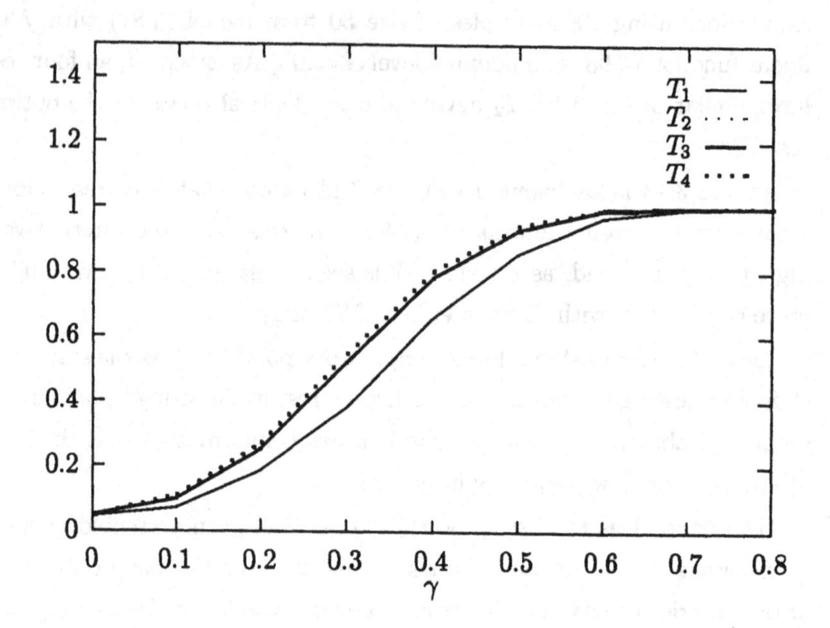

Figure 3.12: Empirical Powers for a Linear Alternative.

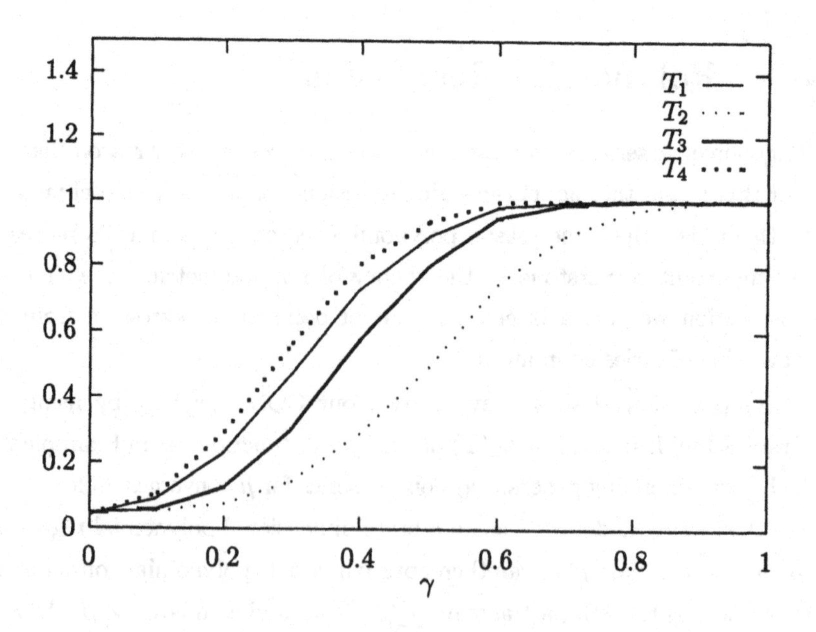

Figure 3.13: Empirical Powers for a Cosine Alternative.

There are a variety of motivations for polynomial regression estimators. One of these is closely related to the frequency motivation for trigonometric series estimators discussed in Section 3.4. The oscillation properties of $\cos(j\pi t)$ that were discussed in that section are reflected in the fact that it has exactly j zeros in (0, 1). A similar property holds for the Jacobi polynomials. It can be shown (see Davis 1975, Chapter X or Szegö 1959, Chapter III) that the jth Jacobi polynomial, x_j, $j = 1, \ldots$, has exactly $j-1$ distinct, real roots in the interior of [0, 1]. Thus, x_j changes sign $j-1$ times in $(0,1)$. By analogy with the cosine functions, one might view this as meaning that x_{j+1} is of higher frequency than x_j and thereby obtain a frequency interpretation for the Fourier coefficients in the orthogonal polynomial expansion for μ. Regression analysis using low order polynomials can then be thought of as an attempt to isolate the low frequency components of the data which should provide the most information about a smooth, slowly varying, regression function.

Another motivation for polynomial regression estimators stems from Taylor's Theorem.

Theorem 3.4 (Taylor's Theorem). *If $\mu \in W_2^m[0,1]$, then there exist coefficients $\theta_1, \ldots, \theta_m$ such that*

$$\mu(t) = \sum_{j=1}^{m} \theta_j t^{j-1} + \int_0^1 \frac{(t-u)_+^{m-1}}{(m-1)!} \mu^{(m)}(u)du \ , \tag{3.100}$$

where

$$(x)_+^r = \begin{cases} x^r, & x \geq 0, \\ 0, & x < 0. \end{cases} \tag{3.101}$$

Proof. Write $\mu(t) = \int_0^1 (t-u)_+^0 \mu'(u)du + \mu(0)$ and integrate by parts. •

Taylor's Theorem suggests that if, for some positive integer λ, the remainder term

$$\text{Rem}_\lambda(t) = [(\lambda-1)!]^{-1} \int_0^1 (t-u)_+^{\lambda-1} \mu^{(\lambda)}(u)du \tag{3.102}$$

is uniformly small then we could write

$$y_i \doteq \sum_{j=1}^{\lambda} \theta_j t_i^{j-1} + \varepsilon_i, \ i = 1, \ldots, n.$$

In other words, the data would follow an approximate polynomial regression model. We could then estimate the polynomial coefficients by least squares or some other method. [Actually, polynomial regression does not give estimates of the θ_j in (3.100). For a precise formulation see Exercise 12.]

Still another motivation for the use of polynomial regression stems from the Weierstrass Approximation Theorem (Davis 1975, Chapter VI) which states that if $\mu \in C[0,1]$ then, given any $\delta > 0$, there exists a polynomial q_λ of sufficiently high order λ such that $|\mu(t) - q_\lambda(t)| \leq \delta$ for all $t \in [0,1]$. Note that in contrast to Taylor's Theorem no differentiability conditions are required for this result to hold. However, the basic implications concerning the possible utility of an approximate polynomial regression model remain unchanged.

Both the Taylor and Weierstrass Theorem arguments for the use of polynomial regression are tantamount to lumping the remainder terms $\mathrm{Rem}_\lambda(t_1), \ldots, \mathrm{Rem}_\lambda(t_n)$ into the random error component of our model. If μ is uniformly well approximated by a polynomial of the order selected and the remainders at the t_i are small relative to the random errors, polynomial regression may work fairly well. If not, problems can arise. Since both the remainder and random errors are unknown there is no way of knowing for certain whether or not the Taylor and Weierstrass Theorem arguments are applicable to a specific choice of λ that is made with any given data set.

In view of the uncertainty about the magnitude of the remainder from polynomial approximation of μ and the random errors it would seem natural to try to modify the polynomial regression estimator to attempt to compensate for the possibililty of large remainder terms. This line of reasoning leads to smoothing and least-squares spline estimators for μ which are the topics of Chapters 5 and 6.

3.5.1 Form of the Estimator

There are a number of useful representations for the polynomial regression estimator. In particular, there is no reason to restrict attention to an orthogonal polynomial basis when conducting polynomial regression analysis. In fact, this is not the usual approach and typically one will use the power basis functions

$$q_j(t) = t^{j-1},\ j = 1, \ldots \tag{3.103}$$

Let $\mathbf{Q}_\lambda$ be the $n \times \lambda$ matrix

$$\mathbf{Q}_\lambda = \{q_j\,(t_i)\}_{i=1,n,j=1,\lambda}\,. \tag{3.104}$$

Then, under the power basis representation, the polynomial regression estimator of $\mu(t)$ is

$$\mu_\lambda(t) = \left(1, t, \ldots, t^{\lambda-1}\right)\left(\mathbf{Q}_\lambda^T\mathbf{Q}_\lambda\right)^{-1}\mathbf{Q}_\lambda^T\mathbf{y}. \tag{3.105}$$

Note that the actual value of $\mu_\lambda(t)$ is not altered by changing from an orthogonal polynomial basis to the power basis. However, the estimated coefficients for the power basis elements, i.e.,

$$\boldsymbol{\theta}_\lambda = (\theta_{\lambda 1}, \ldots, \theta_{\lambda\lambda})^T = \left(\mathbf{Q}_\lambda^T\mathbf{Q}_\lambda\right)^{-1}\mathbf{Q}_\lambda^T\mathbf{y}, \tag{3.106}$$

will not coincide with the estimated Fourier coefficients under an orthogonal basis.

There is another form for the polynomial regression estimator that is instructive. Rather than use the power basis or Legendre basis suppose we define basis functions by

$$x_{jn}(t) = \sum_{k=1}^{\lambda} a_{kj}t^{k-1}, j = 1, \ldots, \lambda,$$

with $\mathbf{A}_\lambda = \{a_{jr}\}$ a nonsingular $\lambda \times \lambda$ matrix satisfying

$$\mathbf{A}_\lambda^T\mathbf{Q}_\lambda^T\mathbf{Q}_\lambda\mathbf{A}_\lambda = n\mathbf{I}.$$

In other words we choose our basis so that

$$n^{-1}\sum_{i=1}^{n} x_{jn}(t_i)x_{kn}(t_i) = \delta_{jk},\ j, k = 1, \ldots, \lambda \tag{3.107}$$

and the x_{jn} have an orthogonality property over the design which mimics that of orthogonal polynomials in $L_2(w)$. The two sets of functions will often be quite similar (Exercises 1-2). There is a large literature in Statistics on the construction of polynomials of this type; see, e.g., Narula (1979).

As a result of (3.107) the polynomial regression estimator now takes on the form

$$\mu_\lambda(t) = \sum_{j=1}^{\lambda} b_{\lambda j} x_{jn}(t)$$

with

$$b_{\lambda j} = n^{-1} \sum_{i=1}^{n} y_i x_{jn}(t_i),\ j = 1, \ldots, \lambda.$$

This can be rewritten as

$$\mu_\lambda(t) = n^{-1} \sum_{i=1}^{n} y_i K_\lambda(t, t_i) \tag{3.108}$$

with

$$K_\lambda(s, t) = \sum_{j=1}^{\lambda} x_{jn}(s) x_{jn}(t). \tag{3.109}$$

The function $K_\lambda(\cdot, \cdot)$ is called a kernel polynomial and plays a parallel role to the weight function (3.45) for the cosine series estimator of Section 3.4. Notice, for example, that the weights corresponding to the kernel polynomial average to one, i.e.,

$$n^{-1} \sum_{i=1}^{n} K_\lambda(t, t_i) = n^{-1} \left(1, t, \ldots, t^{\lambda-1}\right) \mathbf{A}_\lambda \mathbf{A}_\lambda^T \mathbf{Q}_\lambda^T \mathbf{1} = 1,$$

where $\mathbf{1}$ is a vector of all unit elements. This follows from the fact that $\mathbf{A}_\lambda$ is nonsingular and, hence, $n^{-1}\mathbf{A}_\lambda^T \mathbf{Q}_\lambda \mathbf{1}$ is the first column of $\mathbf{A}_\lambda^{-1}$.

The preceeding discussion shows that the polynomial regression estimator is a weighted average with weights provided by a kernel function. There is, of course, nothing unique about polynomial estimators in this respect and it should now be clear that parallels of (3.108) and (3.109) will hold for any series type estimator. This viewpoint leads us into the study of kernel estimators in the next chapter.

3.5.2 An Example

To illustrate the use of polynomial regression estimators we fitted a set of data concerning the DC output (y) produced by a windmill as a function of wind velocity (t). The data consisted of $n = 25$ observations that are given in Exercise 24. A plot of the data is shown in Figure 3.14.

Table 3.3 gives values of the GCV and unbiased risk criteria (using the GSJS variance estimator) for the windmill data that suggests using a polynomial of degree 4 or order 5. The resulting fit is shown in Figure 3.14. For this data the GSJS estimator is $\widehat{\sigma} = .00797$ while the more traditional analysis of variance estimator (3.28) is found to be $\sigma_5^2 = \text{RSS}(5)/(n-5) = .0093$.

Table 3.3: RSS, GCV and $\widehat{P}$ for the Windmill Data

λ	RSS(λ)	GCV(λ)	$\widehat{P}$
1	10.211	.4432	.4091
2	1.282	.0606	.0525
3	0.331	0.0171	.0152
4	0.235	0.0133	.0120
5	0.186	0.0116	.0106
6	0.183	0.0127	.0111
7	0.181	0.0139	.0117
8	0.177	0.0156	.0123

Diagnostic analysis of our fit can be conducted similarly to that for the voltage drop data in Section 3.4.2. For example, Cook's distance measures from the polynomial fit are shown in Figure 3.15. It appears from this that only the observation corresponding to the smallest wind velocity is particularly influential and its Cook's distance does not exceed the rough bound of .463 for the 10th percentile of an F-distribution with 5 numerator and 20 denominator degrees-of-freedom.

The estimated polynomial coefficients (3.106) for μ_5 are given in Table 3.4 along with their standardized values using the GSJS estimator. All

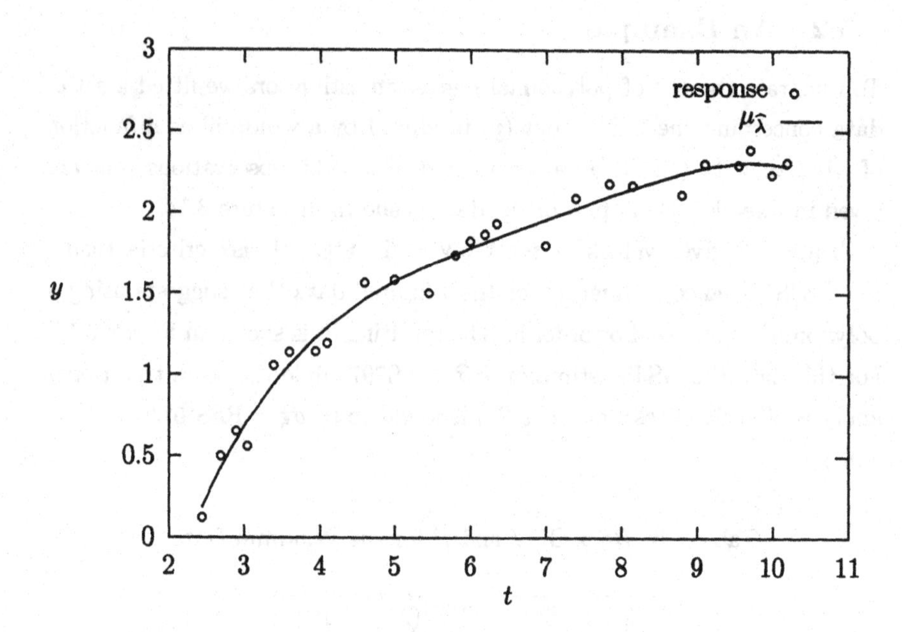

Figure 3.14: Polynomial Fit for the Windmill Data.

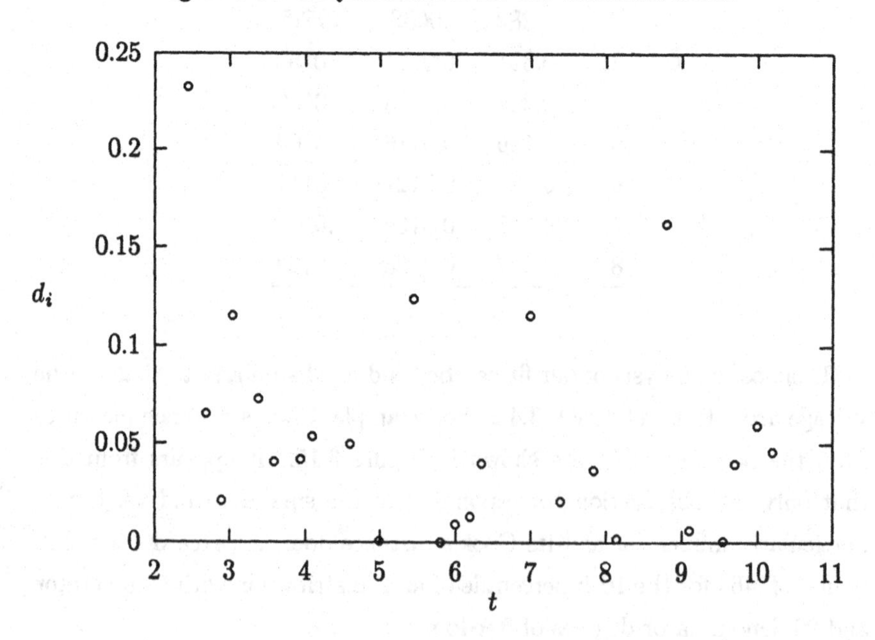

Figure 3.15: Cook's Distances for the Windmill Data

the standardized coefficients exceed 2 in magnitude which suggests that all the fitted terms are of some value to the fit. Of course the magnitudes of the estimated quartic and quintic coefficients are so small that any actual utility they provide is necessarily limited and one might consider excluding them.

Table 3.4: Coefficient Estimates for the Windmill Data

Coefficient	Estimate	Estimate/(Standard Error)
θ_1	-4.715	-4.632
θ_2	3.277	4.238
θ_3	-.662	-3.229
θ_4	.063	2.760
θ_5	-.002	-2.485

3.5.3 Large Sample Properties of the Estimator

Polynomial regression is undoubtedly the most widely used nonparametric regression technique. However, the classical approach has been to treat the polynomial as being a parametric model for the regression function after an implicit invocation of Taylor's Theorem or some related argument. Such practices are usually not justified since there is typically no way to know whether the remainder term in (3.102) can be ignored to an extent that warrants the assumption of a finite dimensional (i.e., parametric) polynomial model for μ. As noted in Chapter 1, the parametric modeling philosophy fails when the actual form of the true regression function is not known and the use of polynomials or other functions to produce *ad hoc* parametric models is a dangerous practice.

It is, in fact, difficult to imagine a case where some finite dimensional polynomial model for μ could be presumed to hold with any degree of certainty. It therefore seems much more sensible to regard a polynomial as only an approximation to the true regression function and then analyze the polynomial regression estimator as what it actually is: a series type nonparametric regression estimator. This means that one must actually deal with, rather than assume away, issues of bias in estimation.

The conclusions concerning the effectiveness of the linear models approach to inference in Section 3.3 for polynomial regression are similar to those for trigonometric series estimators discussed in Section 3.4. While standard t-test methodology can be expected to work quite well for inference about polynomial coefficients, the usual linear models type interval estimators (3.32) are likely to fail due to estimator bias when the number of polynomial terms is selected by automated methods such as GCV. As in the trigonometric regression setting, some protection against these bias problem can be obtained by using Bonferroni intervals where α is replaced by α/n in (3.32).

The properties of GCV and related order selection techniques in the polynomial regression setting have been studied by Shibata (1981), Li (1987), Eubank and Jayasuriya (1993) and Hurvich and Tsai (1995). In particular, Hurvich and Tsai (1995) show that a version of (3.37) holds for certain polynomial regression estimators. These results suggest that the order selection methods discussed in Section 2.2 should provide effective means of selecting λ in polynomial regression.

To analyze the large sample risk behavior of polynomial regression we will, for simplicity, deal only with the case of a uniform design where $w(t) \equiv 1$ in (3.21). Thus, we take x_j to be the jth Legendre polynomial in (3.8)–(3.9) and define

$$\beta_j = \int_0^1 \mu(t)x_j(t)dt, j = 1, \ldots, \tag{3.110}$$

as the jth Legendre Fourier coefficient. Then, one can show the following result (Exercise 21).

Theorem 3.5. *Assume that $\mu \in W_2^1[0,1]$ and $\lambda, n \to \infty$ with $\lambda/n \to 0$. Then the polynomial regression estimator μ_λ satisfies*

$$R_n(\lambda) = n^{-1}\sum_{i=1}^{n} \mathrm{E}(\mu_\lambda(t_i) - \mu(t_i))^2 \le \frac{\sigma^2\lambda}{n} + \sum_{j=\lambda+1}^{\infty} \beta_j^2 + r_n \tag{3.111}$$

with

$$r_n = O\left(\frac{\lambda^{5/2}}{n}[\sum_{j=\lambda+1}^{\infty} \beta_j^2]^{1/2}\sum_{j=1}^{\lambda}|\beta_j|\right). \tag{3.112}$$

It can be shown (Eubank and Jayasuriya 1993) that

$$\beta_j = \frac{2}{\sqrt{2j+1}}\{\frac{\beta'_{j-1}}{\sqrt{2j-1}} - \frac{\beta'_{j+1}}{\sqrt{2j+3}}\} \tag{3.113}$$

with β'_j the jth Legendre Fourier coefficient for μ'. Since $|\beta'_j| \leq ||\mu'||$, it follows that

$$|\beta_j| \leq \frac{2}{\sqrt{(2j+1)(2j-1)}}||\mu'|| \leq j^{-1}||\mu'||.$$

Repeating this process and continuing by induction we arrive at the conclusion that

$$|\beta_j| \leq j^{-m}||\mu^{(m)}|| \tag{3.114}$$

if $\mu \in W_2^m[0,1]$. This has the implication that $\sum_{j=1}^{\lambda}|\beta_j|$ is summable when $\mu \in W_2^m[0,1]$ for $m \geq 2$. This fact along with (3.113)-(3.114) give the following result.

Corollary 3.2. *If $\mu \in W_2^m[0,1]$ for $m \geq 2$, then*

$$R_n(\lambda) \leq \frac{\sigma^2\lambda}{n} + \lambda^{-2m}||\mu^{(m)}||^2 + O(\frac{\lambda^{-(2m-5)/2}}{n}). \tag{3.115}$$

As a result of Corollary 3.2 we see that any sequence of orders λ_n growing at the rate $n^{1/(2m+1)}$ will result in $R_n(\lambda_n) = O(n^{-2m/(2m+1)})$ and, as a result, polynomial regression can achieve the optimal convergence rate for estimation of a $W_2^m[0,1]$ function in the case of a uniform design. This result is actually true for general choices of w in (3.21) as shown in Cox (1988), Eubank and Jayasuriya (1993) and Exercise 22.

Just as for trigonometric series estimators, the convergence properties of the risk for polynomial regression can be tied directly to the rate of decay for the Fourier coefficients in the squared bias term $\sum_{j=\lambda+1}^{\infty}\beta_j^2$ in (3.111). For very smooth (i.e., infinitely differentiable) regression functions the Legendre Fourier coefficients can be shown to decay exponentially fast in which case a choice of $\lambda_n \sim C\log(n)$ for some constant C produces $R_n(\lambda_n) = O(\frac{\log(n)}{n})$ which is almost a parametric rate of convergence.

Our asymptotic analysis of the risk shows that polynomial regression has an "adaptive" feature in that it is capable of producing convergence rates that conform to whatever smoothness qualities the regression function may have that are reflected in its Legendre Fourier coefficients. This feature is not shared by the estimators in subsequent chapters which are capable of converging no faster than some pre-specified rate irregardless of any additional smoothness that μ might possess. However, the practical consequences of this "adaptive" quality for polynomial regression are somewhat circumspect. The values of a polynomial in any small interval determine the function globally which has the consequence that polynomial fits tend to be somewhat inflexible and are prone to exhibit local wiggles that are not features of the data. This quality tends to make them less esthetically desirable than estimators which use only local information around some point of estimation, such as the kernel estimators in Chapter 4.

3.6 Polynomial-Trigonometric Regression

One of the nice features of polynomial regression is that it is free of the boundary bias problems that plague trigonometric series estimators. However, the power function basis (3.103), which is usually used in estimation, is ill-conditioned causing substantial numerical difficulties in the computation of estimators even for moderate order polynomials. In contrast, the design matrices for trigonometric series estimators tend to be well conditioned, at least for approximately uniform designs.

In this section we derive an alternative to cosine series and polynomial regression that can be viewed as a combination of the two estimators. The resulting estimator is called a polynomial-cosine regression estimator and is a series type estimator that includes both polynomial and cosine functions. One can think of this estimator as resulting from the use of polynomials to remove the boundary bias of cosine series estimators or, alternatively, as the use of cosines to ameliorate the conditioning problems of polynomial estimators.

The basic idea is easiest to illustrate for estimation of a regression function $\mu \in W_2^2[0,1]$ with the cosine series estimator of Section 3.4 when the design is uniform as in (3.38). In that case we know that the cosine series estimator μ_λ in (3.43) can attain the optimal $n^{-4/5}$ convergence rate for the risk by choosing λ to be of order $n^{1/5}$ provided that μ satisfies the boundary conditions $\mu'(0) = \mu'(1) = 0$. Such conditions cannot be expected to hold for general μ . However, we can always write

$$\mu = \mu^* + \mu'(0)q_0 + \mu'(1)q_1, \tag{3.116}$$

where $\mu^* = \mu - \mu'(0)q_0 - \mu'(1)q_1$ and q_0, q_1 are the quadratic polynomials

$$q_0(t) = t - \frac{t^2}{2} \tag{3.117}$$

and

$$q_1(t) = \frac{t^2}{2}. \tag{3.118}$$

Note that μ^* does, in fact, have the requisite boundary properties that allow it to be estimated at the optimal rate using a cosine series estimator. This suggests that we try estimating μ through a fit to the data using the constant function, q_0, q_1 and elements of the cosine sequence (3.7). One might hope that by doing this the polynomial functions would account for the polynomial term in (3.116) while leaving the cosine functions free to take care of μ^*. We will show that this strategy works through an application of the following lemma.

Lemma 3.3 (Bias Reduction Lemma). *Let $\boldsymbol{\theta}$ be an n-vector of unknown parameters and let $\widehat{\boldsymbol{\theta}}$ be an estimator of $\boldsymbol{\theta}$ with the property that*

$$\mathrm{E}\widehat{\boldsymbol{\theta}} = \boldsymbol{\theta} + \mathbf{Gc} + \mathbf{r} \tag{3.119}$$

for $\mathbf{G}$ a known $n \times p$ matrix and $\mathbf{c}$ and $\mathbf{r}$ unknown vectors of parameters. Let $\bar{\boldsymbol{\theta}}$ be an unbiased estimator of $\boldsymbol{\theta}$ and define the projection matrix $\mathbf{P}_G = \mathbf{G}(\mathbf{G}^T\mathbf{G})^-\mathbf{G}^T$ with $(\mathbf{G}^T\mathbf{G})^-$ a generalized inverse of $\mathbf{G}^T\mathbf{G}$. Then,

$$\widetilde{\boldsymbol{\theta}} = \widehat{\boldsymbol{\theta}} + \mathbf{P}_G(\bar{\boldsymbol{\theta}} - \widehat{\boldsymbol{\theta}}) \tag{3.120}$$

satisfies

$$(\mathrm{E}\widetilde{\boldsymbol{\theta}}-\boldsymbol{\theta})^T(\mathrm{E}\widetilde{\boldsymbol{\theta}}-\boldsymbol{\theta}) \leq \mathbf{r}^T\mathbf{r} \tag{3.121}$$

and

$$\mathrm{tr}\mathrm{Var}(\widetilde{\boldsymbol{\theta}}) \leq \mathrm{tr}\mathrm{Var}(\widehat{\boldsymbol{\theta}}) + \mathrm{rank}(\mathbf{G})\mathrm{ch}_{\max}(\mathrm{Var}(\bar{\boldsymbol{\theta}})), \tag{3.122}$$

where $\mathrm{rank}(\mathbf{G})$ *is the rank of* $\mathbf{G}$ *and* $\mathrm{ch}_{\max}(\mathrm{Var}(\bar{\boldsymbol{\theta}}))$ *is the largest characteristic root or eigenvalue of the variance-covariance matrix for* $\bar{\boldsymbol{\theta}}$.

Proof. To prove (3.121) observe that

$$(\mathrm{E}\widetilde{\boldsymbol{\theta}}-\boldsymbol{\theta})^T(\mathrm{E}\widetilde{\boldsymbol{\theta}}-\boldsymbol{\theta}) = \mathbf{r}^T(\mathbf{I}-\mathbf{P}_G)\mathbf{r} \leq \mathbf{r}^T\mathbf{r}$$

due to (3.119), the unbiasedness of $\bar{\boldsymbol{\theta}}$ and the fact that $\mathbf{I}-\mathbf{P}_G$ is a projection matrix. For (3.122) first note that $\mathrm{tr}\mathrm{Var}(\widetilde{\boldsymbol{\theta}}) = \mathrm{tr}(\mathbf{I}-\mathbf{P}_G)\mathrm{Var}(\widehat{\boldsymbol{\theta}}) + \mathrm{tr}\mathbf{P}_G\mathrm{Var}(\bar{\boldsymbol{\theta}})$. Then write $\mathrm{tr}(\mathbf{I}-\mathbf{P}_G)\mathrm{Var}(\widehat{\boldsymbol{\theta}}) = \sum_{j=1}^{n}\delta_j\mathbf{v}_j^T(\mathbf{I}-\mathbf{P}_G)\mathbf{v}_j$ for δ_j and $\mathbf{v}_j$ the eigenvalues and eigenvectors of $\mathrm{Var}(\widehat{\boldsymbol{\theta}})$. Since $\mathbf{v}_j^T(\mathbf{I}-\mathbf{P}_G)\mathbf{v}_j \leq \mathbf{v}_j^T\mathbf{v}_j$, the first part of the variance bound has been established. For the second part, proceed similarly to obtain $\mathrm{tr}\mathbf{P}_G\mathrm{Var}(\bar{\boldsymbol{\theta}}) = \sum_{j=1}^{\mathrm{rank}(\mathbf{G})}\mathbf{u}_j^T\mathrm{Var}(\bar{\boldsymbol{\theta}})\mathbf{u}_j$ with the $\mathbf{u}_j$ being the eigenvectors corresponding to the unit eigenvalues of $\mathbf{P}_G$. The proof is completed by using the fact that $\sup_{\mathbf{u}^T\mathbf{u}=1}\mathbf{u}^T\mathrm{Var}(\bar{\boldsymbol{\theta}})\mathbf{u} = \mathrm{ch}_{\max}(\mathrm{Var}(\bar{\boldsymbol{\theta}}))$. •

Lemma 3.3 is designed for use in situations where there is a biased estimator (i.e., $\widehat{\boldsymbol{\theta}}$) with good variance properties and an unbiased estimator (i.e., $\bar{\boldsymbol{\theta}}$) whose variance behavior may not be satisfactory. It states that in such cases the two estimators may be combined using (3.120) to produce the new estimator $\widetilde{\boldsymbol{\theta}}$ that attains the target bias $\mathbf{r}$ in (3.119) while increasing the total variance by at most $\mathrm{rank}(\mathbf{G})\mathrm{ch}_{\max}(\mathrm{Var}(\bar{\boldsymbol{\theta}}))$.

To apply Lemma 3.3 to our function estimation problem, we first make the identifications $\boldsymbol{\theta} = \boldsymbol{\mu} = (\mu(t_1),\ldots,\mu(t_n))^T$, $\bar{\boldsymbol{\theta}} = \mathbf{y} = (y_1,\ldots,y_n)^T$ and $\widehat{\boldsymbol{\theta}} = \boldsymbol{\mu}_\lambda = (\mu_\lambda(t_1),\ldots,\mu_\lambda(t_n))^T$ for μ_λ the cosine series estimator of order λ in (3.43). Now set $\mathbf{Q} = \{q_j(t_i)\}_{i=1,n,j=0,1}$, $\mathbf{c} = (\mu'(0),\mu'(1))^T$ and write

$$\boldsymbol{\mu} = \boldsymbol{\mu}^* + \mathbf{Q}\mathbf{c} \tag{3.123}$$

for $\boldsymbol{\mu}^* = (\mu^*(t_1), \ldots, \mu^*(t_n))^T$. Then one can readily check that the cosine series estimator satisfies (3.119) with

$$\mathbf{G} = -(\mathbf{I} - \mathbf{S}_\lambda)\mathbf{Q} \tag{3.124}$$

and

$$\mathbf{r} = -(\mathbf{I} - \mathbf{S}_\lambda)\boldsymbol{\mu}^* \tag{3.125}$$

for $\mathbf{S}_\lambda = n^{-1}\mathbf{X}_\lambda\mathbf{X}_\lambda^T$ the cosine series smoother or hat matrix that corresponds to $\mathbf{X}_\lambda$ in (3.39). The new, bias reduced, estimator (3.120) is then given by

$$\widetilde{\boldsymbol{\mu}}_\lambda = (\widetilde{\mu}_\lambda(t_1), \ldots, \widetilde{\mu}_\lambda(t_n))^T = \boldsymbol{\mu}_\lambda + \mathbf{P}_G(\mathbf{y} - \boldsymbol{\mu}_\lambda) \tag{3.126}$$

with $\mathbf{P}_G = \mathbf{G}(\mathbf{G}^T\mathbf{G})^-\mathbf{G}^T$ for $\mathbf{G}$ in (3.124).

At first (3.126) looks somewhat formidable. However, it is easy to see (Exercise 23) that $\widetilde{\boldsymbol{\mu}}_\lambda$ is nothing more than the result of standard least-squares regression using the functions $1, t, t^2$ and $\{\cos(j\pi t)\}_{j=1}^{\lambda-1}$. That is,

$$\widetilde{\mu}_\lambda(t) = d_{\lambda 0}q_0(t) + d_{\lambda 1}q_1(t) + d_{\lambda 2} + \sum_{j=1}^{\lambda-1} d_{\lambda(j+2)}\cos(j\pi t), \tag{3.127}$$

where the $d_{\lambda j}$ are the minimizers of

$$\sum_{i=1}^{n}(y_i - d_0q_0(t_i) - d_1q_1(t_i) - d_2 - \sum_{j=1}^{\lambda-1} d_{j+2}\cos(j\pi t_i))^2,$$

with respect to $d_j, j = 0, \ldots, \lambda+1$. The resulting hat or smoother matrix for $\widetilde{\boldsymbol{\mu}}_\lambda$ is therefore given by

$$\mathbf{S}_\lambda = \mathbf{X}_\lambda(\mathbf{X}_\lambda^T\mathbf{X}_\lambda)^{-1}\mathbf{X}_\lambda^T \tag{3.128}$$

for $\mathbf{X}_\lambda$ the $n \times (\lambda+2)$ partitioned matrix

$$\mathbf{X}_\lambda = [\mathbf{1}|\mathbf{Q}|\{\cos(j\pi t_i)\}_{j=1,\lambda-1,i=1,n}]. \tag{3.129}$$

Consequently, a fit using $\widetilde{\mu}_\lambda$ can be obtained using any standard linear regression software package. We refer to the estimator in (3.127) as a polynomial-cosine series estimator.

As a result of (3.55), (3.59) and (3.121) the average squared bias for $\widetilde{\mu}_\lambda$ is bounded by

$$\begin{aligned} n^{-1}\sum_{i=1}^{n}(\mathrm{E}\widetilde{\mu}_\lambda(t_i) - \mu(t_i))^2 &\leq n^{-1}(\boldsymbol{\mu}^*)^T(\mathbf{I} - \mathbf{S}_\lambda)\boldsymbol{\mu}^* \\ &\leq \frac{1}{\pi^4\lambda^4}\|(\mu^*)''\|^2 + o((n\lambda)^{-1}) \quad (3.130) \end{aligned}$$

because the derivative of μ^* vanishes at 0 and 1. For the variance of $\widetilde{\mu}_\lambda$ we have $\mathrm{Var}(\bar{\boldsymbol{\theta}}) = \sigma^2\mathbf{I}$ so that $\mathrm{ch}_{\max}\mathrm{Var}(\bar{\boldsymbol{\theta}}) = \sigma^2$ and $\mathrm{rank}(\mathbf{G}) = 2$ in (3.122). Thus,

$$n^{-1}\sum_{i=1}^{n}\mathrm{Var}(\widetilde{\mu}_\lambda(t_i)) \leq \sigma^2\frac{\lambda+2}{n} \quad (3.131)$$

and one may check that this bound actually holds with equality here. Combining (3.130)–(3.131) leads to the conclusion that the risk for $\widetilde{\mu}_\lambda$ will decay at the optimal $n^{-4/5}$ rate for estimation of a $W_2^2[0,1]$ function if we choose λ to grow at the rate $n^{1/5}$. Consequently, we have effectively removed the boundary bias from the cosine series estimator.

An illustration of the results that can be obtained from polynomial-cosine regression can be seen in Figure 3.16. Here we show the polynomial-cosine fit to the voltage drop data of Section 3.4.2 that corresponds to the GCV and unbiased risk choice of $\widehat{\lambda} = 3$. Notice that this fit seems to adequately handle the boundary behavior of the data while producing a more visually satisfying resolution of the peak around $t = .5$ with far fewer terms than were needed for the cosine series estimator in Figure 3.2. The estimated coefficients for q_0 and q_1 are $d_{30} = -24.37$ and $d_{31} = 1.09$ both of which have estimated standard errors (using the GSJS variance estimator) of 2.35. Since d_{30} and d_{31} can be viewed as estimating $\mu'(0)$ and $\mu'(1)$, respectively, this suggests that the derivative of the true regression curve for this data may vanish at 1. Thus, only boundary adjustments at 0 may actually be necessary.

While we have focused on estimating a $W_2^2[0,1]$ function under a uniform design, the approach we have used for boundary correction is actually quite general and can be extended to a variety of other problems using similar

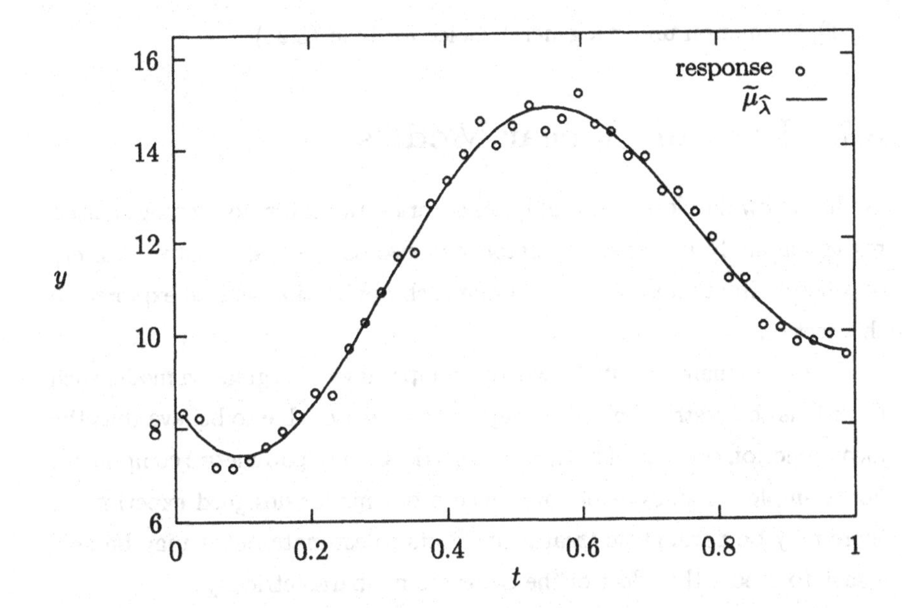

Figure 3.16: Polynomial-Cosine Estimator for the Voltage Drop Data.

arguments. For example, if $\mu \in W_2^{2m}[0,1]$ we write $\mu = \mu^* + p$ for p a polynomial of order $2m+1$ and μ^* a function whose $(2k-1)$st derivative vanishes at 0 and 1 for $k = 1, \ldots, m$. An application of Lemma 3.3 then shows that the least-squares estimator of μ constructed from the functions $1, \ldots, t^{2m}$ and $\{\cos(j\pi t)\}_{j=1}^{\lambda}$ will attain the optimal convergence rate $n^{-4m/(4m+1)}$ for the risk if λ grows at the rate $n^{1/(4m+1)}$. If $\mu \in W_2^{2m-1}[0,1]$ only polynomials of order $2m-1$ are needed to attain the optimal risk convergence rate.

There is nothing special about our use of the cosine series. Lemma 3.3 can be employed to produce boundary corrected ***polynomial-trigonometric*** regression estimators using either the sine functions alone or both the sines and cosines (Exercises 26–29). With a little extra work the assumption of a uniform design can also be relaxed (Eubank and Speckman 1990, 1991). This gives the conclusion that polynomial-trigonometric regression estimators are competitive with polynomial regression in the sense of being capable of obtaining the optimal risk convergence rate for estimation of any

$W_2^m[0,1]$ function under a general choice for w in (3.21).

3.7 Partially Linear Models

We have now developed several types of series estimators for nonparametric regression analysis. Series estimators can also be adapted to solve a variety of other estimation problems and one such application will be explored in this section.

There are many situations where a nonparametric regression model such as (3.1) is too restrictive and it may be more reasonable to believe that the mean function consists of both a parametric and nonparametric component. For example, in analysis of covariance problems for designed experiments there may be parametric treatment effects to estimate but it may be necessary to model the effect of the covariate nonparametrically.

One regression model that allows for parametric and nonparametric components is the *partially linear model*. In this case the responses satisfy

$$y_i = \mathbf{u}_i^T\gamma + f(t_i) + \varepsilon_i,\ i = 1, \ldots, n, \tag{3.132}$$

with $\mathbf{u}_i = (u_{i1}, \ldots, u_{ip})^T, i = 1, \ldots, n$, known p-vectors, $\boldsymbol{\gamma} = (\gamma_1, \ldots, \gamma_p)^T$ an unknown parameter vector, $f \in W_2^m[0,1]$ an unknown function and $\boldsymbol{\varepsilon} = (\varepsilon_1, \ldots, \varepsilon_n)^T$ a vector of zero mean random variables having variance-covariance matrix $\sigma^2\mathbf{I}$. In vector and matrix form we can write (3.132) as

$$\mathbf{y} = \mathbf{U}\boldsymbol{\gamma} + \mathbf{f} + \boldsymbol{\varepsilon}, \tag{3.133}$$

where $\mathbf{U} = \{u_{ij}\}_{i=1,n,j=1,p}$ and $\boldsymbol{f} = (f(t_1), \ldots, f(t_n))^T$. This model generalizes both the parametric linear regression and nonparametric regression models which correspond to the cases $\mathbf{f} = \mathbf{0}$ and $\boldsymbol{\gamma} = \mathbf{0}$, respectively. It is of interest to estimate both the mean vector $\boldsymbol{\mu} = \mathbf{U}\boldsymbol{\gamma} + \mathbf{f}$ and the parameter vector $\boldsymbol{\gamma}$ in this setting.

One approach to estimation in (3.133) can be obtained from the Bias Reduction Lemma of the previous section. To give a somewhat general development, suppose that we choose to fit the data using a generic series estimator of order λ having the associated smoother matrix

$$\mathbf{S}_\lambda = \mathbf{X}_\lambda(\mathbf{X}_\lambda^T\mathbf{X}_\lambda)^{-1}\mathbf{X}_\lambda^T$$

for $\mathbf{X}_\lambda$ defined as in (3.25). Let us further suppose that our series smoother is mth order in the sense that

$$n^{-1}\mathbf{f}^T(\mathbf{I}-\mathbf{S}_\lambda)\mathbf{f} = O(\lambda^{-2m}) \tag{3.134}$$

when $n, \lambda \to \infty$ with $\lambda/n \to 0$. This is true, for example, if we construct $\mathbf{S}_\lambda$ from a polynomial or suitable polynomial-trigonometric regression estimator.

Given our choice for an estimator one might naively attempt to estimate the mean vector for the data using

$$\boldsymbol{\mu}_\lambda = \mathbf{S}_\lambda\mathbf{y}.$$

Of course this estimator is unsatisfactory since it does not allow for dependence of the mean vector on $\mathbf{u}_i, i = 1, \ldots, n$. However,

$$\begin{aligned} \mathrm{E}\boldsymbol{\mu}_\lambda &= \boldsymbol{\mu} - (\mathbf{I}-\mathbf{S}_\lambda)\mathbf{U}\boldsymbol{\gamma} - (\mathbf{I}-\mathbf{S}_\lambda)\mathbf{f} \\ &= \boldsymbol{\mu} + \mathbf{G}\boldsymbol{\gamma} + \mathbf{r} \end{aligned}$$

with $\mathbf{G} = -(\mathbf{I}-\mathbf{S}_\lambda)\mathbf{U}$ and $\mathbf{r} = -(\mathbf{I}-\mathbf{S}_\lambda)\mathbf{f}$. This is of the form (3.119) that is needed for application of Lemma 3.3. Thus, upon choosing $\bar{\boldsymbol{\theta}} = \mathbf{y}$ for our unbiased estimator of $\boldsymbol{\theta} = \boldsymbol{\mu}$ we obtain the "bias corrected" estimator

$$\begin{aligned} \tilde{\boldsymbol{\mu}}_\lambda &= \mathbf{S}_\lambda\mathbf{y} + \mathbf{G}(\mathbf{G}^T\mathbf{G})^-\mathbf{G}^T(\mathbf{y}-\mathbf{S}_\lambda\mathbf{y}) \\ &= \mathbf{f}_\lambda + \mathbf{U}\boldsymbol{\gamma}_\lambda \end{aligned} \tag{3.135}$$

with

$$\boldsymbol{\gamma}_\lambda = (\mathbf{U}^T(\mathbf{I}-\mathbf{S}_\lambda)\mathbf{U})^-\mathbf{U}^T(\mathbf{I}-\mathbf{S}_\lambda)\mathbf{y} \tag{3.136}$$

and

$$\mathbf{f}_\lambda = \mathbf{S}_\lambda(\mathbf{y}-\mathbf{U}\boldsymbol{\gamma}_\lambda) \tag{3.137}$$

being the natural estimators of $\boldsymbol{\gamma}$ and $\mathbf{f}$ that come out of this formulation.

Again the estimation prescription (3.135) produced by the Bias Reduction Lemma appears to be quite complicated. However, it is easy enough to check (Exercise 31) that this estimator is the same as would be produced from least squares by minimizing

$$\sum_{i=1}^{n}(y_i - \sum_{j=1}^{p} u_{ij}a_j - \sum_{j=1}^{\lambda} x_j(t_i)c_j)^2$$

with respect to $a_1, \ldots, a_p$ and $c_1, \ldots, c_\lambda$. Thus, this type of estimator can be readily computed for data using standard linear regression software.

An application of (3.121)–(3.122) along with (3.134) shows that

$$\begin{aligned} n^{-1}\mathrm{E}\,(\tilde{\boldsymbol{\mu}}_\lambda - \boldsymbol{\mu})^T(\tilde{\boldsymbol{\mu}}_\lambda - \boldsymbol{\mu}) &\leq n^{-1}\mathbf{f}(\mathbf{I} - \mathbf{S}_\lambda)\mathbf{f} + n^{-1}\sigma^2(\lambda + \mathrm{rank}(\mathbf{U})) \\ &\leq O(\lambda^{-2m}) + n^{-1}\sigma^2(\lambda + p). \end{aligned} \tag{3.138}$$

Thus, a choice of λ that grows at the rate $n^{1/(2m+1)}$ will ensure that $\boldsymbol{\mu}$ can be estimated as efficiently as if the model consisted of only a nonparametric component. That is, the additional difficulty of estimating the parameter vector γ has not affected how well we can estimate the mean vector.

While (3.138) shows that the entire mean vector can be estimated as efficiently as in the totally nonparametric case, one might hope that the parameter vector γ could be estimated somewhat more efficiently. Specifically, it is of interest to know whether or not

$$\sqrt{n}(\gamma_\lambda - \gamma) = O_p(1). \tag{3.139}$$

A result of this nature would have the consequence that we could actually estimate the parametric part of the model at the usual parametric rate irregardless of the presence of a nonparametric component.

Result (3.139) is actually true under fairly general conditions (Speckman 1988). Here we show that it is satisfied in a somewhat simplified setting that allows us to present the basic ideas without the need for too much additional mathematical machinery.

Let us now specialize to the case of $p = 1$ under which model (3.132) reduces to

$$y_i = u_i\gamma + f(t_i) + \varepsilon_i,\ i = 1, \ldots, n, \tag{3.140}$$

with $u_i, i = 1, \ldots, n$, known values, γ an unknown parameter and $f \in W_2^m[0,1]$ an unknown function. We will also now assume that the $\varepsilon_i, i = 1, \ldots, n$, are *iid* with $E\varepsilon_1 = E\varepsilon_1^3 = 0$ and $E\varepsilon_1^4 < \infty$.

The estimator of γ that is obtained from (3.136) is

$$\begin{aligned} \gamma_\lambda &= (\mathbf{u}^T(\mathbf{I} - \mathbf{S}_\lambda)\mathbf{u})^{-1}\mathbf{u}^T(\mathbf{I} - \mathbf{S}_\lambda)\mathbf{y} \\ &= \gamma + (\mathbf{u}^T(\mathbf{I} - \mathbf{S}_\lambda)\mathbf{u})^{-1}\mathbf{u}^T(\mathbf{I} - \mathbf{S}_\lambda)(\mathbf{f} + \boldsymbol{\varepsilon}) \end{aligned} \tag{3.141}$$

for $\mathbf{u} = (u_1, \ldots, u_n)^T$. To analyze its large sample properties we need some suitable way of modeling the interrelationship between the u and t variables. One simple, mathematically tractable, approach is to assume that

$$u_i = g(t_i) + \eta_i,\ i = 1, \ldots, n, \tag{3.142}$$

for $g \in W_2^m[0,1]$. Here we take $\boldsymbol{\eta} = (\eta_1, \ldots, \eta_n)^T$ to be a vector of *iid* random variables such that $E\eta_1 = E\eta_1^3 = 0$, $E\eta_1^2 = \sigma_\eta^2$ and $E\eta_1^4 < \infty$. The η_i are also assumed to be independent of $\varepsilon_1, \ldots, \varepsilon_n$ from model (3.140). Formulation (3.142) allows the u_i to depend on t through a deterministic, unknown trend function g with the remainders $u_i - g(t_i)$ being treated as "random errors."

We will begin by analyzing the $\mathbf{u}^T(\mathbf{I} - \mathbf{S}_\lambda)\mathbf{u}$ term in (3.141). We may write this quantity as

$$\mathbf{u}^T(\mathbf{I} - \mathbf{S}_\lambda)\mathbf{u} = \mathbf{g}^T(\mathbf{I} - \mathbf{S}_\lambda)\mathbf{g} + 2\mathbf{g}^T(\mathbf{I} - \mathbf{S}_\lambda)\boldsymbol{\eta} + \boldsymbol{\eta}^T(\mathbf{I} - \mathbf{S}_\lambda)\boldsymbol{\eta}, \tag{3.143}$$

where $\mathbf{g} = (g(t_1), \ldots, g(t_n))^T$. We already know from (3.134) that

$$\mathbf{g}^T(\mathbf{I} - \mathbf{S}_\lambda)\mathbf{g} = O(n\lambda^{-2m})$$

and, since $E\mathbf{g}^T(\mathbf{I} - \mathbf{S}_\lambda)\boldsymbol{\eta} = 0$, Markov's inequality gives

$$P(|\mathbf{g}^T(\mathbf{I} - \mathbf{S}_\lambda)\boldsymbol{\eta}| > \delta) \le \sigma_\eta^2\delta^{-2}\mathbf{g}^T(\mathbf{I} - \mathbf{S}_\lambda)\mathbf{g} = O(n\lambda^{-2m}).$$

Similarly, results in Section 2.4 give $E\boldsymbol{\eta}^T\mathbf{S}_\lambda\boldsymbol{\eta} = \sigma_\eta^2\lambda$ so that $\boldsymbol{\eta}^T\mathbf{S}_\lambda\boldsymbol{\eta} = O_p(\lambda)$. Combining all these bounds with the fact that $n^{-1}\boldsymbol{\eta}^T\boldsymbol{\eta} \xrightarrow{P} \sigma_\eta^2$ due to the weak law of large numbers gives the conclusion that

$$n^{-1}\mathbf{u}^T(\mathbf{I} - \mathbf{S}_\lambda)\mathbf{u} \xrightarrow{P} \sigma_\eta^2 \tag{3.144}$$

if $n, \lambda \to \infty$ and $\lambda/n \to 0$.

We have now shown that

$$\gamma_\lambda - \gamma = \frac{1}{n\sigma_\eta^2}\mathbf{u}^T(\mathbf{I} - \mathbf{S}_\lambda)(\mathbf{f} + \boldsymbol{\varepsilon})(1 + o_p(1)).$$

Thus, we will have shown (3.139) if we can show that

$$\begin{aligned} n^{-1/2}\mathbf{u}^T(\mathbf{I} - \mathbf{S}_\lambda)(\mathbf{f} + \boldsymbol{\varepsilon}) &= n^{-1/2}[\mathbf{g}^T(\mathbf{I} - \mathbf{S}_\lambda)\mathbf{f} + \mathbf{g}^T(\mathbf{I} - \mathbf{S}_\lambda)\boldsymbol{\varepsilon} \\ &+ \boldsymbol{\eta}^T(\mathbf{I} - \mathbf{S}_\lambda)\mathbf{f} + \boldsymbol{\eta}^T(\mathbf{I} - \mathbf{S}_\lambda)\boldsymbol{\varepsilon}] \end{aligned}$$

has a limiting normal distribution. The first three terms on the right hand side of this expression can be handled similarly to our treatment of $\mathbf{u}^T(\mathbf{I} - \mathbf{S}_\lambda)\mathbf{u}$ in (3.143) using the Cauchy-Schwarz and Markov inequalities. The only new difficulty stems from the quantity $\boldsymbol{\eta}^T(\mathbf{I} - \mathbf{S}_\lambda)\boldsymbol{\varepsilon} = \boldsymbol{\eta}^T\boldsymbol{\varepsilon} - \boldsymbol{\eta}^T\mathbf{S}_\lambda\boldsymbol{\varepsilon}$. However, $E\boldsymbol{\eta}^T\mathbf{S}_\lambda\boldsymbol{\varepsilon} = 0$ and $\mathrm{Var}(\boldsymbol{\eta}^T\mathbf{S}_\lambda\boldsymbol{\varepsilon}) = \sigma^2\sigma_\eta^2\lambda$ while $\boldsymbol{\eta}^T\boldsymbol{\varepsilon} = \sum_{i=1}^n \eta_i\varepsilon_i$ is a sum of n independent random variables all identically distributed as $\eta_1\varepsilon_1$. We therefore have that if $n, \lambda \to \infty$ in such a way that $\lambda/n \to 0$, then

$$n^{-1/2}\mathbf{u}^T(\mathbf{I} - \mathbf{S}_\lambda)(\mathbf{f} + \boldsymbol{\varepsilon}) = n^{-1/2}\boldsymbol{\eta}^T\boldsymbol{\varepsilon} + o_p(1) \xrightarrow{d} \sigma\sigma_\eta Z \tag{3.145}$$

for Z having a standard normal distribution.

Results such as (3.145) have the important consequence that tests of hypotheses and interval estimates concerning γ that are obtained for model (3.140) using standard linear regression techniques will be effective, at least in an asymptotic sense. In particular, it follows that the interval

$$\gamma_\lambda \pm z_{\alpha/2}\sigma(\mathbf{u}^T(\mathbf{I} - \mathbf{S}_\lambda)\mathbf{u})^{-1/2} \tag{3.146}$$

will have asymptotic probability $1 - \alpha$ of containing γ. Notice that this holds under rather mild conditions on the order sequence λ that allows for λ to grow at the optimal $n^{1/(2m+1)}$ rate for estimation of the purely nonparametric part of the model.

To illustrate the use of estimators such as (3.135) in a model setting such as (3.133) we consider the Rothamsted mildew control data from Draper and Guttman (1980) that is shown in Figure 3.17. This data comes from a study of four mildew control treatments for wheat: none, early spring, late spring and repeated application. The experiment was designed in a single

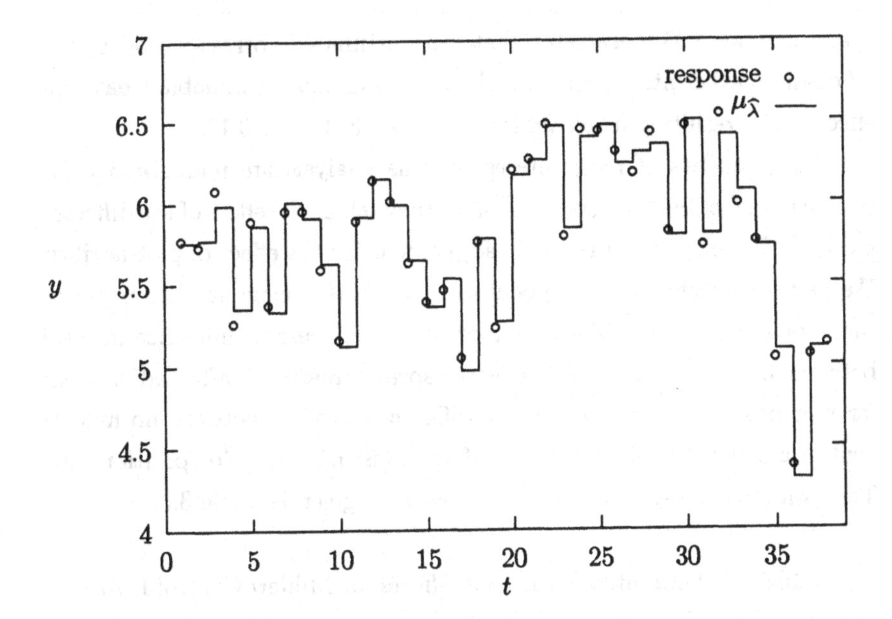

Figure 3.17: Polynomial-Cosine Fit to Rothamstead Data.

column of 38 plots in nine blocks of four plots each with an extra plot on each end. The four treatments were randomly assigned to the plots within each block.

We will treat this data as coming from a model of the form (3.133) with the response variable being the plot yield. The matrix $\mathbf{U}$ is just the incidence matrix that indicates which treatment was applied to each plot and γ represents the treatment effects. The covariate or t variable is plot position and one can then think of $\mathbf{f}$ as representing the effect on yield of changes in fertility as one moves from plot to plot. In this case t is discrete so that the components of $\mathbf{f}_\lambda$ in (3.137) cannot be viewed as estimating a $W_2^m[0,1]$ function but rather as estimating values of a function on the integers. Nevertheless, $\mathbf{f}$ represents a nonparametric component in the model and our estimator (3.135) remains applicable in this setting.

The polynomial-cosine smoother matrix (3.128)-(3.129) was used for $\mathbf{S}_\lambda$ in (3.135) and the order of the estimator was selected by minimizing the GCV criterion. The resulting choice for λ was $\hat{\lambda} = 12$. This means that there were 17 parameter that were estimated corresponding to the 12 cosine coefficients, 2 polynomial coefficients and 3 estimable treatment effects. The resulting fitted values are shown in Figure 3.17.

The quantities of most interest in this analysis are undoubtedly the treatment effects $\gamma = (\gamma_1, \gamma_2, \gamma_3)^T$ since they tell us the effect of the different methods for mildew control after adjustment for the effect of plot fertility. We have parameterized the model so that all treatments are compared to the "none" treatment. Thus, γ_1 represents the average difference in yield between no mildew control and early spring treatment after adjustment for plot position, γ_2 is the average difference in yield between no mildew control and late spring treatment after adjustment for plot position, etc. The estimated treatment effects $\gamma_{\hat{\lambda}} = \gamma_{12}$ are given is Table 3.5.

Table 3.5: Estimated Treatment Effects for Mildew Control Data

Coefficient	Estimate	Estimate/(Standard Error)
γ_1	.552	10.81
γ_2	.665	12.06
γ_3	.682	12.50

The variance-covariance matrix for γ_λ in (3.136) is

$$\mathrm{Var}(\gamma_\lambda) = \sigma^2(\mathbf{U}^T(\mathbf{I} - \mathbf{S}_\lambda)\mathbf{U})^{-1} \tag{3.147}$$

which can be used for the construction of asymptotically valid confidence intervals and tests about the components of γ for a non-stochastic sequence of orders. For our case, as in most cases in practice, λ is estimated. However, we will still use this formula to provide an idea of the variability of the estimated treatment effects with the cautionary note that this may not be strictly valid with estimated orders.

To estimate σ^2 in (3.147) we will use the linear regression type variance estimator

$$\sigma^2_{\hat{\lambda}} = \sum_{i=1}^{n}(y_i - \tilde{\mu}_{\lambda i})^2/(n - p - \lambda - 2),$$

with $\widetilde{\mu}_{\lambda i}, i = 1, \ldots, n$, the elements of $\widetilde{\boldsymbol{\mu}}_\lambda$. For the mildew data this becomes $\sigma^2_{\widetilde{\lambda}} = \sigma^2_{12} = .0082$ at the GCV choice of λ. The corresponding matrix in (3.147) is found to be

$$(\mathbf{U}^T(\mathbf{I} - \mathbf{S}_{12})\mathbf{U})^{-1} = \begin{pmatrix} .318 & .173 & .157 \\ .173 & .371 & .178 \\ .157 & .178 & .363 \end{pmatrix}.$$

These two quantities can now be used to provide rough ideas concerning the statistical significance of any linear combination of the treatment effects. For example, the standardized coefficient values in Table 3.5 were obtained in this fashion from which it appears that all three methods of treatment are "significantly" better than no mildew control at all. An F-ratio type statistic can also be computed for this data to obtain

$$\frac{\gamma_{12}^T \mathbf{U}^T(\mathbf{I} - \mathbf{S}_{12})\mathbf{U}\gamma_{12}}{3\sigma^2_{12}} = 55.29$$

which suggests that the treatment effects are "highly significant."

3.8 Appendix

In this section we derive certain orthogonality results concerning the cosine and sine functions over uniform designs. We begin with the case of $t_i = (2i-1)/2n, i = 1, \ldots, n$, and study the quantities

$$\begin{aligned} d_{jr} &= 2n^{-1}\sum_{i=1}^{n}\cos(j\pi t_i)\cos(r\pi t_i) \\ &= n^{-1}\sum_{i=1}^{n}\cos((r-j)\pi t_i) + n^{-1}\sum_{i=1}^{n}\cos((r+j)\pi t_i) \\ &= n^{-1}\sum_{i=1}^{n}\cos((2i-1)\frac{(r-j)\pi}{2n}) \\ &+ n^{-1}\sum_{i=1}^{n}\cos((2i-1)\frac{(r+j)\pi}{2n}) \end{aligned} \qquad (3.148)$$

that appear in the Euclidean norms and inner products involving the cosine series estimator of Section 3.4. In this regard, we have the following result.

Lemma 3.4. *For* $j = 1, \ldots, n-1$,

$$d_{jr} = \begin{cases} 1, & \text{if } r - j = 2kn \text{ or } r + j = 2kn \text{ for } k \text{ even}, \\ -1, & \text{if } r - j = 2kn \text{ or } r + j = 2kn \text{ for } k \text{ odd}, \\ 0, & \text{otherwise.} \end{cases}$$

Proof. Formula (4) from Section 1.342 of Gradshteyn and Ryzhik (1980) gives

$$\sum_{k=1}^{n} \cos((2k-1)x) = \frac{1}{2}\frac{\sin(2nx)}{\sin(x)} \tag{3.149}$$

provided $\sin(x) \neq 0$. An application of this result to the sums in (3.148) for $x = (r-j)\pi/2n$ and $x = (r+j)\pi/2n$ produces

$$d_{jr} = \frac{1}{2n}\left[\frac{\sin((r-j)\pi)}{\sin(\frac{(r-j)\pi}{2n})} + \frac{\sin((r+j)\pi)}{\sin(\frac{(r+j)\pi}{2n})}\right] \tag{3.150}$$

provided the denominators in the fractions in (3.150) do not vanish. Thus, in particular the d_{jr} vanish when $r \leq n$ and $r \neq j$. If we allow $r = j$ then

$$\begin{aligned} d_{jj} &= 2n^{-1}\sum_{i=1}^{n} \cos^2(j\pi t_i) = 1 + n^{-1}\sum_{i=1}^{n} \cos(j\pi(2i-1)/n) \\ &= 1 + \frac{1}{2n}\frac{\sin(2j\pi)}{\sin(\frac{j\pi}{n})} \end{aligned}$$

and we conclude that $d_{jj} = 1$ for $j = 1, \ldots, n-1$.

Now observe that our previous formula for the d_{jr} with $r \neq j$ and $r \leq n$ can also be used to show that $d_{jr} = 0$ for $r > n$ provided that $r - j \neq 2kn$ or $r + j \neq 2kn$ for any $k = 1, \ldots$ To handle the remaining cases suppose, for example, that $r - j = 2kn$ for k odd. Then,

$$d_{jr} = 2n^{-1}\sum_{i=1}^{n} \cos(j\pi t_i)\cos(j\pi t_i + k\pi(2i-1)) = -2n^{-1}\sum_{i=1}^{n} \cos^2(j\pi t_i)$$

since $k(2i-1)$ being odd entails that $\cos(j\pi t_i + k\pi(2i-1)) = -\cos(j\pi t_i)$, $i = 1, \ldots, n$. The remaining three cases may be handled similarly and one finds that $d_{jr} = -1$ when $r + j = 2kn$ for k odd while $d_{jr} = 1$ for either $r - j = 2kn$ or $r + j = 2kn$ when k is even. •

In a similar manner one can establish a parallel result for the sine functions. Specifically, we have the following lemma whose proof is left as an exercise.

Lemma 3.5. *Let $t_i = i/n, i = 1,\ldots,n$, and define*

$$d_{jr} = 2n^{-1}\sum_{i=1}^{n}\sin(j\pi t_i)\sin(r\pi t_i).$$

For $j = 1,\ldots,n$,

$$d_{jr} = \begin{cases} 1, & \text{if } r-j=2kn, \\ -1, & \text{if } r+j=2kn, \\ 0, & \text{otherwise.} \end{cases}$$

3.9 Exercises

1. Compute the first three Legendre polynomials using formulas (3.8)-(3.9).

2. Compute the first three polynomials which are orthonormal on $t_i = (i-.5)/n$, $i = 1,\ldots,n$, i.e., construct functions x_1, x_2, x_3 from the functions, $1, t, t^2$ which satisfy (3.107). Compare these to the first three Legendre polynomials you computed in Exercise 1. [It may help to recall that $\sum_{j=0}^{n-1} j = n(n-1)/2$, $\sum_{j=0}^{n-1} j^2 = (2n-1)n(n-1)/6$, $\sum_{j=0}^{n-1} j^3 = n^2(n-1)^2/4$ and $\sum_{j=0}^{n-1} j^4 = (n^2-3n-1)(2n-1)n(n-1)/30$.]

3. Let $L_2[0,1]\times[0,1]$ denote the collection of all square integrable functions on $[0,1]\times[0,1]$, i.e., the set of all functions $\mu(s,t)$ such that

$$\int_0^1\int_0^1 \mu^2(s,t)dsdt < \infty.$$

A well known result from functional analysis has the consequence that if $\{x_j\}_{j=1}^{\infty}$ is a CONS for$L_2[0,1]$ then the product sequence

$$\{x_j x_r\}_{j,r=1}^{\infty}$$

is a CONS for $L_2[0,1]\times[0,1]$. Verify the orthonormality property of the product sequence and give a formula for the generalized Fourier series of a square integrable bivariate regression function. Discuss the implications of this for series estimation in the case of bivariate dependent variables.

4. Show that for generalized Fourier series estimators the ordinary cross-validation criterion is

$$\mathrm{CV}(\lambda) = n^{-1}\sum_{i=1}^{n}\left(y_i - \mu_\lambda(t_i)\right)^2 / (1 - s_i)^2,$$

where s_i is the ith diagonal element of $\mathbf{S}_\lambda = \mathbf{X}_\lambda(\mathbf{X}_\lambda^T\mathbf{X}_\lambda)^{-1}\mathbf{X}_\lambda^T$. Discuss the relationship between CV and GCV in this case.

5. Show that $\inf_{\lambda\le n} L_n(\lambda) \xrightarrow{P} 0$ if $L_n(\lambda_n) \xrightarrow{P} 0$ for any sequence λ_n. Conclude from this (Why?) that $\inf_{\lambda\le n} R_n(\lambda) \to 0$ implies that $\inf_{\lambda\le n} L_n(\lambda) \xrightarrow{P} 0$. Discuss any implication this might have for data driven methods of selecting λ. [Hint: Use Markov's inequality.]

6. Establish a version of Theorem 3.1 that allows for the order to be stochastic. In particular, suppose that ε_1 has a normal distribution and that $\tilde{\lambda}$ is a random variable satisfying $\tilde{\lambda}/a_n = 1 + O_p(n^{-\delta})$ for a divergent, deterministic sequence of positive intergers a_n and some $0 < \delta \le 1/2$. Then, obtain conditions which ensure that for a fixed $t \in (0,1)$

$$\sqrt{2n}\frac{\sum_{j=2}^{\tilde{\lambda}}(b_j - \mathrm{E}b_j)\cos(j\pi t)}{\sigma\sqrt{\tilde{\lambda}-1}} \xrightarrow{d} Z$$

for Z a standard normal random variable. [Hint: Use the fact that the $b_j - \mathrm{E}b_j, j = 1,\ldots,n$, are *iid* normal random variables in this case and then find conditions under which $\sum_{j=2}^{\tilde{\lambda}}(b_j - \mathrm{E}b_j)\cos(j\pi t) - \sum_{j=2}^{a_n}(b_j - \mathrm{E}b_j)\cos(j\pi t) = o_p(\sqrt{a_n})$.]

7. Prove (3.78) and discuss its implications for testing the hypothesis that β_j is zero.

8. Consider the data in Table 1.1.

   (a) Fit a cosine series estimator to these data using GCV to select λ. Through examination of the estimated coefficients, diagnostics, etc., determine any adjustments which should be made to your estimator.

(b) Compare the estimated Fourier coefficients to the true Fourier coefficients for μ in this case.

(c) Compute the GSJS variance estimator and estimator (3.28) using the GCV choice of λ. Compare these estimators to the true value of $\sigma = .2$ for this data.

(d) Construct interval estimates of μ at the design points using your estimators of σ^2 in part (c) and discuss their coverage performance.

(e) Compute the loss $L_n(\lambda)$ for various values of λ and compare the optimal λ to the value selected by GCV. Also compare the values of GCV(λ) and $L_n(\lambda)$ and discuss the magnitude of their difference.

(f) How do your results change if you use a polynomial-cosine estimator? What are the coefficients of q_0 and q_1 in (3.117)-(3.118) estimating in this case and how do the actual estimators compare to these true parameter values?

9. The data given in Table 3.6 represents the voltage drop in the battery of a guided missile motor during its time of flight.

   (a) Fit a polynomial regression estimator to this data using automated methods to select the degree of the polynomial. Through examination of estimated coefficients, diagnostics, etc., determine any adjustments which should be made to the estimator.

   (b) Obtain approximate interval estimates for the true regression function at the design points using the interval estimates discussed in Section 3.3.

10. Suppose that $\mu \in C^1[0,1]$ and the ε_i are independent and identically distributed. Show that $\hat{\sigma}_R^2$ in (3.29) is a consistent estimator of σ^2.

Table 3.6: Voltage Drop Data

t	y	t	y	t	y
.0000	8.33	.0244	8.23	.0488	7.17
.0732	7.14	.0976	7.31	.1220	7.60
.1463	7.94	.1707	8.30	.1951	8.76
.2195	8.71	.2439	9.71	.2683	10.26
.2927	10.91	.3171	11.67	.3415	11.76
.3659	12.81	.3902	13.30	.4146	13.88
.4390	14.59	.4634	14.05	.4878	14.48
.5122	14.92	.5366	14.37	.5610	14.63
.5854	15.18	.6098	14.51	.6341	14.34
.6585	13.81	.6829	13.79	.7073	13.05
.7317	13.04	.7561	12.6	.7805	12.05
.8049	11.15	.8293	11.15	.8537	10.14
.8780	10.08	.9024	9.78	.9268	9.80
.9512	9.95	.9756	9.51		

Source: Montgomery and Peck (1982).

11. Let $\{x_j\}_{j=1}^{\infty}$ be a CONS with $x_1(t) \equiv 1$. Show that a generalized Fourier series estimator based on x_j can be written as

$$\mu_\lambda(t) = n^{-1}\sum_{i=1}^{n} K_\lambda(t, t_i)y_i$$

for some function $K_\lambda(\cdot, \cdot)$ with $\sum_{i=1}^{n} K_\lambda(t, t_i) = n$.

12. The λth partial sum for the Legendre series for $\mu \in L_2[0,1]$ is $\sum_{j=1}^{\lambda} \beta_j x_j$, where $\beta_j = \int_0^1 \mu(t)x_j(t)dt$ and the x_j are defined by (3.8)–(3.9). Thus, an expression for μ obtained from Proposition 3.1 is

$$\mu(t) = \sum_{j=1}^{\lambda} \beta_j x_j(t) + E_\lambda(t),$$

with $E_\lambda(t) = \mu(t) - \sum_{j=1}^{\lambda} \beta_j x_j(t)$. In contrast, Taylor's Theorem gives

$$\mu(t) = \sum_{j=1}^{\lambda} \eta_j x_j(t) + \text{Rem}_\lambda(t)$$

for coefficients $\eta_1, \ldots, \eta_\lambda$ and $\text{Rem}_\lambda(\cdot)$ defined by (3.102). Relate the coefficients and remainders in these two expressions and draw conclusions about the utility of the least-squares estimators $b_{\lambda 1}, \ldots, b_{\lambda\lambda}$ of the β_j as estimators of the η_j.

13. The estimated Fourier coefficients obtained from least squares are optimal in the sense that they minimize the residual sum of squares. Show that the true Fourier coefficients have a similar optimality property. Specifically, show that $\beta_1, \ldots, \beta_\lambda$ with

$$\beta_j = \int_0^1 \mu(t) x_j(t) w(t) dt$$

minimize

$$||\mu - \sum_{j=1}^{\lambda} c_j x_j||_w$$

over all possible values of $c_1, \ldots, c_\lambda$.

14. Derive the form of the test statistic T_4 in (3.97) and show it has a limiting chi-squared distribution with one degree-of-freedom under the no effect hypothesis.

15. Show that maximization of the criterion $\widetilde{M}$ in (3.88) is equivalent to minimizing the Schwarz criterion (2.40).

16. Prove Theorem 3.2. Also, show that the test based on T_1 in (3.89) is consistent against any alternative to the no effect model having a nonzero cosine Fourier coefficient β_j for some $j \geq 2$. [Hint: Use Bonferroni's inequality along with the result that if χ_k^2 is a central chi-squared random variable with k degrees-of-freedom then, for $k > 2$, $\ln(n) \geq e/2$ and $v^2 \geq \max\{(k-1)\ln(n), k-1\}$,

$$P(\chi_k^2 \geq v^2) \leq 4\; 2^{-\frac{(k-2)}{2}} e^{-v^2(1-\delta_n)/2}$$

for some $0 \leq \delta_n \leq [3\ln(\ln(n^2)) + 4]/2\ln(n)$.]

17. Show that maximization of M in (3.90) with $C = 2$ is equivalent to minimization of the unbiased risk criterion $\widehat{P}$ for a cosine series regression estimator.

18. Prove Theorem 3.3.

19. Establish the alternative form (3.95) for the order selection based test statistic for the no effect hypothesis and use this to establish consistency for the corresponding test.

20. Verify that the Bayesian test statistic T_2 for the no effect hypothesis has the same limiting distribution as that of the Cramér-von Mises statistic and that the test obtained by rejecting H_0 for "large" values of T_2 is consistent against any fixed alternative to the null model.

21. Prove Theorem 3.5. [Hint: Use a parallel of the argument that was used to establish this type of result for the cosine series estimator along with the result that

$$\max_{1 \le j \le \lambda} \max_{0 \le t \le 1} |x_j'(t)| = O(\lambda^{5/2})$$

for x_j the jth Legendre polynomial of Example 3.2.]

22. Generalize Theorem 3.5 and Corollary 3.2 to include the case where w is not the unit function in (3.21). [Hint: Bound the average squared bias of the estimator by that of the average squared bias for the Legendre polynomial estimator in the uniform design case.]

23. Show that the polynomial-cosine regression estimator of Section 3.6 is equivalent to a least-squares fit to the data using the functions $1, t, t^2$ and $\cos(j\pi t), j = 1, \ldots, \lambda - 1$.

24. The data given in Table 3.7 represents the DC output as a function of wind velocity for a windmill.

    (a) Fit this data using the polynomial-cosine estimator of Section 3.6. Through examination of estimated coefficients, diagnostics, etc., determine any adjustments which should be made to the

Table 3.7: Windmill Data

t	y	t	y	t	y
2.45	.123	2.70	.500	2.90	.653
3.05	.558	3.40	1.057	3.60	1.137
3.95	1.144	4.10	1.194	4.60	1.562
5.00	1.582	5.45	1.501	5.80	1.737
6.00	1.822	6.20	1.866	6.35	1.930
7.00	1.800	7.40	2.088	7.85	2.179
8.15	2.166	8.80	2.112	9.10	2.303
9.55	2.294	9.70	2.386	10.00	2.236
10.20	2.310				

Source: Montgomery and Peck (1982).

estimator. [Note: The t values for this data do not fall in the interval [0, 1] and must be adjusted accordingly.]

(b) Obtain approximate interval estimates for the true regression function at the t_i using the interval estimates discussed in Section 3.3.

25. Discuss the convergence properties of the other trigonometric series estimators for the case of the sine or sine and cosine CONS. Specifically, assume a uniform design $t_i = i/n, i = 1, \ldots, n$, and derive upper bounds for the risks associated with series estimators obtained from the functions $\sqrt{2}\sin(j\pi t), j = 1, \ldots, \lambda$, or 1 and $(\sqrt{2}\cos(2j\pi t)$, $\sqrt{2}\sin(2j\pi t))$, $j = 1, \ldots, \lambda$. What types of boundary behavior are required for μ in order to be able to obtain the optimal rates of convergence with these estimators?

26. Construct a boundary corrected sine series estimator that is guaranteed to be second order irregardless of the boundary behavior of μ. Under what conditions might this estimator produce a better asymptotic rate of convergence for its risk than can be obtained with the boundary corrected cosine series estimator of Section 3.6?

Table 3.8: Mouthwash Data

3 Week SBI	Baseline SBI	u	3 Week SBI	Baseline SBI	u
0.39	0.25	1	0.19	0.25	0
0.30	0.30	1	0.15	0.33	0
0.14	0.34	0	0.15	0.38	1
0.17	0.40	0	0.19	0.45	1
0.18	0.46	1	0.33	0.48	0
0.29	0.54	0	0.45	0.55	1
0.41	0.57	1	0.29	0.59	1
0.09	0.60	0	0.65	0.63	1
0.45	0.63	1	0.15	0.63	1
0.51	0.65	0	0.44	0.65	1
0.50	0.66	1	0.18	0.69	1
0.54	0.71	0	0.47	0.71	0
0.51	0.75	1	0.42	0.99	0
0.42	0.99	0	0.69	1.32	0
0.57	1.42	0	0.31	1.72	0

Source: Speckman (1988).

27. Implement your boundary corrected sine series estimator of Exercise 26 and use it to fit the voltage drop data of Exercise 9.

28. Construct a boundary corrected version of the sine and cosine series estimator that is guaranteed to be second order irregardless of the boundary behavior of μ. Under what conditions might this estimator produce a better asymptotic rate of convergence for its risk than can be obtained with a boundary corrected cosine series estimator of Section 3.6?

29. Implement your boundary corrected sine and cosine series estimator of Exercise 28 and use it to fit the windmill data of Exercise 24.

30. The data in Table 3.8 is from an experiment to determine whether the use of a particular mouthwash is effective in treating gum disease. For this study 30 subjects were randomly assigned to control and treatment groups ($u_i = 0$ for the control and $u_i = 1$ otherwise). An initial assessment of each subject's baseline SBI (a measurement of gum shrinkage) was made and this was followed up by a further assessment after a 3 week period. Conduct an analysis of this data to determine if the mouthwash is effective in reducing gum disease.

31. Show that estimator (3.135) for the partially linear model (3.133) can be obtained by ordinary least-squares regression using the "design" matrix $\mathbf{Z}_\lambda = [\mathbf{U}|\mathbf{X}_\lambda]$. That is, $\tilde{\boldsymbol{\mu}}_\lambda = \mathbf{Z}_\lambda(\mathbf{Z}_\lambda^T\mathbf{Z}_\lambda)^{-1}\mathbf{Z}_\lambda^T\mathbf{y}$.

32. Prove Lemma 3.5. [Hint: Use the fact that $\sum_{k=0}^{n}\cos(kx) = \cos(\frac{nx}{2}) \sin(\frac{(n+1)x}{2})/\sin(\frac{x}{2})$.]

Chapter 4

Kernel Estimators

4.1 Introduction

Suppose, as before, that we have n observations $(t_1, y_1), \ldots, (t_n, y_n)$ which follow the model

$$y_i = \mu(t_i) + \varepsilon_i,\ i = 1, \ldots, n. \tag{4.1}$$

The $\varepsilon_i, i = 1, \ldots, n$, are zero mean, uncorrelated random variables having common variance σ^2 and the $t_i, i = 1, \ldots, n$, are nonstochastic design points assumed to satisfy

$$0 \leq t_1 < t_2 < \cdots < t_n \leq 1.$$

The assumption of a distinct and scalar valued design is for ease of presentation and the results which follow typically continue to hold even when this condition is not satisfied (see, e.g., Exercise 1).

In this chapter we discuss what are known as kernel estimators of μ in model (4.1). Like series estimators, kernel estimators are linear estimators in the sense described in Chapter 1. They arise from a particularly simple method of weighting the response values.

Our study of kernel estimators begins with a discussion of the motivations and forms for these estimators in the next section. Sections 4.3, 4.6 and 4.8 contain treatments of the asymptotic theory for kernel estimators while Sections 4.4 and 4.5 detail methods for choosing the kernel (or weight) function and bandwidth (or smoothing) parameter. Extensions of the basic kernel smoothing methodology are described in Sections 4.7, 4.9 and 4.11. Section 4.10 summarizes some of what is known about the behavior of kernel estimators when the independent variable, t, is also random.

4.2 Kernel Estimators

Kernel estimators are linear estimators in the sense that we can express the value of the estimator at any point t as a weighted sum of the responses. The weights in this sum all derive from a common kernel function that is independent of the design. The use of such an estimator can be motivated, for example, by referring back to our work in Chapter 3 on cosine series estimators.

We saw in Section 3.4 that a λth order cosine series estimator could be written as $n^{-1}\sum_{i=1}^{n} y_i K(t, t_i; \lambda)$ for a function K that has its peak at the point of estimation t and produces weights that average to one over the design. The plots of this function in Figure 3.1 show that it has the displeasing property of oscillating as we move away from t. This has the effect of giving larger weights than one might desire to responses whose corresponding t_i are far away from the point of estimation. The obvious way to remove this feature is to replace the weight function for the estimator by another that decays smoothly to zero as we move away from t.

One simple method of producing a weight function with the desired properties is to use weights of the form

$$K(t, t_i; \lambda) = \frac{1}{\lambda} K\left(\frac{t - t_i}{\lambda}\right) \tag{4.2}$$

with K a function supported on $[-1, 1]$ that has a maximum at zero; the latter property insures that y_i contributes most to its own prediction. Also we would like our function K to satisfy the moment conditions

$$\int_{-1}^{1} K(u)du = 1, \tag{4.3}$$

$$\int_{-1}^{1} uK(u)du = 0, \tag{4.4}$$

$$M_2 = \int_{-1}^{1} u^2K(u)du \neq 0 \tag{4.5}$$

and

$$V = \int_{-1}^{1} K(u)^2du < \infty. \tag{4.6}$$

Condition (4.3) is roughly equivalent to having the weights sum to one while (4.4) is a type of symmetry condition that is automatically satisfied if K is symmetric about zero. The reasons for condition (4.5)–(4.6) will be seen shortly. A function K with these properties will hereafter be referred to as a *second order kernel* or simply a *kernel* if the context is clear. Some commonly used examples of second order kernels are given in Table 4.1. In subsequent discussions we will have occasion to consider kernels of higher order than two which arise from imposing additional moment conditions on the function K. We will also use the term kernel for these functions when the specification of order is immaterial to the discussion.

Table 4.1: Examples of Second Order Kernels

Kernel	K	V	M_2
Uniform	$K(u) = \frac{1}{2}I_{[-1,1]}(u)$	$\frac{1}{2}$	$\frac{1}{3}$
Quadratic	$K(u) = \frac{3}{4}(1-u^2)I_{[-1,1]}(u)$	$\frac{3}{5}$	$\frac{1}{5}$
Biweight	$K(u) = \frac{15}{16}(1-u^2)^2I_{[-1,1]}(u)$	$\frac{5}{7}$	$\frac{1}{7}$

The parameter λ in (4.2) is called the *bandwidth* or smoothing parameter and is roughly the equivalent of the reciprocal of the number of terms in a series estimator. However, there is no reason to restrict $1/\lambda$ to be an integer in the present context. Accordingly, we allow λ to be any nonnegative number.

Note that if $K(u)$ has support on [-1, 1] then $K(u/\lambda)$ has support on $[-\lambda, \lambda]$, so the bandwidth determines, among other things, how far away observations are allowed to be from t and still contribute to the estimation of $\mu(t)$. The bandwidth also governs the peakedness of the weight function and, hence, the degree of dependence of the estimator on information near t. Small values of λ will result in rougher (wigglier) estimators that rely heavily on the data near t. In contrast, larger λ's allow more averaging to occur and thereby give smoother estimators. If K is nonnegative then, as a result of (4.3), it is a probability density on [-1,1] and, in that case, λ is a scale parameter for the density.

Figure 4.1 shows kernel estimator fits to data on the level of a chemical in blood sampled from pregnant women on day t of their pregnancy (see Exercise 19). The estimators are Gasser-Müller kernel estimators discussed below which use the quadratic kernel from Table 4.1. The three fits in the figure correspond to different values of the bandwidth and are quite different ranging from very wiggly to nearly flat. Thus, a good choice of the bandwidth appears to be rather crucial in obtaining a satisfactory fit to data using a kernel estimator. We explore this problem in more detail in Section 4.5.

There is no loss in assuming that K has support on [-1,1]. This is because any kernel with finite support can be rescaled to have support on [-1,1]. It should be noted, however, that kernels which do not have finite support are sometimes used. As seen from Exercise 3 the use of kernels with support on the entire line results in estimators with global bias difficulties which are localized to the boundaries when K has finite support. Thus, kernels with finite support seem preferable.

The estimator we obtain from (4.2) is

$$\mu_\lambda(t) = (n\lambda)^{-1} \sum_{i=1}^{n} K\left(\lambda^{-1}(t - t_i)\right) y_i. \tag{4.7}$$

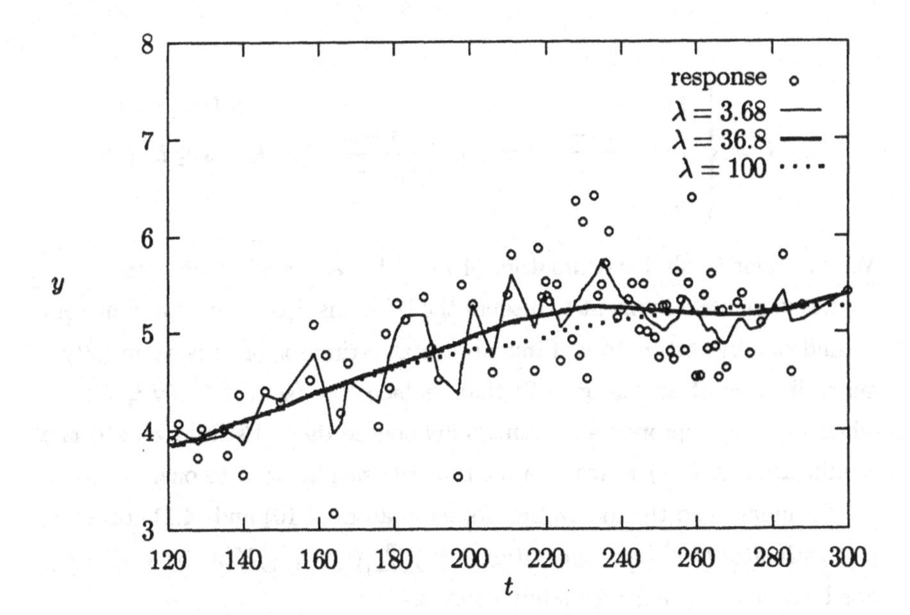

Figure 4.1: Kernel Fits to Assay Data.

However, there are several variants on this theme which include

$$\mu_\lambda(t) = \sum_{i=1}^{n}(t_i - t_{i-1})\,\lambda^{-1}K\left(\lambda^{-1}(t-t_i)\right)y_i, \text{ (with } t_0 = 0) \quad (4.8)$$

$$\mu_\lambda(t) = \sum_{i=1}^{n} K\left(\lambda^{-1}(t-t_i)\right)y_i \Big/ \sum_{j=1}^{n} K\left(\lambda^{-1}(t-t_j)\right), \quad (4.9)$$

$$\mu_\lambda(t) = \sum_{i=1}^{n}\left(\lambda^{-1}\int_{s_{i-1}}^{s_i} K\left(\lambda^{-1}(t-s)\right)ds\right)y_i, \quad (4.10)$$

where

$$s_0 = 0,\ s_{i-1} \le t_i \le s_i,\ i = 1, \ldots, n-1,\ s_n = 1,$$

and

$$\mu_\lambda(t) = \int_0^1 \lambda^{-1}K\left(\lambda^{-1}(t-s)\right)h_y(s)ds, \quad (4.11)$$

for

$$h_y(s) = \begin{cases} y_1, & s \le t_1, \\ y_i\left(\frac{t_{i+1}-s}{t_{i+1}-t_i}\right) + y_{i+1}\left(\frac{s-t_i}{t_{i+1}-t_i}\right), & t_i < s \le t_{i+1}, \\ y_n, & s > t_n. \end{cases}$$

We will refer to all five estimators (4.7)–(4.11) as kernel estimators.

Let us pause a moment to discuss the relationships between and history behind our five estimators. First note that estimator (4.8) is essentially a generalization of estimator (4.7) that replaces the factor n^{-1} by $t_i - t_{i-1}$, which is more appropriate for unequally spaced data. Estimator (4.9) is a modification of (4.7) which ensures that the weights sum to one.

To understand the motivation for estimators (4.10) and (4.11) observe, e.g., estimator (4.8) has expectation $\lambda^{-1}\sum_{i=1}^{n}(t_i-t_{i-1})K(\lambda^{-1}(t-t_i))\mu(t_i)$. For large n this is approximately equal to

$$\lambda^{-1}\int_0^1 \mu(s)K\left(\lambda^{-1}(t-s)\right)ds.$$

Thus, estimator (4.10) is in the spirit of (4.7)–(4.8) but can viewed as using $\lambda^{-1}y_i\int_{s_{i-1}}^{s_i} K\left(\lambda^{-1}(t-s)\right)ds$ instead of, e.g., $\lambda^{-1}y_i(t_i-t_{i-1})K(\lambda^{-1}(t-t_i))$ to estimate the quantity $\lambda^{-1}\int_{s_{i-1}}^{s_i}\mu(s)K(\lambda^{-1}(t-s))ds$. Estimator (4.11) is similar to (4.10). The difference is that a piecewise linear approximation, $h_y(s)$, is used to estimate $\mu(s)$ for $t_i < s \le t_{i+1}$ rather than the piecewise constant approximation of y_i, for $t_i < s \le t_{i+1}$, used by estimator (4.10).

Estimators (4.7)-(4.8) stem from the work of Priestley and Chao (1972). The use of estimator (4.9) with nonstochastic t_i was initially proposed by Benedetti (1975). Estimator (4.9) is often used in the case where the t_i are also random. In this case its use was suggested by Nadaraya (1964) and Watson (1964). For this reason it is often called the Nadaraya-Watson estimator.

Estimator (4.10) was originally studied by Gasser and Müller (1979) and Cheng and Lin (1981). The closely related estimator (4.11) was proposed by Clark (1977) and has also been studied by Cheng and Lin (1981).

All five estimators (4.7)–(4.11) have very similar properties for uniformly spaced t_i. However, estimator (4.10) has received more attention in the literature and is also more appropriate for unequally spaced design points than (4.7) and (4.9). For these reasons we will focus primarily on (4.10) in this book. However, it should be noted that similar developments are also possible for the other estimators.

In the next section we will study the asymptotic properties of estimator (4.10). Before doing so, however, it will be useful to complete the circle and relate kernel estimators back to the cosine series estimators which were used to motivate their study. By doing this we will gain some insight into the way that kernel estimators actually smooth the data.

To avoid some technical problems occurring at the boundaries of [0, 1], let us assume that we are in the uniform design setting with $t_i = (2i - 1)/2n, i = 1, \ldots, n$, and are estimating μ at some specific design point $t_k \in [\lambda + (2n)^{-1}, 1 - \lambda - (2n)^{-1}]$ using the estimator in (4.7). Then we can write

$$\mu_\lambda(t_k) = (n\lambda)^{-1}\mathbf{k}_\lambda^T\mathbf{y}$$

for $\mathbf{y}$ the response vector and

$$\mathbf{k}_\lambda = (\underbrace{0, \ldots, 0}_{k-[n\lambda]}, K(-\frac{[n\lambda]}{n\lambda}), \ldots, K(\frac{[n\lambda]}{n\lambda}), \underbrace{0, \ldots, 0}_{n-k-[n\lambda]})^T.$$

If we now define the $n \times n$ matrix $\mathbf{X} = [\mathbf{1}|\sqrt{2}\{\cos(j\pi t_i)\}_{i=1,n,j=1,n-1}]$, we have $\mathbf{X}^T\mathbf{X} = \mathbf{X}\mathbf{X}^T = n\mathbf{I}$ and, hence,

$$\begin{aligned} \mu_\lambda(t_k) &= (n^2\lambda)^{-1}\mathbf{k}_\lambda^T\mathbf{X}\mathbf{X}^T\mathbf{y} \\ &= \mathbf{f}_\lambda^T\mathbf{b} \end{aligned}$$

for $\mathbf{b} = (b_1, \ldots, b_n)^T$ the sample cosine Fourier coefficients in (3.42) and

$$\mathbf{f}_\lambda = (n\lambda)^{-1}\mathbf{X}^T\mathbf{k}_\lambda.$$

Assuming that K is symmetric about zero we can use the identity $\cos(x + y) = \cos(x)\cos(y) - \sin(x)\sin(y)$ to see that $\mathbf{f}_\lambda$ has typical element

$$\begin{aligned} \frac{\sqrt{2}}{n\lambda}\sum_{r=1}^{n} K(\frac{t_k - t_r}{\lambda})\cos(j\pi t_r) &= \frac{\sqrt{2}}{n\lambda}\sum_{r=-[n\lambda]}^{[n\lambda]} K(\frac{r}{n\lambda})\cos(j\pi\frac{r}{n} + j\pi t_k) \\ &= \sqrt{2}\cos(j\pi t_k)f_{jn}(\lambda) \end{aligned}$$

with

$$f_{jn}(\lambda) = \frac{1}{n} \sum_{r=-[n\lambda]}^{[n\lambda]} \frac{1}{\lambda} K(\frac{r}{n\lambda}) \cos(j\pi \frac{r}{n}) .$$

Combining all our calculations then gives

$$\mu_\lambda(t_k) = f_{1n}(\lambda) b_1 + \sum_{j=1}^{n-1} f_{jn}(\lambda) b_{j+1} \sqrt{2} \cos(j\pi t_k) \tag{4.12}$$

for $f_{1n}(\lambda) = (n\lambda)^{-1} \sum_{r=-[n\lambda]}^{[n\lambda]} K(r/(n\lambda))$.

Equation (4.12) has the implication that μ_λ is really a type of cosine series estimator. The estimator uses all n of the possible cosine basis functions which can be fitted to the data and multiplies the jth cosine function by the coefficient $b_{j+1} f_{jn}(\lambda)$. The multipliers $f_{jn}(\lambda)$ tend to work as damping factors. To see this note that

$$\begin{aligned} f_{jn}(\lambda) &\doteq \int_{-1}^{1} \frac{1}{\lambda} K\left(\frac{x}{\lambda}\right) \cos(j\pi x) dx \\ &= \int_{-1/\lambda}^{1/\lambda} K(u) \cos(j\pi\lambda u) du \\ &= \int_{-1}^{1} K(u) \cos(j\pi\lambda u) du = f(j\pi\lambda) \end{aligned}$$

for λ sufficiently small and n large. Thus, $f_{jn}(\lambda)$ is approximately the Fourier cosine transform f of K evaluated at the point $j\pi\lambda$.

Fourier cosine transforms for the kernels in Table 4.1 are shown in Figure 4.2. If we now think about evaluating these functions at points $j\pi\lambda$ the effect, for any fixed λ, will be that $f(j\pi\lambda)$ tends to grow smaller as j increases in magnitude. Thus, $f_{jn}(\lambda)$ tends to decrease as $|j|$ increases meaning that we are damping out the contributions of the higher frequency Fourier coefficients.

As a result of the above analysis it now becomes apparent that the choice of a kernel is tantamount to selecting a particular way of weighting Fourier coefficients in the Fourier representation (4.12) of the kernel estimator. The converse is also true. Thus, for example, we recognize that the kernel estimator based on the uniform kernel and the cosine series estimator

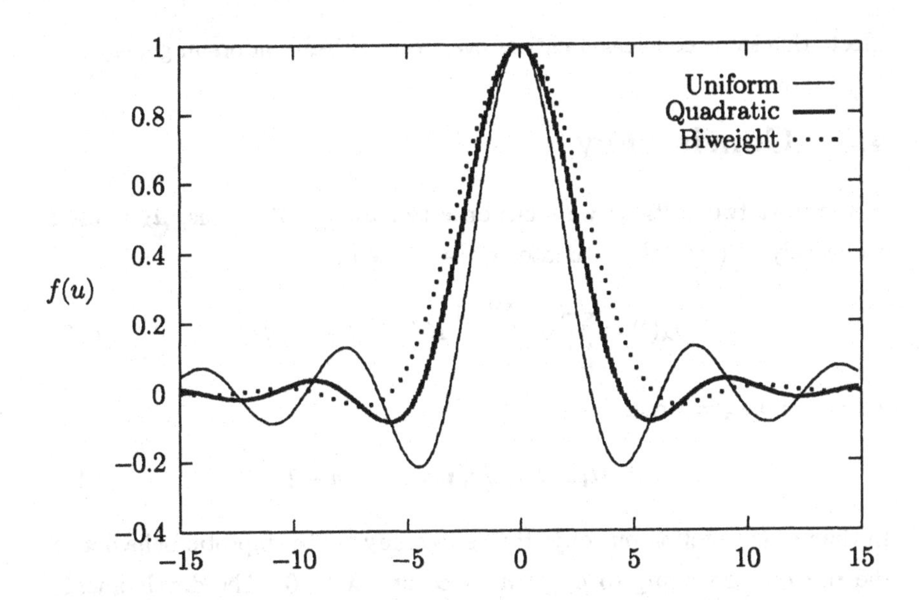

Figure 4.2: Fourier Cosine Transforms of the Kernels in Table 4.1.

studied in Section 3.4 exhibit a dual nature giving uniform weights to observations and sample Fourier coefficients respectively. One might conclude from this that a kernel estimator which uses the uniform kernel does the same thing in the t or "time" domain that a cosine series estimator does in the frequency domain.

As noted in Chapter 3 the fact that μ is smooth makes us think that most of the information about the regression function should be captured in the lower frequency Fourier coefficients while the coefficients for higher frequencies will contain mostly information about the random noise. When looked at from this viewpoint the kernel estimator seems to smooth in a rather natural way. It uses information from all the sample Fourier coefficients and hence from all allowable frequencies. Then the value of the bandwidth governs how much influence the high frequency content of the data is allowed to have on the fit. If λ is large, the estimator contains primarily information from lower frequencies while smaller bandwidths allow for the use of more high frequency information. Apart from boundary effects this same conclusion carries over to kernel estimators in general.

4.3 Consistency

In the next two sections we focus on estimator (4.10). Thus, μ_λ is used exclusively to mean the estimator whose value at t is

$$\mu_\lambda(t) = \sum_{i=1}^{n} y_i \int_{s_{i-1}}^{s_i} \frac{1}{\lambda} K\left(\lambda^{-1}(t-s)\right) ds \tag{4.13}$$

with $s_0 = 0, s_n = 1$ and

$$s_i = (t_{i+1} + t_i)/2, \, i = 1, \ldots, n-1. \tag{4.14}$$

In the present section our objective is to study the asymptotic behavior of the risk corresponding to μ_λ as $n \to \infty$ and $\lambda \to 0$. The developments follow those in Gasser and Müller (1979) and Müller (1988). Initially we will assume that the design is uniform with

$$t_i = (2i-1)/2n, \, i = 1, \ldots, n,$$

and that $K \in C^1[-1,1]$ satisfies (4.3)-(4.6).

Eventually we want to obtain an asymptotic expression for

$$R_n(\lambda) = n^{-1} \sum_{i=1}^{n} \mathrm{E}\left(\mu(t_i) - \mu_\lambda(t_i)\right)^2. \tag{4.15}$$

It is difficult to do so directly. Consequently, we begin with the somewhat more modest undertaking of obtaining an asymptotic form for the risk at a fixed point $t \in (0,1)$, defined by

$$R_n(t;\lambda) = \mathrm{E}\left(\mu(t) - \mu_\lambda(t)\right)^2.$$

Under certain conditions it will be possible to sum asymptotic expressions for $R_n(t_i;\lambda), i = 1, \ldots, n$, across the design to obtain the desired form for $R_n(\lambda)$.

To study $R_n(t;\lambda)$ let us first note that

$$R_n(t;\lambda) = \left(\mu(t) - \mathrm{E}\,\mu_\lambda(t)\right)^2 + \mathrm{Var}\left(\mu_\lambda(t)\right).$$

Thus, we can study the asymptotic behavior of the bias, $\mu(t) - \mathrm{E}\,\mu_\lambda(t)$, and the variance of $\mu_\lambda(t)$ separately and then combine our results to obtain an approximation for $R_n(t;\lambda)$. An initial step in this regard is provided by the following lemma.

Lemma 4.1.

$$\operatorname{Var}(\mu_\lambda(t)) = \frac{\sigma^2}{n\lambda^2}\int_0^1 K^2\left(\lambda^{-1}(t-u)\right)du + O\left((n\lambda)^{-2}\right). \tag{4.16}$$

Proof. First note that the exact variance for $\mu_\lambda(t)$ is

$$\operatorname{Var}(\mu_\lambda(t)) = \frac{\sigma^2}{\lambda^2}\sum_{i=1}^{n}\left[\int_{s_{i-1}}^{s_i} K\left(\lambda^{-1}(t-s)\right)ds\right]^2. \tag{4.17}$$

Let us therefore consider the difference between (4.17) and the first term in (4.16).

Using the mean value theorem for integrals we find that

$$\sum_{i=1}^{n}\left(\int_{s_{i-1}}^{s_i} K\left(\frac{t-s}{\lambda}\right)ds\right)^2 - n^{-1}\int_0^1 K^2\left(\frac{t-s}{\lambda}\right)ds$$
$$= \sum_{i=1}^{n}(s_i - s_{i-1})\left[(s_i - s_{i-1})K^2\left(\frac{t-\theta_i}{\lambda}\right)^2 - n^{-1}K^2\left(\frac{t-\xi_i}{\lambda}\right)\right]$$

for mean values θ_i, ξ_i satisfying $s_{i-1} \le \theta_i, \xi_i \le s_i, i = 1,\ldots,n$. Since the design is uniform the last expression is n^{-2} times

$$\left|\sum_{i=1}^{n}\left[K^2\left(\frac{t-\theta_i}{\lambda}\right) - K^2\left(\frac{t-\xi_i}{\lambda}\right)\right]\right| \le \sum_{i=1}^{n}\left|K^2(u_i) - K^2(v_i)\right| \tag{4.18}$$

with $u_i = (t-\theta_i)/\lambda$ and $v_i = (t-\xi_i)/\lambda, i = 1,\ldots,n$. Now use the fact that $K \in C^1[-1,1]$ to see that $|K(u_i)^2 - K(v_i)^2| \le C|u_i - v_i|$ for some positive constant C. Hence (4.18) is bounded by a constant multiple of

$$\sum_{u_i, v_i \in [-1,1]} |u_i - v_i| \le (n\lambda)^{-1}O(n\lambda) = O(1).$$

The last inequality makes use of the fact that the cardinality of the set of u_i, v_i which fall in [-1,1] is $O(n\lambda)$. As a result of our last bound the disparity between (4.17) and the integral in (4.16) is $O\left((\lambda n)^{-2}\right)$. •

Lemma 4.1 gives the expression we need for the variance of $\mu_\lambda(t)$. By making a change of variable we see that for any fixed t

$$\mathrm{Var}\,(\mu_\lambda(t)) = \frac{\sigma^2 V}{n\lambda} + O\left((n\lambda)^{-2}\right), \tag{4.19}$$

where V is defined in (4.6) This indicates that $\mu_\lambda(t)$ behaves, as expected, like a weighted average of $n\lambda$ terms. It also has the implication that for $R_n(t;\lambda)$ to decay to zero we not only need that $\lambda \to 0$ as $n \to \infty$ but also that $n\lambda \to \infty$. Recall that the same type of condition was needed for Fourier series estimators. In that case we required $n/p \to \infty$ with p the number of terms included in the estimator.

The next lemma provides a companion result to Lemma 4.1 for the expectation of $\mu_\lambda(t)$.

Lemma 4.2. *If* $\mu \in C^1[0,1]$ *then*

$$\mathrm{E}\,\mu_\lambda(t) = \frac{1}{\lambda}\int_0^1 K\left(\lambda^{-1}(t-s)\right)\mu(s)ds + O\left(n^{-1}\right). \tag{4.20}$$

Proof. The proof rests on the following string of equalities and inequalities:

$$\begin{aligned}
&\left|\sum_{i=1}^n \mu(t_i)\int_{s_{i-1}}^{s_i} K\left(\lambda^{-1}(t-s)\right)ds - \int_0^1 K\left(\lambda^{-1}(t-s)\right)\mu(s)ds\right| \\
=\;& \left|\sum_{i=1}^n \left(\mu(t_i) - \mu(\xi_i)\right)\int_{s_{i-1}}^{s_i} K\left(\lambda^{-1}(t-s)\right)ds\right| \\
\le\;& \lambda n^{-1}\max_{t\in[0,1]}|\mu'(t)|\sum_{i=1}^n\left|\int_{u_{i-1}}^{u_i} K(u)du\right| \\
\le\;& \lambda n^{-1}\max_{t\in[0,1]}|\mu'(t)|\max_{u\in[-1,1]}|K(u)|\sum_{u_i\in[-1,1]}|u_i - u_{i-1}| = O(\lambda/n),
\end{aligned}$$

where the ξ_i are mean values with $s_{i-1} \le \xi_i \le s_i$ and $u_i = (t-s_i)/\lambda$. •

Now suppose that $\mu \in C^2[0,1]$. A change of variable and a Taylor expansion about t along with the continuity of μ'' allows us to write the integral in (4.20) as

$$\begin{aligned}
\frac{1}{\lambda}\int_0^1 K\left(\lambda^{-1}(t-s)\right)\mu(s)ds &= \int_{(t-1)/\lambda}^{t/\lambda} K(u)\mu(t-\lambda u)du \\
&= \mu(t)M_0(t) - \lambda\mu'(t)M_1(t) \\
&+ \frac{\lambda^2}{2}\mu''(t)M_2(t) + o(\lambda^2)
\end{aligned}$$

with

$$M_j(t) = \int_{(t-1)/\lambda}^{t/\lambda} u^j K(u)du, \; j = 0, 1, 2. \tag{4.21}$$

For λ sufficiently small, $[(t-1)/\lambda, t/\lambda] \supset [-1, 1]$ so that conditions (4.3)-(4.5) give $M_0(t) = 1, M_1(t) = 0$ and $M_2(t) = M_2$ for M_2 in (4.5). We have just proved the following corollary.

Corollary 4.1. *If* $\mu \in C^2[0,1]$ *then*

$$\mathrm{E}\,\mu_\lambda(t) - \mu(t) = \frac{\lambda^2}{2}\mu''(t)M_2 + o(\lambda^2) + O(n^{-1}). \tag{4.22}$$

It is now possible to see the import of conditions (4.3)–(4.5) required for K. Condition (4.3) assures us that the leading term in $\mathrm{E}\,\mu_\lambda(t)$ is $\mu(t)$. Condition (4.4) is a restriction which eliminates the term of order λ in the bias thereby assuring that the bias is $O(\lambda^2 + n^{-1})$. Requirement (4.5) has the implication that K is matched to a $C^2[0,1]$ function in the sense that the bias is made to decay as rapidly as possible for a regression function that is known to have only two continuous derivatives. If μ has more than two derivatives other moments of K could be required to vanish and a bias of still smaller order could be obtained. This is discussed in Section 4.6. Finally, requirement (4.6) is needed to insure that $\mu_\lambda(t)$ has finite asymptotic variance.

Corollary 4.1 provides an indication of where the bias of μ_λ will tend to be largest: namely, where $\mu''(t)$ is largest in magnitude. So, μ_λ will tend to be more biased in areas where the derivative (or slope) of μ is changing rapidly. This means, as one might expect, that the estimator does not perform as well in areas where the regression function is very nonlinear.

If the bias $\mathrm{E}\,\mu_\lambda(t) - \mu(t)$ is positive, then $\mu_\lambda(t)$ tends to overestimate $\mu(t)$ while if it is negative $\mu_\lambda(t)$ tends to underestimate $\mu(t)$. Assuming that $M_2 > 0$ in (4.22) we see that μ_λ will tend to overestimate where μ'' is positive and underestimate when μ'' is negative. When μ'' is positive this means that μ is concave upwards and if μ'' is negative this implies that μ is concave downwards. The implication of this is that μ_λ will tend to overestimate the valleys (or troughs) of μ and underestimate the peaks

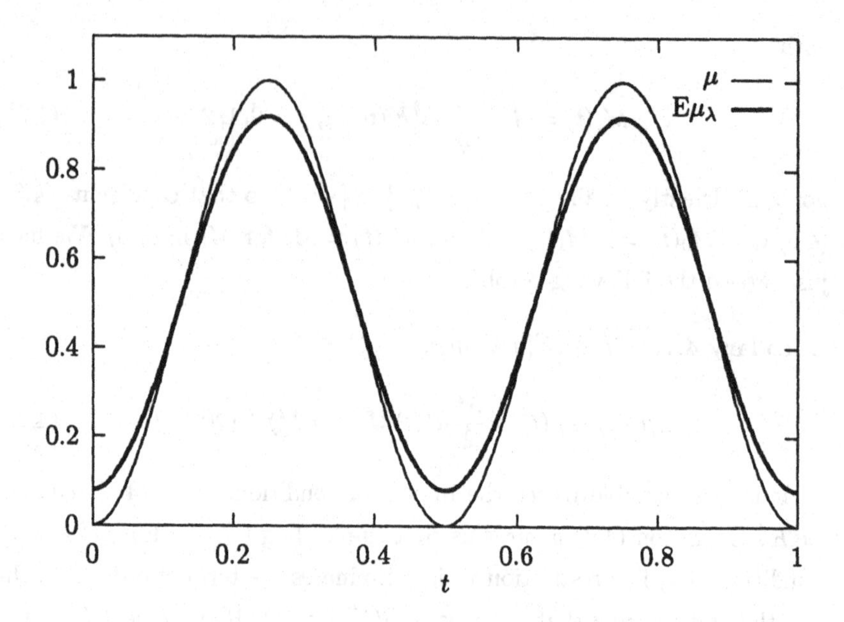

Figure 4.3: Expectation of a Kernel Estimator.

of the function. Since peaks and valleys can often be detected by visual inspection of the data, we will frequently have some idea of whether μ_λ will tend to under or overestimate μ. This is not as much help as it might seem since the actual bias and its magnitude will be unknown. However, the important point is that inspection of the data may suggest areas where the estimator may not behave as well as in others and also the direction in which it will miss μ in such regions.

An illustration of the type of bias behavior discussed above is provided in Figure 4.3. There we have plotted $\mu(t) = .5(1 - \cos(4\pi t))$ along with $\mu(t) + \lambda^2\mu''(t)/10$ for $\lambda = .1$ which is an approximation to $\mathrm{E}\,\mu_\lambda(t)$ for this bandwidth when the quadratic kernel in Table 4.1 is used. We see, for example, that μ_λ will tend to overestimate μ on [0,1/8] where μ is concave upwards and underestimate on [1/8,3/8] where μ is concave downwards, etc.

By combining (4.16) and (4.22) it is possible to establish the asymptotic properties of $R_n(t;\lambda)$. These are stated formally as follows.

Theorem 4.1. *Suppose that $\mu \in C^2[0,1]$ and (4.3)–(4.6) hold. Then, if $n \to \infty$, $\lambda \to 0$ in such a way that $n\lambda \to \infty$*

$$R_n(t;\lambda) \sim \frac{\sigma^2 V}{n\lambda} + \frac{\lambda^4}{4}\mu''(t)^2 M_2^2. \tag{4.23}$$

The asymptotically optimal bandwidth is

$$\lambda_{opt}(t) = n^{-1/5}\left\{\frac{\sigma^2 V}{\mu''(t)^2 M_2^2}\right\}^{1/5}, \tag{4.24}$$

where $\mu''(t) \neq 0$, and

$$R_n\left(t;\lambda_{opt}(t)\right) \sim \frac{1.25}{n^{4/5}}\left\{\left|\mu''(t)\right|\sigma^4 M_2 V^2\right\}^{2/5}. \tag{4.25}$$

Proof. Equation (4.23) is immediate from (4.16) and (4.22). Result (4.24) follows from differentiation of (4.23), and (4.25) is obtained by substituting (4.24) into (4.23). •

Theorem 4.1 has the consequence that the optimal risk is larger where $|\mu''(t)|$ is larger and the optimal bandwidth is larger where $|\mu''(t)|$ is smaller. The first result is a natural consequence of the fact that $\mu(t)$ is more difficult to estimate if μ is very wiggly near t. To understand the second result note, for example, that if μ'' is approximately zero this means that μ is nearly linear around t. But if μ is locally linear then all the observations near t will provide information about the behavior of μ. Consequently the bandwidth should be widened to allow more averaging and thereby reduce the estimator's variance. Conversely, if $|\mu''(t)|$ is large this means that μ is changing rapidly near t so that only the data obtained at points very close to t will provide useful information about $\mu(t)$. The bandwidth should then be narrowed to reduce bias.

While Theorem 4.1 tells us the optimal bandwidth and risk for any particular point it does not tell us how to choose λ to obtain optimal global behavior in terms of minimizing (4.15). Since $R_n(\lambda) = n^{-1}\sum_{i=1}^n R_n(t_i;\lambda)$ it is tempting to use (4.23) to obtain approximations for the $R_n(t_i;\lambda)$ and substitute them into the formula for $R_n(\lambda)$. This approach fails because if t_i is within a bandwidth of the boundary Theorem 4.1 is no longer valid.

To appreciate the problem suppose we wish to estimate μ at a point $t = \lambda q$ for $0 \le q < 1$. That is, suppose we are estimating at some point in the lower boundary region. One may check that $\mathrm{Var}(\mu_\lambda(t)) = O((n\lambda)^{-1})$ so that the variance behaves the same as for estimation at an interior point in terms of its order. However, the expectation of the estimator is

$$\mathrm{E}\,\mu_\lambda(t) = \mu(t)M_0(t) - \lambda\mu'(t)M_1(t) + \frac{\lambda^2}{2}\mu''(t)M_2(t) + o(\lambda^2) + O(n^{-1}),$$

and, in contrast to the case where t is an interior point we no longer have $M_0(t) = 1$ even for large n. Thus, the lead term in the approximation to $\mathrm{E}\,\mu_\lambda(t)$ is no longer $\mu(t)$. In an attempt to correct this boundary bias problem we might try to modify the estimator as

$$\mu_{\lambda c}(t) = \sum_{i=1}^{n} y_i \int_{s_{i-1}}^{s_i} K\left(\lambda^{-1}(t-s)\right) ds \Big/ \sum_{i=1}^{n} \int_{s_{i-1}}^{s_i} K\left(\lambda^{-1}(t-s)\right) ds. \tag{4.26}$$

This is often called the cut-and-normalize estimator and has the property that $\mu_{\lambda c}(t) = \mu_\lambda(t)$ for $t \in [\lambda, 1-\lambda]$. However, in the boundary region $\mu_{\lambda c}$ is less biased than the unmodified estimator. If we now average the pointwise risks for the cut-and-normalize estimator across the design we find that the global risk for the estimator is $O(\lambda^3) + O((n\lambda)^{-1} + \lambda^4)$ as $n \to \infty$, $\lambda \to 0$ and $n\lambda \to \infty$. The $O(\lambda^3)$ portion of the risk comes from summing over the $O(n\lambda)$ squared bias terms of order λ^2 that are contained in the boundary regions. The $O(\lambda^3)$ term dominates $O(\lambda^4)$ and the best rate of decay is only $n^{-3/4}$ rather than the $n^{-4/5}$ rate we would like to obtain.

If μ is smoothly periodic a modified version of μ_λ can be constructed that avoids boundary bias difficulties. For this estimator the optimal bandwidth for minimization of $R_n(\lambda)$ is of exact order $n^{-1/5}$ and the optimal global risk is $O(n^{-4/5})$ (Exercise 4). Another case where the boundary bias issue does not arise is when $\mu'(0) = \mu'(1) = 0$. In this case $\mathrm{E}\,\mu_{\lambda c}(t) - \mu(t) = O(\lambda^2)$ uniformly in t. Since $\mu_{\lambda c}(t)$ agrees with $\mu_\lambda(t)$ in $[\lambda, 1-\lambda]$ we can use Theorem 4.12 to see that if $n \to \infty, \lambda \to 0$ and $n\lambda \to \infty$,

$$n^{-1}\sum_{i=1}^{n} \mathrm{E}\,(\mu_{\lambda c}(t_i) - \mu(t_i))^2 = \frac{\sigma^2 V}{n\lambda} + \frac{\lambda^4}{4} J_2(\mu) M_2^2 + o((n\lambda)^{-1} + \lambda^4), \tag{4.27}$$

where

$$J_2(\mu) = \int_0^1 \mu''(t)^2 dt. \tag{4.28}$$

However, this is only valid when the derivative of the regression function vanishes at 0 and 1 which will not generally be true. So, we need some method of removing the boundary bias from μ_λ that will allow it to be used for estimation of general, unknown regression functions. We will discuss the boundary kernel method for edge effect removal here and then return to this issue in Section 4.4 where other approaches are presented.

It suffices to show how to alter the estimator for one of the boundaries, so we choose the lower region $[0, \lambda)$. If $t \in [0, \lambda)$, then it can always be written as $t = q\lambda$ for some $0 \leq q < 1$. For any such point assume it is possible to construct a modification of the original kernel, K_q, which satisfies

$$\int_{-1}^{q} K_q(u)du = 1, \tag{4.29}$$

$$\int_{-1}^{q} uK_q(u)du = 0, \tag{4.30}$$

$$\int_{-1}^{q} u^2 K_q(u)du = M_{2q} \neq 0 \tag{4.31}$$

and

$$\int_{-1}^{q} K_q(u)^2 du = V_q < \infty. \tag{4.32}$$

Also assume that K_q is differentiable in u and continuous in q with

$$K_q \to K \text{ as } q \to 1. \tag{4.33}$$

A kernel satisfying (4.29)–(4.33) is called a *boundary kernel*. Many kernels of interest can be modified to obtain boundary kernels. Examples of boundary kernels corresponding to the second order kernels in Table 4.1 are shown in Table 4.2. Other examples can be found in Gasser and Müller (1979), Gasser, Müller and Mammitzsch (1985), Müller (1988, 1991, 1993) and Müller and Wang (1994).

Table 4.2: Examples of Lower Boundary Kernels

Kernel	$K_q(u)$
Uniform	$\frac{2}{(1+q)^3}\left\{3(1-q)u+2(1-q+q^2)\right\} I_{[-1,q]}(u)$
Quadratic	$\frac{12(u+1)}{(1+q)^4}\left\{(1-2q)u+.5(3q^2-2q+1)\right\} I_{[-1,q]}(u)$
Biweight	$\frac{15(1+u)^2(q-u)}{(1+q)^5}\left\{2u\left(5\frac{1-q}{1+q}-1\right)+(3q-1)+5\frac{(1-q)^2}{(1+q)}\right\} I_{[-1,q]}(u)$

Kernels for the upper boundary can be obtained from those used in the lower boundary. To see this suppose we are at a point $t \in (1-\lambda, 1]$ and write $1-t=q\lambda$ for $q \in [0,1)$. Then, we require that

$$\int_{-q}^{1} u^j K_q(u)du = \begin{cases} 1, & j=0, \\ 0, & j=1, \end{cases}$$

with $\int_{-q}^{1} u^2 K_q(u)du \neq 0$ and $\int_{-q}^{1} K_q(u)^2 du < \infty$. A kernel satisfying these conditions can be obtained by using the lower boundary kernel with argument $-u$.

For estimation in the boundary region we will now use a modified version of the Gasser-Müller estimator μ_λ obtained by replacing K with a boundary kernel K_q whenever an estimation point t is within a bandwidth of the boundary. Tracing back through the arguments which led to Lemma 4.1 and Corollary 4.1 reveals that this modified estimator at, e.g., a lower boundary point $t=q\lambda$, will have

$$\mathrm{Var}\,(\mu_\lambda(q\lambda)) \leq \frac{\sigma^2}{n\lambda} \max_{0\leq q\leq 1} V_q + o((n\lambda)^{-1})$$

and

$$\mathrm{E}\,\mu_\lambda(q\lambda) - \mu(q\lambda) = \mu''(0)\frac{\lambda^2}{2}M_{2q} + o(\lambda^2) + O(n^{-1}).$$

The latter expression has the implication that if $n \to \infty$ and $\lambda \to 0$

$$(\mathrm{E}\,\mu_\lambda(q\lambda) - \mu(q\lambda))^2 = O((\lambda^2+n^{-1})^2)$$

uniformly in $q \in [0,1]$. Letting $B = \{t_j : t_j < \lambda \text{ or } t_j > 1-\lambda\}$ it then follows that if $n \to \infty$, $\lambda \to 0$ and $n\lambda \to \infty$,

$$\begin{aligned} R_n(\lambda) &= \frac{1}{n}\sum_{t_i \in B} \mathrm{E}\,(\mu_\lambda(t_i) - \mu(t_i))^2 + \frac{1}{n}\sum_{t_i \notin B} \mathrm{E}\,(\mu_\lambda(t_i) - \mu(t_i))^2 \\ &\sim O(\lambda)[O((n\lambda)^{-1} + \lambda^4)] + \frac{\sigma^2}{n\lambda}\int_{-1}^{1} K(u)^2 du \\ &+ \lambda^4 \left(\frac{1}{4n}\sum_{t_i \notin B} \mu''(t_i)^2\right)\left(\int_{-1}^{1} u^2 K(u) du\right)^2 \\ &\sim \frac{\sigma^2 V}{n\lambda} + \frac{\lambda^4}{4} J_2(\mu) M_2^2. \end{aligned} \tag{4.34}$$

In obtaining this we have employed the continuity of μ'' to allow us to replace $n^{-1}\sum_{i=1}^{n} \mu''(t_i)^2$ by an integral. The conclusions drawn from (4.34) can be summarized as follows.

Theorem 4.2. *If $\mu \in C^2[0,1]$, (4.3)-(4.6) hold and second order boundary kernels are used for estimation in the boundary regions, the asymptotic minimizer of $R_n(\lambda)$ is*

$$\lambda_{opt} = n^{-1/5}\left\{\frac{\sigma^2 V}{J_2(\mu) M_2^2}\right\}^{1/5} \tag{4.35}$$

and the asymptotically optimal risk is

$$R_n(\lambda_{opt}) \sim \frac{1.25}{n^{4/5}} J_2(\mu)^{1/5}\left\{\sigma^4 M_2 V^2\right\}^{2/5}. \tag{4.36}$$

Theorem 4.2 has the interpretation that if μ_λ is altered appropriately at the boundaries it is possible to obtain a global rate of $n^{-4/5}$ for $R_n(\lambda)$. Notice that this is the rate one would have anticipated from our discussion in Chapter 1, as $n^{-4/5}$ is the best uniform rate of convergence for any linear estimator of a function in the set $W_2^2[0,1]$ that contains $C^2[0,1]$.

Theorem 4.2 is, unfortunately, not directly useful for selection of λ since the asymptotically optimal bandwidth λ_{opt} involves the unknown regression function and variance. It is possible to estimate μ'' using, for example, the approach in Section 4.6, and thereby obtain an estimate of $J_2(\mu)$ in (4.35) and consistent estimators for σ^2 were discussed in Section 3.3. Thus, in theory it is possible to estimate the unknown quantities in λ_{opt} to obtain a data driven choice for the bandwidth. The practical utility of this approach remains somewhat uncertain for a variety of reasons that we will explore in Section 4.5.

It is interesting to compare Theorem 4.2 with parallel results for series estimators such as Theorems 3.5. This latter result has the implication that a polynomial regression estimator can obtain the optimal $n^{-4/5}$ rate of decay for the risk of a $W_2^2[0,1]$ regression function by choosing the number of terms in the estimator to grow at the rate $n^{1/5}$. However, if the regression function possesses additional smoothness that produces a faster rate of decay for its (Legendre polynomial) Fourier coefficients, then the risk for the estimator can be made to decay much faster. In particular, the rate $O(\log n/n)$ is attainable when μ is infinitely differentiable.

In contrast, for a second order kernel estimator the condition $M_2 \neq 0$ means that the leading term in the bias of the estimator is of order λ^2 irregardless of how many additional derivatives the regression function might have. Thus, the best rate of decay for the risk of a second order kernel estimator is always $n^{-4/5}$ even if μ is infinitely differentiable.

It is unclear to what extent the theoretical ability of series estimators to adapt to additional smoothness is ever realized in practice. In any case, the local nature of kernel type estimators tends to produce more visually appealing fits to data that typically exhibit fewer extraneous wiggles than for series estimators. This feature alone may be sufficient to compensate for any theoretical shortcomings of the kernel method.

Perhaps the most important point to be gleaned from our comparison of series and kernel smoothers is that the two estimators represent distinctly different estimation methods. Unlike other estimators that will be discussed subsequently which have asymptotic representations as kernel type smoother, series estimators are not equivalent to kernel estimators in any asymptotic sense and conversely.

Theorem 4.2 has the implication that a suitably boundary corrected kernel estimator gives a globally consistent estimator of μ in the case of a uniform design provided only that $n \to \infty$, $\lambda \to 0$ and $n\lambda \to \infty$. There is nothing particularly special about uniform designs and similar results apply to the nonuniform case. Suppose, for example, that the design is generated as in (3.21) from a positive, differentiable design density w through the relation

$$\int_0^{t_i} w(t)dt = (2i-1)/2n,\ i = 1, \ldots, n. \tag{4.37}$$

Then, one finds that expression (4.22) for the bias remains valid in this case while the variance becomes

$$\text{Var}\,(\mu_\lambda(t)) = \frac{\sigma^2 V}{w(t)n\lambda} + O\left((n\lambda)^{-2}\right) \tag{4.38}$$

for any fixed interior point t. This produces optimized pointwise bandwidths and risks of the form

$$\lambda_{opt}(t) = n^{-1/5}\left\{\frac{\sigma^2 V}{w(t)\mu''(t)^2 M_2^2}\right\}^{1/5} \tag{4.39}$$

and

$$R_n\left(t; \lambda_{opt}(t)\right) \sim \frac{1.25}{n^{4/5}}\left[\frac{\sigma^2 V}{w(t)}\right]^{4/5}\left[\mu''(t)^2 M_2^2\right]^{1/5}. \tag{4.40}$$

If we then boundary correct the estimator using boundary kernels we obtain the globally asymptotically optimal bandwidths and risks

$$\lambda_{opt} = n^{-1/5}\left\{\frac{\sigma^2 V}{J_2(\mu) M_2^2}\right\}^{1/5} \tag{4.41}$$

and

$$R_n(\lambda) \sim \frac{1.25}{n^{4/5}}\left[\sigma^2 V\right]^{4/5}\left[J_2(\mu) M_2^2\right]^{1/5}, \tag{4.42}$$

where now

$$J_2(\mu) = \int_0^1 \mu''(t)^2 w(t)dt. \tag{4.43}$$

The proof of (4.39)-(4.42) is left as an exercise (Exercise 5).

4.4 Selecting the Kernel Functions

In order to use a kernel estimator it is necessary to select both interior and boundary kernel functions which raises the question of whether there is some best or optimal way of making these choices. This section gives a partial answer to that question.

Let us first deal with the problem of choosing the interior kernel K. In the spirit of Chapter 2, we might try to choose a kernel that minimizes the risk of the estimator once an optimal choice has been made for λ. (See, e.g., Blyth 1993 for an alternative approach.) As a result of (4.42) we see this is asymptotically equivalent to minimizing $R(K) = M_2^{2/5}V^{4/5}$, where V and M_2 are defined in (4.5) and (4.6), respectively. Note that $R(K)$ also appears in (4.40) so a K which is optimal for $R_n(\lambda)$ will also be optimal for $R_n(t;\lambda)$.

The problem, therefore, is to find a kernel K which minimizes $R(K)$ subject to restrictions (4.3)-(4.6). Results in Gasser and Müller (1979) reveal that the solution is a quadratic kernel with two sign changes in [-1,1]. To avoid such oscillations, Gasser and Müller (1979) impose the restriction that the kernel be positive on [-1,1]. The result is the quadratic kernel in Table 4.1. This kernel is well known in the density estimation literature due to the work of Epanechnikov (1969). Benedetti (1977) also pointed out the optimality of the quadratic kernel for nonparametric regression.

Alternatively, we can think of smoothing as merely a way of obtaining an estimator of $\mu(t_i)$ with smaller variance than y_i. From this viewpoint smoothing is really a variance reduction technique and bias becomes something that must simply be tolerated. By taking this perspective one is instead led to select K to minimize the variance of μ_λ. Asymptotically this is equivalent to choosing a kernel that minimizes V subject to (4.3) and (4.4). To obtain the minimizer first note that (4.3) and (4.4) determine a second order (i.e., linear) polynomial, $\theta_0 + \theta_1 u$, as the solution to

$$\int_{-1}^{1} (\theta_0 + \theta_1 u)\, du = 1$$

and

$$\int_{-1}^{1} u(\theta_0 + \theta_1 u) du = 0.$$

We see that $\theta_0 = 1/2$, $\theta_1 = 0$ is the solution which gives the rectangular kernel in Table 4.1.

Let $K_r(u)$ be the rectangular kernel. Then any other kernel K on [-1,1] can be written as $K_r + \delta K$ with δK a function satisfying

$$\int_{-1}^{1} \delta K(u)du = \int_{-1}^{1} u\delta K(u)du = 0.$$

Since

$$\begin{aligned}\int_{-1}^{1} K^2(u)du &= \int_{-1}^{1} K_r^2(u)du + 2\int_{-1}^{1} K_r(u)\delta K(u)du + \int_{-1}^{1} \delta K^2(u)du \\ &= \int_{-1}^{1} K_r^2(u)du + \int_{-1}^{1} \delta K^2(u)du > \int_{-1}^{1} K_r^2(u)du\end{aligned}$$

the rectangular kernel must be the asymptotically minimum variance kernel.

Table 4.1 gives the values of M_2 and V for the quadratic, rectangular and biweight kernels. We see that not much is lost, in terms of $R(K)$, by use of the rectangular over the optimal quadratic kernel. On the other hand, both the quadratic and rectangular kernels have very similar values for V. One might conclude from all this that the actual choice of a kernel, once λ has been optimized, is not very important. This appears to be essentially true and a good choice of λ is what seems to be the most crucial factor in kernel smoothing.

It is also possible to consider the use of optimal boundary kernels. Comprehensive developments of this subject can be found in Müller (1991, 1993) and Müller and Wang (1994). In particular, the kernels in Table 4.2 are the boundary kernels obtained by Müller and Wang (1994) that correspond to the interior kernels in Table 4.1.

If one views boundary effects as being merely a nuisance that must be eliminated for efficient estimation, then optimal selection of a boundary kernel becomes less of an issue and there are a number of other methods that can effectively reduce boundary bias. For example, a very simple approach to constructing a bounday kernel is to start with a kernel of the form

$$K_q(u) = (a_q + b_q u)K(u), \tag{4.44}$$

for K the interior kernel, and then choose a_q and b_q so that K_q satisfies the boundary kernel moment conditions (4.3)-(4.5). A quick calculation

(Exercise 6) reveals that if $t = \lambda q$ with $q < 1$

$$\begin{aligned} a_q &= M_2(\lambda q)/(M_2(\lambda q)M_0(\lambda q) - M_1^2(\lambda q)), \\ b_q &= M_1(\lambda q)/(M_2(\lambda q)M_0(\lambda q) - M_1^2(\lambda q)), \end{aligned}$$

for the $M_j(\cdot), j = 0, 1, 2$, defined in (4.21), and the corresponding kernel is

$$K_q(u) = \frac{M_2(\lambda q) - uM_1(\lambda q)}{M_2(\lambda q)M_0(\lambda q) - M_1^2(\lambda q)} K(u). \tag{4.45}$$

This boundary kernel necessarily satisfies the requisite moment conditions and agrees with K in the interior region $[\lambda, 1 - \lambda]$.

Another approach to boundary correcting a kernel estimator can be developed from the Bias Reduction Lemma in Section 3.6. The idea is essentially the same as the one used to derive the polynomial-cosine estimator of that section. We know that there is no boundary problem if we use the cut-and-normalize kernel estimator when the derivative of the regression function vanishes at zero and one. Thus, write $\mu = \mu^* + \mu'(0)q_0 + \mu'(1)q_1$ for q_0, q_1 the quadratic polynomials in (3.117)-(3.118) and μ^* a function whose derivative vanishes at 0 and 1. Then, arguing as in Section 3.6 we obtain a boundary modified estimator $\widetilde{\boldsymbol{\mu}}_\lambda = \boldsymbol{\mu}_{\lambda c} + \mathbf{P}(\mathbf{y} - \boldsymbol{\mu}_{\lambda c})$ with $\boldsymbol{\mu}_{\lambda c}$ the vector of fitted values for the cut-and-normalize kernel estimator, $\mathbf{P} = \mathbf{G}(\mathbf{G}^T\mathbf{G})^-\mathbf{G}^T$ and $\mathbf{G} = (\mathbf{I} - \mathbf{S}_{\lambda c})\mathbf{Q}$ for $\mathbf{S}_{\lambda c}$ the smoother matrix for the cut-and-normalize kernel estimator and $\mathbf{Q} = \{q_j(t_i)\}_{i=1,n,j=0,1}$. The boundary correction that is made here is global so that, strictly speaking, this estimator is no longer a kernel estimator. When the design is uniform and boundary corrections are necessary, the new estimator has smaller asymptotic risk than the original kernel estimator (Exercise 7). For other discussions of boundary correction see Rice (1984b), Hall and Wehrly (1991) and Hart and Wehrly (1992).

4.5 Selecting the Bandwidth

Perhaps the most important problem associated with the use of a kernel estimator is the selection of a good value of λ. In this section we discuss the use of data driven techniques for smoothing parameter selection.

For the Gasser-Müller kernel estimator (4.13) the vector of fitted values can be written as $\boldsymbol{\mu}_\lambda = \mathbf{S}_\lambda \mathbf{y}$, where $\mathbf{S}_\lambda$ has typical element

$$\lambda^{-1} \int_{s_{j-1}}^{s_j} K(\lambda^{-1}(t_i - s))ds \tag{4.46}$$

with K replaced by a boundary kernel in (4.46) when t_i is in the boundary region. Thus, μ_λ is a linear estimator and we could consider choosing the bandwidth to minimize the GCV criterion

$$\mathrm{GCV}(\lambda) = n^{-1} \sum_{i=1}^{n} (y_i - \mu_\lambda(t_i))^2 / \left[1 - n^{-1}\mathrm{tr}\mathbf{S}_\lambda\right]^2 \tag{4.47}$$

or possibly the unbiased risk criterion

$$\widehat{P}(\lambda) = n^{-1} \sum_{i=1}^{n} (y_i - \mu_\lambda(t_i))^2 + 2\frac{\widehat{\sigma}^2}{n}\mathrm{tr}\mathbf{S}_\lambda. \tag{4.48}$$

We will refer to these two bandwidth estimators as $\widehat{\lambda}_G$ and $\widehat{\lambda}_P$ throughout the remainder of this section.

The fit to the data in Figure 4.1 that used $\lambda = 36.76$ corresponds to the GCV choice of λ for that data. Another GCV fit, this time for simulated data, is shown in Figure 4.4. The data in this figure was generated from model (4.1) using $n = 50$, a uniform design, normal random errors with $\sigma = .15$ and $\mu(t) = t + .5\exp\{-50(t-.5)^2\}$. The fitted curve is a boundary modified Gasser-Müller estimator with the quadratic kernel that seems to do a good job of capturing the features of the true regression curve for the data. For this data we found $\widehat{\lambda}_G = .1202$ and $\widehat{\lambda}_P = .1161$ which can be compared to the asymptotically optimal value (4.35) which is found to be $\lambda_{opt} = .115$ for this regression function (Exercise 9).

Since the true regression function is known we can actually compute the loss for this example. We have done this for various values of λ and plotted the results in Figure 4.5. Also plotted are the corresponding values of the GCV and unbiased risk (UBR) criteria (4.47)-(4.48). Notice how GCV(λ) and $\widehat{P}(\lambda)$ track the loss with the disparity between the selection criteria and the loss being roughly the value of the variance $\sigma^2 = .0225$. The loss for this data was minimized at $\lambda = .1291$.

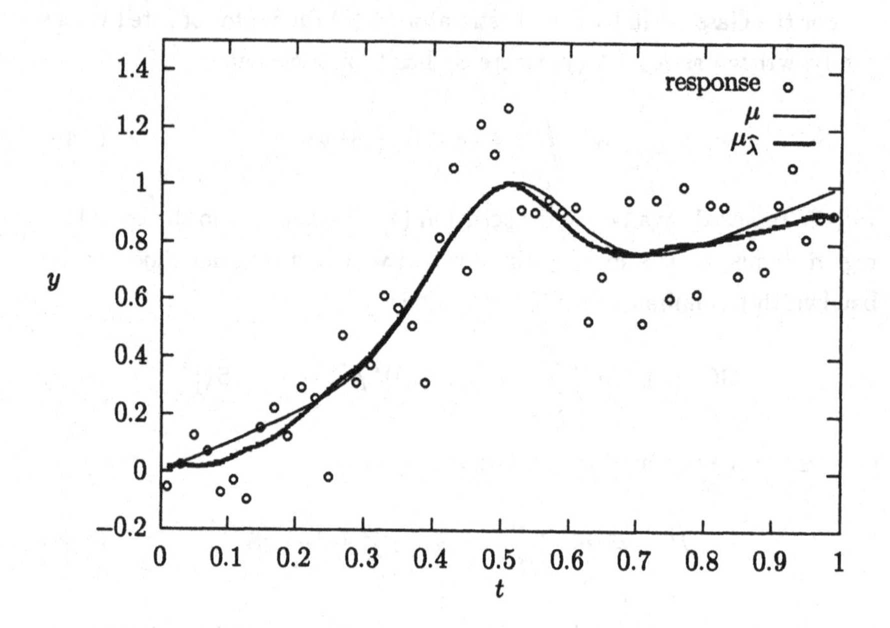

Figure 4.4: Kernel Fit to Simulated Data.

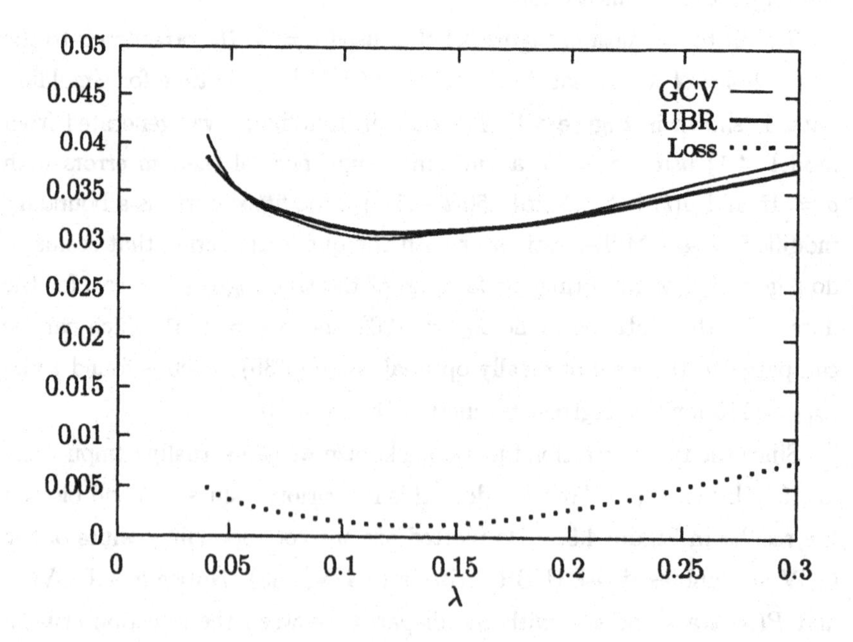

Figure 4.5: GCV, $\widehat{P}$ and Loss for Simulated Data.

There are now a number of results available concerning the consistency and optimality properties of various data driven bandwidth selection procedures. Pioneering work by Wong (1983) and Rice (1984a) has now been advanced and refined to the extent that a relatively complete picture is beginning to emerge concerning the properties of automatic bandwidth selectors.

To discuss the properties of bandwidth estimators we must first decide what we actually wish to estimate. One object of interest is the unknown minimizer $\widehat{\lambda}_{opt}$ of the loss

$$L_n(\lambda) = n^{-1}\sum_{i=1}^{n}(\mu(t_i) - \mu_\lambda(t_i))^2$$

which corresponds to the kernel estimator that is closest to μ in terms of average Euclidean distance across the design.

Härdle, Hall and Marron (1988) showed that if $\widehat{\lambda}$ is obtained from any of the standard bandwidth estimators (e.g., GCV or $\widehat{P}$) then

$$n^{1/10}\frac{\widehat{\lambda} - \widehat{\lambda}_{opt}}{\widehat{\lambda}_{opt}} \xrightarrow{d} v_1 Z, \tag{4.49}$$

where Z is a standard normal random variable and v_1^2 is a variance term that depends on K, σ^2 and μ. This result indicates that standard bandwidth selection methods based on minimizing the usual estimators of the risk or loss provide consistent "estimators" of $\widehat{\lambda}_{opt}$, but the rate of convergence at $n^{-1/10}$ is very slow. One anticipates from this that $\widehat{\lambda}$ will tend to be quite variable as an estimator of $\widehat{\lambda}_{opt}$ and indeed this quality has been well documented in empirical studies.

Figure 4.6 shows the results of a simulation corresponding to 300 replications of the basic experiment that generated the data for Figure 4.4. We computed the values of $\widehat{\lambda}_G, \widehat{\lambda}_P$ as well as $\widehat{\lambda}_{opt}$ for all 300 samples and then computed the standardized bandwidths $n^{1/10}(\widehat{\lambda}_G - \widehat{\lambda}_{opt})/\widehat{\lambda}_{opt}$ and $n^{1/10}(\widehat{\lambda}_P - \widehat{\lambda}_{opt})/\widehat{\lambda}_{opt}$. The empirical densities for the 300 standardized bandwidths corresponding to the two bandwidth estimators are shown in Figure 4.6. Both densities are roughly centered at zero but exhibit substantial positive skewness. In addition, the linear correlations of $\widehat{\lambda}_{opt}$ with

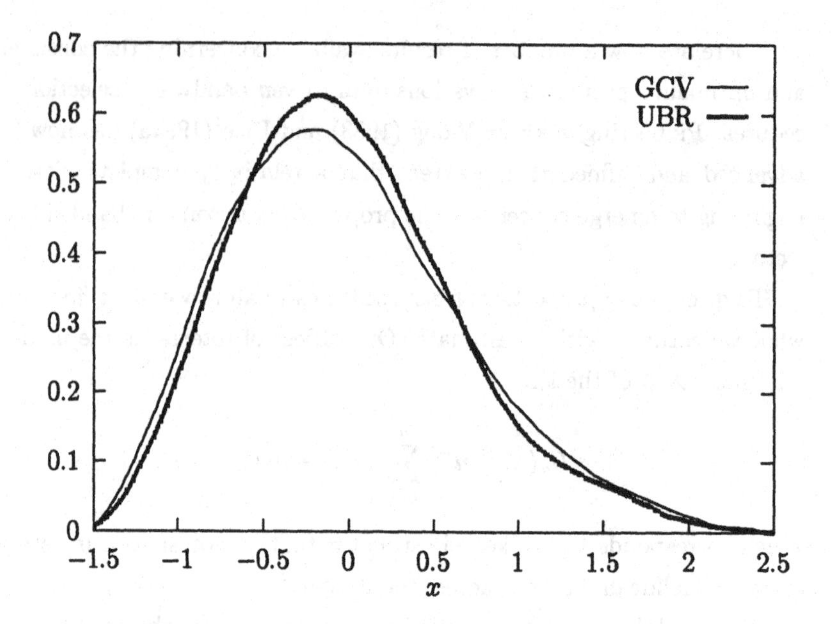

Figure 4.6: Empirical Bandwidth Densities.

$\widehat{\lambda}_G$ and $\widehat{\lambda}_P$ were found to be -.245 and -.253, respectively. One can argue that this negative correlation is what has caused the positive skewness that is seen in the densities.

The less than stellar performance by the bandwidth estimators in the simulation and result (4.49) raise many questions concerning whether there are problems with these selection methods and whether or not they can be improved. At least in terms of convergence rates, the answer to the latter question seems to be negative. Hall and Johnstone (1992) have demonstrated that in the absence of boundary effects and with a uniform design there are statistics $\widehat{A}_1, \widehat{A}_2$ (i.e., the values of $\widehat{A}_1, \widehat{A}_2$ are known once the data has been observed) such that

$$\widehat{\lambda}_{opt} = \widehat{A}_1 + n^{1/5}\widehat{A}_2 J_1 + o_p(n^{-3/10}) \qquad (4.50)$$

with $J_1 = \int_0^1 \mu'(t)^2 dt$ and $\widehat{A}_2 \xrightarrow{P} A_2 < 0$. Consequently, for any estimator $\widehat{\lambda}$ of $\widehat{\lambda}_{opt}$ we can always write

$$\widehat{\lambda} - \widehat{\lambda}_{opt} = n^{1/5}\widehat{A}_2(\widehat{J}_1 - J_1) + o_p(n^{-3/10}) \qquad (4.51)$$

with $\widehat{J}_1 = (\widehat{\lambda} - \widehat{A}_1)/n^{1/5}\widehat{A}_2$. Since $\widehat{J}_1 - J_1$ can converge to zero no faster than $n^{-1/2}$ and $\widehat{\lambda}_{opt}$ converges to zero at the exact rate $n^{-1/5}$ (Härdle, Hall and Marron 1988), it follows that the best rate of convergence for any estimator of $\widehat{\lambda}_{opt}$ is only the $n^{-1/10}$ that is obtained by cross-validation and unbiased risk type estimators.

Optimal convergence rates not withstanding, there still appear to be ways to improve on GCV and related methods for selecting λ. It turns out that the negative correlation between the bandwidth estimators and $\widehat{\lambda}_{opt}$ that was seen in the simulation is a real feature of these bandwidth selectors (Chiu and Marron 1990, Hall and Johnstone 1992 and Zhang 1995) and cross-validation and related methods have larger than optimal variance among the class of estimators satisfying (4.49). These problems and various attempts to solve them are discussed in Scott and Terrell (1987), Chiu (1990, 1991), Park and Marron (1990) and Hart and Yi (1998), for example. In particular, Hall and Johnstone (1992) produce a bandwidth estimator of the form $\widehat{\lambda} = \widehat{A}_1 + n^{1/5}\widehat{A}_2\widehat{J}_1$ which has asymptotic minimum variance and is positively correlated with $\widehat{\lambda}_{opt}$. While this result is promising its practical implications remain unclear since it relies on still developing methodology for producing $\sqrt{n}$-consistent estimators of functionals such as J_1.

Our previous conclusions change substantially if we alter our definition of optimality. Suppose now that we wish to estimate the minimizer λ_{opt} of the risk $R_n(\lambda) = \mathrm{E}\, L_n(\lambda)$. Then, Härdle, Hall and Marron (1992), for example, have shown that it is possible to produce an estimator $\widetilde{\lambda}$ such that

$$n^{1/2}\frac{\widetilde{\lambda} - \lambda_{opt}}{\lambda_{opt}} \xrightarrow{d} v_2 Z \tag{4.52}$$

for Z a standard normal random variable and v_2^2 a variance term similar to that in (4.49). The implication of (4.52) is that we can actually estimate λ_{opt} with a parametric (relative) rate of consistency.

An intuitive understanding of why (4.52) should be true can be obtained by re-examining the asymptotic form for λ_{opt} in (4.41). The unknowns in this quantity are σ^2 and $J_2(\mu) = \int_0^1 \mu''(t)^2 w(t)dt$. We already know that σ^2 can be estimated at a parametric rate so the question of how well we can estimate λ_{opt} revolves around rates of convergence for estimation of $J_2(\mu)$.

Estimators of $J_2(\mu)$ that are $\sqrt{n}$-consistent under various conditions are given in Hall and Johnstone (1992) and Goldstein and Messer (1992), for example. Thus, λ_{opt} can be estimated with relative $\sqrt{n}$-consistency at least in certain cases by plugging in $\sqrt{n}$-consistent estimators of $J_2(\mu)$ and σ^2 into formula (4.41). A practical way of estimating λ_{opt} that achieves the desired parametric rate of convergence has been developed by Gasser, Kneip and Köhler (1991) using the kernel estimators of μ'' that are discussed in Section 4.6.

A final question concerns how well $\widehat{\lambda}_G$ and $\widehat{\lambda}_P$ fare as estimators of λ_{opt}. The answer is essentially the same as before in that $n^{1/10}(\widehat{\lambda} - \lambda_{opt})/\lambda_{opt}$ has a limiting normal distribution for $\widehat{\lambda}$ being either of $\widehat{\lambda}_G$ or $\widehat{\lambda}_P$. Thus, neither the GCV nor unbiased risk criteria provide satisfactory estimators of λ_{opt}. However, it is also true (Härdle, Hall and Marron 1988) that $n^{1/10}(\widehat{\lambda}_{opt} - \lambda_{opt})/\lambda_{opt}$ has a limiting normal distribution. Thus, $\widehat{\lambda}_{opt}$ is also a poor "estimator" of λ_{opt} and, in this sense, we can see that the problems of estimating $\widehat{\lambda}_{opt}$ and λ_{opt} have a distinctly different nature.

To this point we have focused on estimating the bandwidth that minimizes the global risk or loss. Such globally optimal bandwidths can perform poorly, however, in areas where the regression function exhibits rapid changes or where the design is sparse. To see this, observe from (4.40) that the ratio of the pointwise risk at the asymptotically optimal pointwise bandwidth (4.39) to the same risk with the asymptotically globally optimal bandwidth (4.42) satisfies (Exercise 11)

$$\frac{R_n(t; \lambda_{opt})}{R_n(t; \lambda_{opt}(t))} \sim \frac{4}{5}[A(\mu, t)^{1/5} + \frac{1}{4}A(\mu, t)^{-4/5}] \tag{4.53}$$

with $A(\mu, t) = J_2(\mu)/\mu''(t)^2 w(t)$ assuming that $\mu''(t) \neq 0$. This ratio becomes quite large when $|\mu''(t)|$ is large or $w(t)$ is near zero, for example.

To produce kernel estimators that are better able to adapt to local features of the data it seems natural to attempt to use estimators whose effective bandwidths change locally as a function of the point of estimation. There are various ways to accomplish this such as the techniques proposed by Jennen-Steinmetz and Gasser (1988) and Messer and Goldstein (1993). However, many of the locally adaptive methods that have been proposed in the literature employ the ordinary kernel estimator along with some estimator of the locally optimal bandwidth $\lambda_{opt}(t)$ that minimizes the pointwise risk $R_n(t;\lambda)$. From (4.39) we can see that estimation of $\lambda_{opt}(t)$ implicitly requires estimation of $\mu''(t)$ rather than $J_2(\mu)$ as was the case for global bandwidth estimation. Unlike $J_2(\mu)$ it is not possible to estimate $\mu''(t)$ with $\sqrt{n}$-consistency. As a result, nonparametric rates of convergence of the form $(\widehat{\lambda}(t) - \lambda_{opt}(t))/\lambda_{opt}(t) = O_p(n^{-a})$ for $a < 1/2$ are the best that can be expected from any local bandwidth estimator $\widehat{\lambda}(t)$.

The local bandwidth estimation problem has been studied by Müller and Stadtmuller (1987b), Müller (1988), Staniswalis (1989b), Hall and Schucany (1989) and Brockmann, Gasser and Herrmann (1993). Both Staniswalis and Brockmann, et al give practical algorithms for choosing a local bandwidth. The Staniswalis method is based on minimization of an estimator of the local risk while the Brockmann, et al approach uses a modified version of the plug-in global bandwidth estimator in Gasser, Kneip and Köhler (1991).

4.6 Higher Order and Derivative Estimators

In this section we broaden the definition of kernel estimators to take advantage of possible additional smoothness that the regression function might possess. This leads to estimators with better asymptotic rates of convergence when μ has more than two derivatives. Similar developments also produce one possible approach to extending kernel methods to the estimation of derivatives of the regression curve.

Suppose it is known that $\mu \in C^m[0,1]$ for some $m > 2$. Then, referring back to the arguments which led to Corollary 4.1, one might expect this knowledge could be exploited to obtain a similar expression to (4.22) except with a better order for the bias. The idea is that the expectation of $\mu_\lambda(t)$ can now be written as

$$\mathrm{E}\,\mu_\lambda(t) = \sum_{j=0}^{m} \frac{(-\lambda)^j}{j!}\mu^{(j)}(t) \int_{-1}^{1} u^j K(u)du + o(\lambda^m) + O(n^{-1}). \qquad (4.54)$$

So, the leading term in the bias will be of order λ^m if the kernel satisfies

(4.6) and

$$\int_{-1}^{1} u^j K(u)du = \begin{cases} 1, & j=0, \\ 0, & j=1,\ldots,m-1, \\ M_m \neq 0, & j=m. \end{cases}$$

A kernel of this type is called an *mth order kernel* and is referred to as a higher order kernel if $m > 2$.

The use of higher order kernels provides one generalization of the estimators in Section 4.2. It is now straightforward to extend the results in Section 4.3 and obtain rates of convergence for the risk of these estimators. For an mth order version of the Gasser-Müller estimator (4.13) one finds that if $n \to \infty$, $\lambda \to 0$ in such a way that $n\lambda \to \infty$ then (Müller 1984a)

$$\mathrm{Var}\,(\mu_\lambda(t)) = \frac{\sigma^2}{n\lambda}\frac{V}{w(t)}(1+o(1)),$$

with V defined as in (4.6) and

$$\mathrm{E}\,\mu_\lambda(t) - \mu(t) = (-1)^m \frac{\lambda^m}{m!}\mu^{(m)}(t)M_m + o(\lambda^m) + O(n^{-1}).$$

Thus, an asymptotic expression for the risk at t is

$$R_n(t;\lambda) \sim \frac{\sigma^2 V}{n\lambda w(t)} + \frac{\lambda^{2m}}{(m!)^2}\mu^{(m)}(t)^2 M_m^2. \tag{4.55}$$

Differentiating (4.55) with respect to λ one finds the asymptotically optimal bandwidth at t to be

$$\lambda_{opt}(t) = \left[\frac{\sigma^2}{n}\frac{(m!)^2}{2m}\frac{V/M_m^2}{w(t)\mu^{(m)}(t)^2}\right]^{1/(2m+1)} \tag{4.56}$$

and

$$R_n(t;\lambda_{opt}(t)) \sim C_m(\sigma^2) n^{-2m/(2m+1)} \left[\frac{V^{2m} M_m^2 \mu^{(m)}(t)^2}{w(t)^{2m}}\right]^{1/(2m+1)}, \tag{4.57}$$

where $C_m(\sigma^2)$ is a constant involving only m and σ^2. Conclusions that follow from this include that i) increased smoothness of μ translates into faster rates of convergence and ii) bandwidths should be made wider in regions where the design is sparse or $|\mu^{(m)}(t)|$ is small.

If (mth order) boundary kernels are used, one can average the approximations (4.55) over the design to assess the global performance of μ_λ. The result is an asymptotic expression for the global risk. Specifically, if $n \to \infty, \lambda \to 0$ and $n\lambda \to \infty$,

$$R_n(\lambda) = \sum_{i=1}^{n} \mathrm{E}\,(\mu_\lambda(t_i) - \mu(t_i))^2 \sim \frac{\sigma^2}{n\lambda}V + \frac{\lambda^{2m}}{(m!)^2}M_m^2 J_m(\mu) \tag{4.58}$$

with

$$J_m(\mu) = \int_0^1 \mu^{(m)}(t)^2 w(t)dt. \tag{4.59}$$

From (4.58) the asymptotically optimal global bandwidth and risk are found to be

$$\lambda_{opt} = \left[\frac{(m!)^2\sigma^2 V/M_m^2}{2mnJ_m(\mu)}\right]^{1/(2m+1)} \tag{4.60}$$

and

$$R_n\left(\lambda_{opt}\right) = C_m(\sigma^2)n^{-2m/(2m+1)}\left[V^{2m}M_m^2 J_2(\mu)\right]^{1/(2m+1)}. \tag{4.61}$$

This gives the conclusion that if $\mu \in C^m[0,1]$ we can use higher order kernels and thereby obtain the optimal $n^{-2m/(2m+1)}$ rate of decay for the risk in this case. Expression (4.61) can be futher optimized through optimal selection of the kernel function and even the design density. Discussions of these issues can be found in Gasser and Müller (1984), Gasser, Müller and Mammitzsch (1985) and Müller (1984b, 1988).

Frequently, it is necessary to estimate some functional of μ. One functional of particular interest is the derivative or, more generally the dth derivative of μ at some point t. One approach to estimating such derivatives using kernel methods can be deduced from further examination of the expansion (4.54) that led us to higher order kernels. Assuming that $d \le m$ we can see from (4.54) that the expected value of $\mu_\lambda(t)$ will be approximately $\lambda^d\mu^{(d)}(t)$ provided that we choose our kernel to satisfy

$$\int_{-1}^{1} u^j K(u)du = \begin{cases} 0, & j = 0,\ldots,m-1, j \ne d, \\ (-1)^d d!, & j = d, \\ M_m \ne 0, & j = m. \end{cases}$$

This suggests that we estimate $\mu^{(d)}(t)$ using

$$\mu_{\lambda d}(t) = \lambda^{-(d+1)} \sum_{i=1}^{n} y_i \int_{s_{i-1}}^{s_i} K(\lambda^{-1}(t-s))ds. \tag{4.62}$$

Note that derivative estimators also suffer from boundary effects so that estimator (4.62) must be altered in the boundary region using boundary kernels such as those in Müller (1991, 1993) and Müller and Wang (1994).

It can be shown that under suitable restrictions on the bandwidth, the variance of $\mu_{\lambda d}(t)$ is of exact order $(n\lambda^{2d+1})^{-1}$ and its bias is of order λ^{m-d} which entails that the optimal rate of decay for $R_{nd}(t;\lambda) = \mathrm{E}\,(\mu_{\lambda d}(t) - \mu^{(d)}(t))^2$ is $n^{-2(m-d)/(2m+1)}$ (See Exercise 14). This rate becomes slower for higher derivatives which indicates, as might be expected, that derivatives are harder to estimate than regression function values. By using a boundary corrected estimator one can develop asymptotic expressions for the global risk and optimal global bandwidth for derivative estimators that are similar to (4.41)-(4.42) (Exercise 15).

Since the optimal bandwidth for estimating μ and $\mu^{(d)}$ are not the same, standard cross-validation methods for estimating μ will not be satisfactory for selecting bandwidths to use in estimating derivatives. The development of parallels of cross-validation techniques for estimation of $\mu^{(d)}$ can be quite difficult. Part of the problem stems from the difficulty in defining residuals that can be used to assess the performance of the fitted derivative values $\mu_{\lambda d}(t_i), i = 1, \ldots, n$, that correspond to any particular bandwidth choice. One approach is to compare the fitted derivative estimator to dth order differences of the responses (Rice 1986a, Müller, Stadtmüller and Schmitt 1987, Müller 1988). However, differencing increases variability so this technique is not altogether satisfactory. Another way to choose the bandwidth is through use of the factor method of Müller, Stadtmüller and Schmitt (1987) and Müller (1988, Section 7.4) which uses a transformed version of a data-selected bandwidth for estimating μ to produce a bandwidth for the derivative estimator. A particular example of this method is explored in Exercise 16.

Another approach to estimating $\mu^{(d)}$ would be to just differentiate the original kernel estimator of the regression function d times. It turns out that this is actually equivalent in most cases to the use of an estimator such as (4.62) as shown in Gasser and Müller (1984).

4.7 Locally Linear Estimators

There are a number of estimators, including smoothing splines in Chapter 5, which have close ties to kernel estimators. A particularly important example of this is the locally linear or, more generally, local polynomial smoother that we will discuss in this section.

To motivate the form of these estimators, recall that the Nadaraya-Watson estimator (4.9) of $\mu(t)$ is $\mu_\lambda(t) = \sum_{i=1}^n y_i K(\frac{t-t_i}{\lambda}) / \sum_{i=1}^n K(\frac{t-t_i}{\lambda})$. A quick calculation reveals that this estimator is the minimizer of the local least-squares criterion

$$\sum_{i=1}^n K\left(\frac{t-t_i}{\lambda}\right)(y_i - \theta)^2 \tag{4.63}$$

with repect to θ. This gives another intuitive motivation for kernel estimators via a Taylor's Theorem argument. Specifically, the y_i means satisfy $\mu(t_i) = \mu(t) + O(|t_i - t|)$ so that if λ is sufficiently small, criterion (4.63) restricts us to estimation using only those observations whose means are "nearly" the same as $\mu(t)$.

Of course this development employs only the constant term in the Taylor expansion for $\mu(t_i)$ and it seems quite reasonable that better results could be produced using higher order expansions. If, for example, we use a second order approximation $\mu(t_i) = \mu(t) + \mu'(t)(t_i - t) + O(|t_i - t|^2)$ this leads to estimation of $\mu(t)$ through minimization of

$$\sum_{i=1}^n K\left(\frac{t-t_i}{\lambda}\right)(y_i - \theta_1 - \theta_2(t_i - t))^2 \tag{4.64}$$

with respect to θ_1 and θ_2 with the resulting value for θ_1 being our estimator of $\mu(t)$. To give an explicit form for the estimator, first define

$$M_{jn}(t) = (\lambda n)^{-1} \sum_{i=1}^n K\left(\frac{t-t_i}{\lambda}\right)(t - t_i)^j,\ j = 0, 1, 2. \tag{4.65}$$

Then one may verify that the resulting estimator of $\mu(t)$ is

$$\mu_\lambda(t) = \sum_{i=1}^{n} y_i w(t, t_i; \lambda) \tag{4.66}$$

with

$$w(t, t_i; \lambda) = \frac{1}{n\lambda} K\left(\frac{t - t_i}{\lambda}\right) \frac{M_{2n}(t) - (\frac{t-t_i}{\lambda}) M_{1n}(t)}{M_{2n}(t) M_{0n}(t) - M_{1n}^2(t)}. \tag{4.67}$$

It is now easy to see the connection between the locally linear estimator obtained from (4.64) and the boundary corrected kernel estimators from Sections 4.3–4.4. Since the $M_{jn}(t)$ in (4.65) are just quadrature approximations to the $M_j(t)$ in (4.21), it follows that the local linear estimator (4.66) is (apart from quadrature error) a boundary corrected kernel estimator that uses the boundary kernel in (4.45). We would therefore expect the estimator (4.66) to have all the same sort of qualities as a second order boundary corrected kenel estimator with, e.g., its asymptotically optimal global bandwidth and risk being given by (4.41)-(4.42). Rigorous verifications of this can be found in Müller (1988, Section 4.6), Wand and Jones (1995, Chapter 5), and Fan and Gijbels (1996, Section 3.2).

Although it may at first seem that there is little to recommend the use of local linear smoothers over the boundary corrected kernel estimators, this is not entirely the case. An important feature of the local linear estimator is that it produces a boundary correction automatically thereby eliminating the problem of explicitly constructing and then using boundary kernels. Perhaps more importantly, the asymptotic equivalence between local linear and second order kernel smoothers no longer holds when the t_i are random rather than deterministic. In that case the local linear smoother has some important advantages over either the Nadaraya-Watson or Gasser-Müller estimators that will be discussed in Section 4.10.

An example of a local linear fit is shown in Figure 4.7. This data corresponds to simulated motorcycle crashes with the response being the head acceleration of a postmortem human test object at t milliseconds after the "crash." (See, e.g., Härdle 1990.) The plot of the data suggests heteroskedastic rather than constant error variances. However, we will ignore this and proceed as if the assumptions for model (4.1) are met.

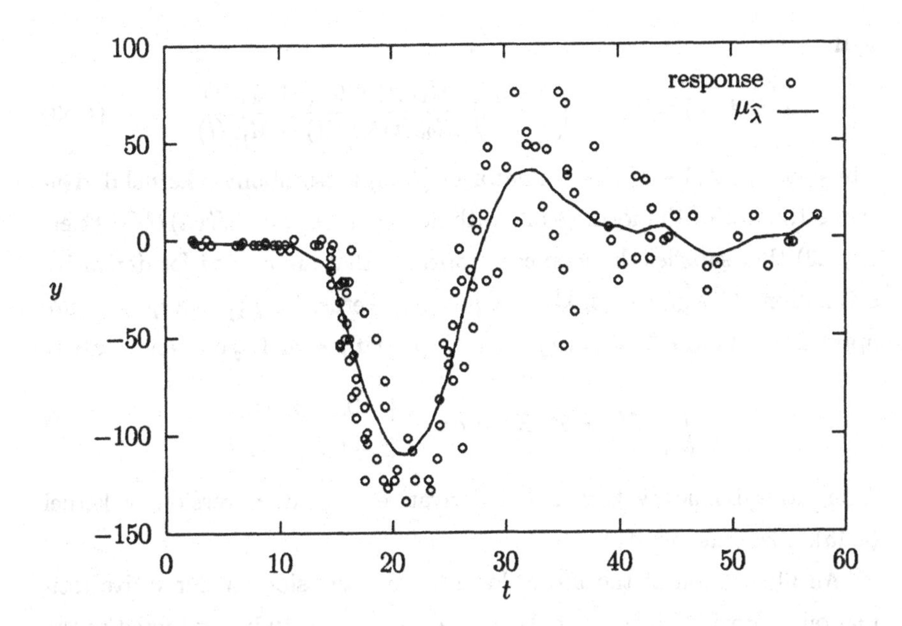

Figure 4.7: Local Linear Regression Fit to the Motorcycle Data.

The fit to the motorcycle data shown in Figure 4.7 is a locally linear estimator that uses the quadratic kernel from Table 4.1 with a bandwidth selected by GCV. Note that the GCV and unbiased risk criteria for a locally linear smoother have the same essential form (4.47)-(4.48) as for a kernel estimator except that $\mathrm{tr}\mathbf{S}_\lambda$ must now represent the trace of the hat or smoother matrix for locally linear regression. This trace is given explicitly by $\sum_{i=1}^n w(t_i, t_i; \lambda)$ with $w(\cdot, \cdot; \lambda)$ the weights in (4.67). The fit in Figure 4.7 appears to do an adequate job of capturing the key features of the data.

To this point we have dealt only with the intercept or θ_1 term obtained from minimizing (4.64). One might expect the slope coefficient to be useful as well and that it could provide an estimator of $\mu'(t)$. To see that this is in fact the case, observe that the value of θ_2 which minimizes (4.64) is

$$\mu^*_{\lambda 1}(t) = \sum_{i=1}^n y_i w^*(t, t_i; \lambda) \tag{4.68}$$

with

$$w^*(t,t_i;\lambda)=\frac{1}{n\lambda^2}K\left(\frac{t-t_i}{\lambda}\right)\frac{M_{1n}(t)-(\frac{t-t_i}{\lambda})M_{0n}(t)}{M_{2n}(t)M_{0n}(t)-M_{1n}^2(t)}. \tag{4.69}$$

Thus, for $t\in[\lambda,1-\lambda]$ the estimator in (4.68) is essentially a kernel derivative estimator of the form (4.62) with kernel $K^*(u)=-uK(u)/M_2$ (Exercise 23) that satisfies the moment properties that are needed for derivative estimation. If, e.g., $t\in[0,\lambda)$ the equivalent kernel for $\mu^*_{\lambda 1}$ is a quadrature approximation to $K_q^*(u)=(a_q+b_qu)K^*(u)$ with a_q and b_q chosen to satisfy

$$\int_{-1}^{q}u^j(a+bu)K^*(u)du=\begin{cases}0, & j=0,\\ -1, & j=1.\end{cases}$$

This latter boundary kernel is a derivative estimation version of kernel (4.45). (See Exercise 17).

An illustration of the use of local linear regression for derivative estimation is provided by growth data on a boy from a study conducted at the University of Zürich. The actual data is shown in Figure 4.8 and represents the growth in stature for a particular boy from the study. In growth studies it is often of more interest to estimate the velocity curve or derivative of the growth curve, rather than the regression function itself, since it provides more direct information about surges or spurts in the rate of growth.

To estimate the velocity curve (4.68) was used with the quadratic kernel. A bandwidth of $\lambda=3$ was chosen to provide a similar estimate to those obtained by Gasser, et. al. (1984) who have undertaken an extensive study of the Zürich growth data. A plot of the resulting estimated velocity curve is shown in Figure 4.9. The estimated curve exhibits two peaks: one at around 13 years of age and another at about 7 years. The peak at 13 years corresponds to the growth spurt that accompanies the advent of puberty. The second, less pronounced, peak at 7 years is due to a so-called mid-growth spurt. Gasser, et. al. (1984) found that popular parametric growth curve models were incapable of fitting this mid-growth spurt which led to biases in estimation of other features of the velocity curve. Note that in this example we have implicitly used the locations of peaks in our derivative estimator to estimate the locations of peaks in the derivative of the true regression function. A theoretical justification for estimating the

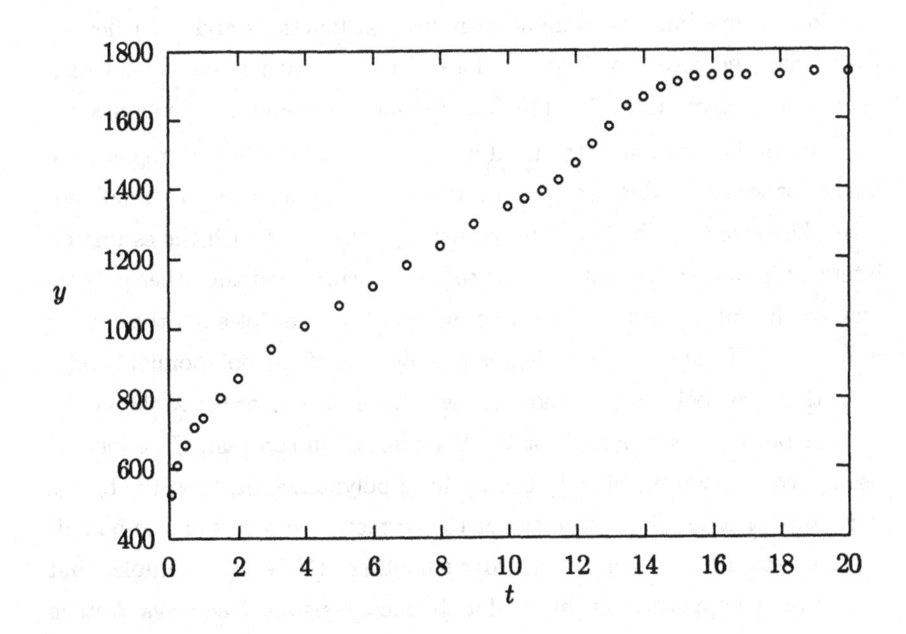

Figure 4.8: Growth in Stature of a Boy.

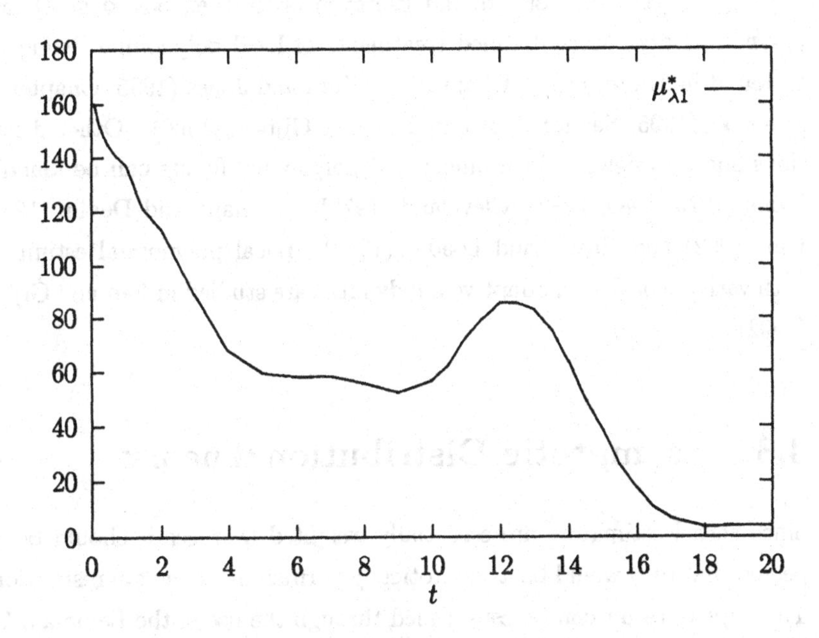

Figure 4.9: An Estimated Growth Velocity Curve.

locations of maxima and minima of regression functions and their derivatives using the corresponding locations of maxima and minima for kernel estimators is given in Müller (1988, 1989) and Härdle (1990, Section 8.2).

A natural extension of the local linear estimator (4.65) is to replace the linear (or second order) polynomial in (4.64) with an mth order polynomial. The result is a higher order kernel type smoother with the estimated intercept giving an estimator of the regression curve and the other polynomial coefficients providing estimators of (constant multiples of) the curve's derivatives. If the order (i.e., degree plus one) m of the polynomial is odd, then the local polynomial smoother will have bias of order λ^{m+1} in the interior but only of order λ^m at the boundaries. In contrast, if we instead use the corresponding $(m+1)$st order local polynomial fit, then the bias is uniformly of order λ^{m+1} and we get the correct type of automatic boundary correction for an $(m+1)$st order smoother. (Note, for example, that this is what happens when $m = 1$ for the local constant Nadaraya-Watson estimator and the local linear regression estimator.) This, as well as other factors, suggests that one should use even order (i.e., odd degree) local polynomial fits. More detailed treatments of local polynomial fitting can be found in Müller (1988, Chapter 4), Wand and Jones (1995, Chapter 5), Simonoff (1995, Section 5.2) and Fan and Gijbels (1996). Other discussions and overviews of local linear and polynomial fitting can be found in Stone (1975, 1980, 1982), Cleveland (1979), Cleveland and Devlin (1988), Fan (1992) and Hastie and Loader (1993). Local polynomial estimators with variable or design adaptive bandwidths are studied in Fan and Gijbels (1992).

4.8 Asymptotic Distribution Theory

Since kernel estimators are essentially weighted averages it should be expected that they would be asymptotically normal under certain restrictions. This type of result can be established through the use of the Lemma 3.1 of Section 3.4.4. For simplicity we will restrict attention to the Gasser-Müller kernel estimator with a second order kernel. The same type of results can be shown to hold for the other estimators in Section 4.2 and kernels of higher order.

For the application of Lemma 3.1 define

$$w_{in}(t;\lambda) = \lambda^{-1}\int_{s_{i-1}}^{s_i} K\left(\lambda^{-1}(t-s)\right)ds, i = 1,\ldots,n. \tag{4.70}$$

Then, for any fixed $t \in (0,1)$ we have from Lemma 4.1 with $n \to \infty, \lambda \to 0, n\lambda \to \infty$ that

$$\sum_{i=1}^{n} w_{in}^2(t;\lambda) \sim \frac{V}{w(t)n\lambda} \tag{4.71}$$

and

$$\max_{1\le i\le n} |w_{in}(t;\lambda)| \le \lambda^{-1} \max_{-1\le u\le 1} |K(u)| \max_{1\le i\le n}(s_i - s_{i-1}) = O((n\lambda)^{-1}).$$

The implications of these two facts are summarized in Theorem 4.3.

Theorem 4.3. *Assume that the ε_i are independent and identically distributed with* $\mathrm{E}\,\varepsilon_1 = 0$ *and* $\mathrm{Var}(\varepsilon_1) = \sigma^2 < \infty$ *and that $n \to \infty$, $\lambda \to 0$ in such a way that $n\lambda \to \infty$. Then, for any fixed $t \in (0,1)$*

$$\frac{\mu_\lambda(t) - \mathrm{E}\,\mu_\lambda(t)}{\sqrt{\mathrm{Var}(\mu_\lambda(t))}} \xrightarrow{d} Z,$$

where Z is a standard normal random variable.

Notice that no smoothness restrictions are required on μ for this to hold. This is because $\mu_\lambda(t) - \mathrm{E}\,\mu_\lambda(t)$ is independent of μ and what has actually been proved is just a theorem about a particular type of weighted average of independent, identically distributed, random variables. Smoothness of μ is needed when we try to obtain a somewhat more useful result concerning asymptotic normality of $\mu_\lambda(t) - \mu(t)$.

To establish asymptotic normality for $\mu_\lambda(t) - \mu(t)$ we can proceed as in Section 3.4.4 and write

$$\frac{\mu_\lambda(t) - \mu(t)}{\sqrt{\mathrm{Var}(\mu_\lambda(t))}} = \frac{\mu_\lambda(t) - \mathrm{E}\,\mu_\lambda(t)}{\sqrt{\mathrm{Var}(\mu_\lambda(t))}} + \frac{\mathrm{E}\,\mu_\lambda(t) - \mu(t)}{\sqrt{\mathrm{Var}(\mu_\lambda(t))}}.$$

The first term on the right hand side of this expression is asymptotically standard normal as a result of Theorem 4.3. Thus, due to (4.71), all that is needed for $(\mu_\lambda(t) - \mu(t))/\sqrt{\mathrm{Var}(\mu_\lambda(t))}$ to be asymptotically normal are conditions under which

$$\sqrt{n\lambda}(\mathrm{E}\,\mu_\lambda(t) - \mu(t)) \to 0.$$

From Corollary 4.1 this condition is satisfied if $n\lambda^5 \to 0$ which proves the following result.

Corollary 4.2. *Suppose that the conditions of Theorem 4.3 are met and that $\mu \in C^2[0,1]$. Then,*

$$\frac{\mu_\lambda(t) - \mu(t)}{\sqrt{\mathrm{Var}(\mu_\lambda(t))}} \xrightarrow{d} Z,$$

where Z has a standard normal distribution, provided $n \to \infty$, $\lambda \to 0$ in such a way that $n\lambda \to \infty$ and $n\lambda^5 \to 0$.

Under the conditions of Corollary 4.2 an approximate $100(1-\alpha)\%$ confidence interval for $\mu(t)$ is provided by

$$\mu_\lambda(t) \pm z_{\alpha/2}\widehat{\sigma}\sqrt{\sum_{i=1}^{n} w_{in}^2(t;\lambda)} \tag{4.72}$$

with z_α the $100(1-\alpha)$th percentage point of the standard normal distribution, $w_{in}(t;\lambda)$ defined in (4.70) and $\widehat{\sigma}$ any $\sqrt{n}$-consistent estimator of σ. Unfortunately, the optimal value of λ in (4.41) does not satisfy the conditions of Corollary 4.2 since $\lambda_{opt}^5 n$ converges to a nonzero constant. This is a consequence of the fact that the optimal λ attempts to balance variance and squared bias with the result that both have the same asymptotic rate of decay.

Since Corollary 4.2 does not hold for λ_{opt} this suggests that (4.72) will not perform well with λ selected by GCV or some other criterion which attempts to estimate the minimizer of $R_n(\lambda)$. Indeed, it follows from the arguments leading to Corollary 4.2 that if we use the globally optimal bandwidth (4.41) in our estimator, $(\mu_\lambda(t) - \mu(t))/\sqrt{\mathrm{Var}(\mu_\lambda(t))}$ will have a limiting normal distribution with variance one and mean $\mu''(t)\sqrt{w(t)}/2\sqrt{J_2(\mu)}$

if $M_2 > 0$ (Exercise 26). As a result, intervals such as (4.72) will be off center asymptotically and give lower than nominal coverage when they are based on bandwidths that behave like the minimizer of the global risk. This means that we cannot expect the intervals (4.72) to work in practice with data selected bandwidths.

To illustrate the problem with using "normal theory" intervals such as (4.72) a small simulation was conducted using the example of data from model (4.1) with $n = 50$, normal errors with $\sigma = .15$ and $\mu(t) = t + .5\exp\{-50(t-.5)^2\}$. A typical data set of this type was shown in Figure 4.4. Each data set in the simulation was fit using a local linear regression estimator with bandwidths selected through the unbiased risk criterion using the GSJS variance estimator $\widehat{\sigma}^2$ from (2.29)–(2.30). Pointwise 95% confidence intervals of the form

$$\mu_{\widehat{\lambda}}(t_i) \pm 1.96\widehat{\sigma}\sqrt{\sum_{i=1}^{n} w^2(t_i, t_i; \widehat{\lambda})}\,,\ i = 1, \ldots, n, \tag{4.73}$$

were then computed across the design with $w(\cdot,\cdot;\lambda)$ the local linear smoother weights in (4.67). This basic experiment was replicated 1000 times and the proportion of times the ith interval in (4.73) contained the value of $\mu(t_i)$ was recorded for each design point with the results being shown in Figure 4.10. The empirical coverages in the figure are seen to depart substantially from the nominal 95% level, particularly in areas where μ is very nonlinear such as around $t = .5$.

To construct confidence intervals that are asymptotically valid and usable in practice with data selected bandwidths, it appears one must take the bias of the estimator into account when constructing the intervals. There are a number of ways this can be accomplished. One direct approach is to actually subtract off an estimator of the bias to produce bias corrected confidence intervals. To be somewhat more specific, note from (4.54) that if μ has, for example, three derivatives then

$$\mathrm{E}\,\mu_\lambda(t) - \mu(t) = \frac{\lambda^2}{2}\mu''(t)M_2 + O(\lambda^3 + n^{-1}). \tag{4.74}$$

Given a choice for λ, the only unknown in the lead term in (4.74) is $\mu''(t)$ which can be estimated using the techniques discussed in Section 4.6. If

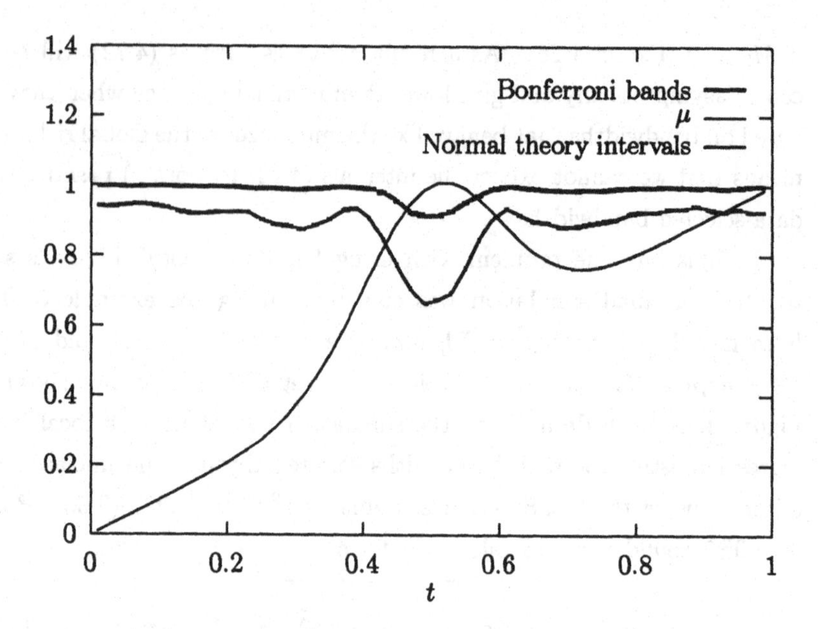

Figure 4.10: Empirical Coverages of Normal and Bonferroni Intervals.

$\mu_{\widetilde{\lambda}2}(t)$ is the kernel estimator (4.62) of $\mu''(t)$ based on an estimator $\widetilde{\lambda}$ of the optimal bandwidth for second derivative estimation, one could then use an interval such as

$$\mu_{\widehat{\lambda}}(t) - \frac{\widehat{\lambda}^2}{2} M_2 \mu_{\widetilde{\lambda}2}(t) \pm z_{\alpha/2} \widehat{\sigma} \sqrt{\sum_{i=1}^{n} w_{in}^2(t;\widehat{\lambda})} \tag{4.75}$$

with $\widehat{\lambda}$, for example, a GCV estimator of λ. Arguments in Eubank and Speckman (1993) can be used to show that this approach actually works in certain special cases with the intervals (4.75) having the desired asymptotic coverage levels $1-\alpha$.

Further inspection of (4.75) reveals that when using these intervals we are, in a sense, no longer using $\mu_{\widehat{\lambda}}(t)$ to estimate $\mu(t)$ but rather the bias corrected estimator

$$\mu^*(t) = \mu_{\widehat{\lambda}}(t) - \frac{\widehat{\lambda}^2}{2} M_2 \mu_{\widetilde{\lambda}2}(t). \tag{4.76}$$

This suggests that we replace the $w_{in}(t;\widehat{\lambda})$ in (4.75) with the weights that are actually being applied to the responses; namely

$$\widehat{\lambda}^{-1}\int_{s_{i-1}}^{s_i} K\left(\widehat{\lambda}^{-1}(t-s)\right)ds - \frac{\widehat{\lambda}^2}{2}M_2\widetilde{\lambda}^{-3}\int_{s_{i-1}}^{s_i} K^*\left(\widetilde{\lambda}^{-1}(t-s)\right)ds \quad (4.77)$$

with K^* the kernel for the derivative estimator.

Notice in (4.77) that if we were to choose $\widetilde{\lambda}$ to be the same as $\widehat{\lambda}$ then the bias corrected estimator (4.76) would be just a kernel estimator whose effective kernel is higher order in the sense that its second moment is zero (Exercise 27). Thus, another method of bias correcting the confidence interval (4.72) is to first fit the data with a second order kernel estimator and obtain the bandwidth for this estimator using GCV or some other related method. Then, use this bandwidth in a kernel estimator that employs a higher (than second) order kernel and use this new estimator to center the confidence intervals. This method also works in the sense of producing the correct asymptotic nominal coverages for the intervals in certain cases.

Both of the above proposals have the somewhat troubling feature that the corresponding confidence intervals are not being centered around the estimator $\mu_{\widehat{\lambda}}$ that is likely to give the most desirable fit to the data. In addition, the intervals only work under additional smoothness conditions on the regression curve which could be exploited (through the use of higher order kernels, for example) to produce a better point estimator of the regression function. Unfortunately, these better estimators have the same types of problems in terms of their associated confidence intervals in that an estimator using an optimal bandwidth will have bias and standard error of the same asymptotic order. Thus, if we use bias correction type methods it appears that the construction of effective interval estimators will require a compromise in the use of a less than "ideal" point estimators to center the intervals. An alternative approach that can avoid this compromise and produce conservative intervals in some cases is considered in Clark (1980) and Müller, Stadtmüller and Schmitt (1987). Bootstrap and related methods for interval estimation in nonparametric regression are discussed in Härdle (1990, Chapter 4), Hall (1992) and Wang(1993), for example. A data driven way to construct asymptotically valid confidence intervals using undersmoothing has been developed by Neumann (1995).

If we think of μ as a point in some function space W or as the "parameter" that we wish to estimate, then we should be attempting to obtain interval estimators of the entire regression function rather than at some fixed point t. The problem then becomes that of constructing a confidence band such that

$$\lim_{n\to\infty} P(\max_{t\in[0,1]} |\widehat{\mu}(t) - \mu(t)| \leq l_{\alpha n}) \geq 1 - \alpha \tag{4.78}$$

with $\widehat{\mu}$ some nonparametric estimator of μ and $l_{\alpha n}$ a suitable upper bound determined from the data. Problems of this nature have been considered by Révész (1979), Bjerve, Doksum and Yandell (1985), Johnston (1982), Stadtmüller (1986), Härdle (1989), Knafl, Sacks, and Ylvisaker (1982, 1985), Hall and Titterington (1988), Härdle and Bowman (1988), Härdle and Marron (1991) and Sun and Loader (1994) among others. Eubank and Speckman (1993) study the limiting distribution of a stochastic process related to the bias corrected estimator in (4.76) under a periodic regression model and a uniform design. This allows them to obtain an exact asymptotic expression for $l_{\alpha n}$ under which (4.78) holds with equality. Perhaps the most practical consequence that can be deduced from their results is that a confidence band satisfying (4.78) can be obtained without bias correction provided that one uses the Bonferroni method to construct the confidence intervals in (4.72). More specifically, let $\widehat{\lambda}$ be any data driven bandwidth such as the GCV or unbiased risk estimator that satisfies (4.49). Then, assuming a uniform design and under certain other restrictions, it can be shown that the corresponding kernel estimator $\mu_{\widehat{\lambda}}$ will have

$$\lim_{n\to\infty} P(\max_{t\in[0,1]} |\mu_{\widehat{\lambda}}(t) - \mu(t)| \leq l_{\alpha n}) \geq 1 - \alpha \tag{4.79}$$

if we take

$$l_{\alpha n} = z_{\alpha/2n}\widehat{\sigma}\sqrt{\frac{V}{n\widehat{\lambda}}}. \tag{4.80}$$

This result has the interesting consequence that a Bonferroni type confidence band derived from the normal theory intervals (4.72) will work even when the individual pointwise intervals fail. The reason this is true is because the factor $z_{\alpha/2n}$ in (4.80) grows with n in a way that expands the interval widths sufficiently to compensate for the bias of the estimator. Results of the form (4.79)–(4.80) have been extended to nonuniform and even random designs by Xia (1999) in the case of local linear regression estimators.

An example of Bonferroni confidence bands is shown in Figure 4.11 for a typical data set from the simulation used to produce Figure 4.10. The 95% Bonferroni bands shown in the figure were obtained by replacing the value of $z_{\alpha/2} = 1.96$ by $z_{\alpha/2n} = 3.29$ in (4.73). The true regression function is shown in the plot and is captured by the bands for this data set.

The pointwise empirical coverage levels that were produced using 95% Bonferroni bands are also shown in Figure 4.10. While the coverage levels are generally quite conservative, they are more satisfactory than the normal theory intervals (4.73) for this case. Of course, the Bonferroni intervals are intended as confidence bands, so that the real measure of their performance is the proportion of our 1000 replicate samples where all the values of the regression function fell inside the bands. For this experiment the coverage level was only 82.2% meaning that there were 188 (rather than the expected 50) of the 1000 samples where a value of the regression curve fell outside of the Bonferroni band. The problem here stems from several factors including estimation of the variance and the relatively small sample size. More extensive simulations that explore this issue further can be found in Eubank and Speckman (1993) and Xia (1999).

One of the most important implications of the simulation is that if we view the Bonferroni intervals from a pointwise, rather than bandwise, perspective, then they appear to provide a simple way of producing bias corrected pointwise confidence intervals for kernel type estimators. They tend to be wider than necessary and thereby overly conservative in areas where μ is nearly linear, but can nonetheless provide an effective method of producing pointwise intervals in some cases.

Bonferroni confidence bands corresponding to the assay data of Figure 4.1 are shown in Figure 4.12 that were produced using a local linear smoother and GCV. These intervals are seen to contain the simple linear regression fit $\widehat{\mu}$ to this data that is also shown in the plot. One might

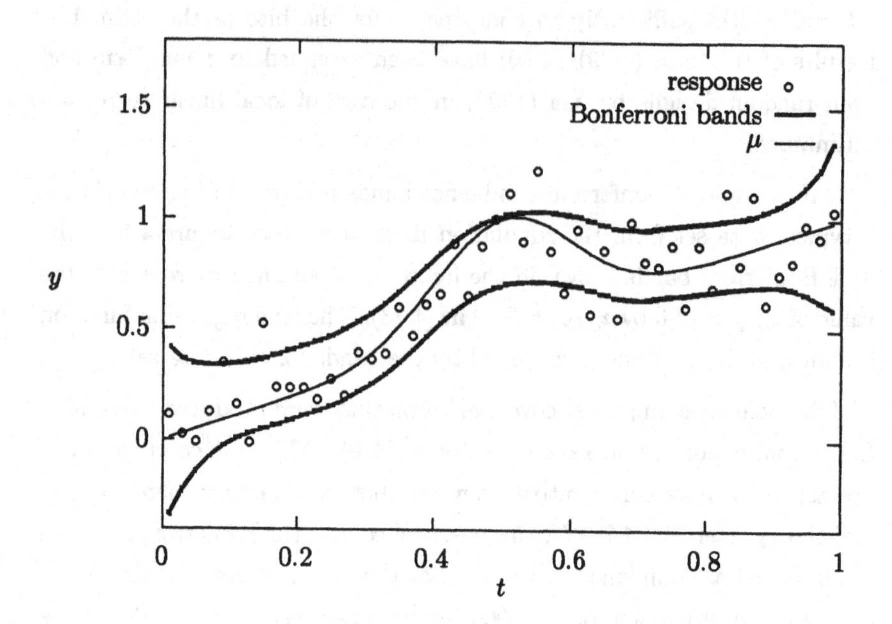

Figure 4.11: Bonferroni Bands for Simulated Data.

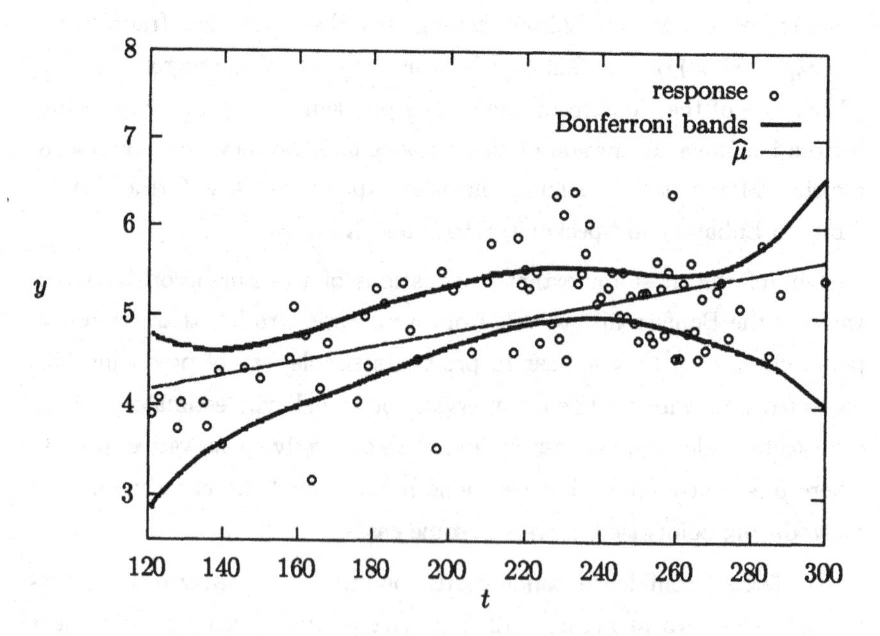

Figure 4.12: Bonferroni Bands for Assay Data.

conclude from this that a linear regression function represents one possible model for the true, unknown, regression function for this data.

4.9 Partially Linear Models

Kernel estimators can be modified in a number of ways to provide solutions to other estimation problems some of which will be discussed in Section 4.11. In this section we explore the use of kernel smoothers for estimation in the context of the partially linear model of Section 3.7.

We will again restrict attention to the simplest case where the responses follow the model

$$y_i = \gamma u_i + f(t_i) + \varepsilon_i,\ i = 1, \ldots, n, \tag{4.81}$$

for $u_i, i = 1, \ldots, n$, known constants, γ an unknown parameter and f an unknown function in $C^2[0,1]$. The $\varepsilon_i, i = 1, \ldots, n$, are assumed to be *iid* random variables with zero first moment and finite variance σ^2. The more general case of vector-valued u_i is treated in Speckman (1988). In vector-matrix form we write model (4.81) as

$$\mathbf{y} = \gamma\mathbf{u} + \mathbf{f} + \boldsymbol{\varepsilon}, \tag{4.82}$$

with $\mathbf{y}$ the response vector, $\mathbf{u} = (u_1, \ldots, u_n)^T, \mathbf{f} = (f(t_1), \ldots, f(t_n))^T$ and $\boldsymbol{\varepsilon} = (\varepsilon_1, \ldots, \varepsilon_n)^T$.

Arguing as in Section 3.7, we find that the Bias Reduction Lemma suggests that we estimate the parameters in model (4.81) by

$$\tilde{\gamma} = (\mathbf{u}^T(\mathbf{I} - \mathbf{S})^T(\mathbf{I} - \mathbf{S})\mathbf{u})^{-1}\mathbf{u}^T(\mathbf{I} - \mathbf{S})^T(\mathbf{I} - \mathbf{S})\mathbf{y} \tag{4.83}$$

and

$$\tilde{\mathbf{f}} = \mathbf{S}(\mathbf{y} - \tilde{\gamma}\mathbf{u}) \tag{4.84}$$

with $\mathbf{S}$ being the smoother or hat matrix corresponding to some suitably boundary corrected kernel type estimator. For example, we can take $\mathbf{S}$ to be the matrix that transforms the responses to the vector of fitted values for the Gasser-Müller estimator with boundary kernel correction or we could use the smoother matrix for the local linear regression estimator of Section 4.7. The estimator of the overall mean vector $\boldsymbol{\mu} = \gamma\mathbf{u} + \mathbf{f}$ obtained from (4.83)-(4.84) is then

$$\tilde{\boldsymbol{\mu}} = \tilde{\gamma}\mathbf{u} + \tilde{\mathbf{f}}. \tag{4.85}$$

Notice that unlike formulas (3.135)-(3.137) for series estimators with model (4.81), our coefficient estimator involves $(\mathbf{I}-\mathbf{S})^T(\mathbf{I}-\mathbf{S})$ rather than just $(\mathbf{I}-\mathbf{S})$. The two estimation prescriptions are essentially the same however because the smoother matrix for series estimators is a projection operator. Thus, (4.83)-(4.85) represent one possible way to extend (3.135)-(3.137) to the case of smoothers with non-symmetric hat matrices.

We know from our work in Sections 4.3, for example, that for a second order kernel type smoother with an optimally chosen bandwidth we will have

$$n^{-1}\mathbf{f}^T(\mathbf{I}-\mathbf{S})^T(\mathbf{I}-\mathbf{S})\mathbf{f} = O(n^{-4/5}) \tag{4.86}$$

and

$$n^{-1}\mathrm{tr}\mathrm{Var}(\mathbf{S}\mathbf{y}) = \sigma^2 n^{-1}\mathrm{tr}\mathbf{S}^T\mathbf{S} = O(n^{-4/5}). \tag{4.87}$$

Thus, from the Bias Reduction Lemma and the fact that the rank of $(\mathbf{I}-\mathbf{S})\mathbf{u}$ is at most one, we have

$$n^{-1}\mathrm{E}\,(\tilde{\boldsymbol{\mu}} - \boldsymbol{\mu})^T(\tilde{\boldsymbol{\mu}} - \boldsymbol{\mu}) = O(n^{-4/5}) + \frac{\sigma^2}{n}.$$

Consequently, we see that the presence of the parametric component in (4.82) does not keep us from obtaining a second order $n^{-4/5}$ rate of decay for the risk in estimating $\boldsymbol{\mu}$.

The next question to be considered is how well we can estimate the parametric regression coefficient γ. To address this issue we proceed as in Section 3.7 and assume that

$$u_i = g(t_i) + \eta_i,\ i = 1, \ldots, n, \tag{4.88}$$

for $g \in C^2[0,1]$ and $\boldsymbol{\eta} = (\eta_1, \ldots, \eta_n)^T$ a vector of *iid* random variables that are independent of the ε_i in (4.81) and satisfy $\mathrm{E}\eta_1 = 0$ and $\mathrm{E}\eta_1^2 = \sigma_\eta^2 < \infty$.

The first step in our analysis is to show that

$$n^{-1}\mathbf{u}^T(\mathbf{I}-\mathbf{S})^T(\mathbf{I}-\mathbf{S})\mathbf{u} \xrightarrow{P} \sigma_\eta^2. \tag{4.89}$$

The argument proceeds along similar lines to that in Section 3.7. We first write

$$\begin{aligned}\mathbf{u}^T(\mathbf{I}-\mathbf{S})^T(\mathbf{I}-\mathbf{S})\mathbf{u} &= \mathbf{g}^T(\mathbf{I}-\mathbf{S})^T(\mathbf{I}-\mathbf{S})\mathbf{g} \\ &+ 2\mathbf{g}^T(\mathbf{I}-\mathbf{S})^T(\mathbf{I}-\mathbf{S})\boldsymbol{\eta} + \boldsymbol{\eta}^T(\mathbf{I}-\mathbf{S})^T(\mathbf{I}-\mathbf{S})\boldsymbol{\eta},\end{aligned}$$

where $\mathbf{g} = (g(t_1), \ldots, g(t_n))^T$. The first term on the right hand side of this last expression is $O(n^{1/5})$ because $g \in C^2[0,1]$ and (4.86) holds for $C^2[0,1]$ functions. For the last term observe that

$$\boldsymbol{\eta}^T(\mathbf{I}-\mathbf{S})^T(\mathbf{I}-\mathbf{S})\boldsymbol{\eta} = \boldsymbol{\eta}^T\boldsymbol{\eta} - \boldsymbol{\eta}^T[\mathbf{S}^T+\mathbf{S}]\boldsymbol{\eta} + \boldsymbol{\eta}^T\mathbf{S}^T\mathbf{S}\boldsymbol{\eta}.$$

By Markov's inequality and (4.87) we have $\boldsymbol{\eta}^T\mathbf{S}^T\mathbf{S}\boldsymbol{\eta} = O_p(n^{1/5})$ and, since $n^{-1}\boldsymbol{\eta}^T\boldsymbol{\eta} \xrightarrow{P} \sigma_\eta^2$, the Cauchy-Schwarz inequality gives $n^{-1}\boldsymbol{\eta}^T(\mathbf{I}-\mathbf{S})^T(\mathbf{I}-\mathbf{S})\boldsymbol{\eta} \xrightarrow{P} \sigma_\eta^2$. Thus, (4.89) now follows through another application of the Cauchy-Schwarz inequality.

We have now shown that

$$\tilde{\gamma} = \gamma + (n\sigma_\eta^2)^{-1}\mathbf{u}^T(\mathbf{I}-\mathbf{S})^T(\mathbf{I}-\mathbf{S})(\mathbf{f}+\boldsymbol{\varepsilon})(1+o_p(1)).$$

The next step is to analyze

$$\begin{aligned}\mathbf{u}^T(\mathbf{I}-\mathbf{S})^T(\mathbf{I}-\mathbf{S})(\mathbf{f}+\boldsymbol{\varepsilon}) &= \mathbf{g}^T(\mathbf{I}-\mathbf{S})^T(\mathbf{I}-\mathbf{S})\mathbf{f} \\ &+ \mathbf{g}^T(\mathbf{I}-\mathbf{S})^T(\mathbf{I}-\mathbf{S})\boldsymbol{\varepsilon} + \boldsymbol{\eta}^T(\mathbf{I}-\mathbf{S})^T(\mathbf{I}-\mathbf{S})\mathbf{f} \\ &+ \boldsymbol{\eta}^T(\mathbf{I}-\mathbf{S})^T(\mathbf{I}-\mathbf{S})\boldsymbol{\varepsilon}.\end{aligned}$$

The $\mathbf{g}^T(\mathbf{I}-\mathbf{S})^T(\mathbf{I}-\mathbf{S})\mathbf{f}$ term is $O(n^{1/5})$ due to (4.86) and the Cauchy-Schwarz inequality. To deal with terms such as $\mathbf{g}^T(\mathbf{I}-\mathbf{S})^T(\mathbf{I}-\mathbf{S})\boldsymbol{\varepsilon}$ and $\boldsymbol{\eta}^T(\mathbf{I}-\mathbf{S})^T(\mathbf{I}-\mathbf{S})\mathbf{f}$ observe, for example, that

$$\mathbf{g}^T(\mathbf{I}-\mathbf{S})^T(\mathbf{I}-\mathbf{S})\boldsymbol{\varepsilon} = \mathbf{g}^T(\mathbf{I}-\mathbf{S})^T\boldsymbol{\varepsilon} - \mathbf{g}^T(\mathbf{I}-\mathbf{S})^T\mathbf{S}\boldsymbol{\varepsilon}. \tag{4.90}$$

The first term on the right hand side of (4.90) has mean zero and variance of order $\mathbf{g}^T(\mathbf{I}-\mathbf{S})^T(\mathbf{I}-\mathbf{S})\mathbf{g} = O(n^{1/5})$ while the second term is $O_p(n^{1/5})$ due to the Cauchy-Schwarz and Markov inequalities.

It only remains to analyze $\boldsymbol{\eta}^T(\mathbf{I}-\mathbf{S})^T(\mathbf{I}-\mathbf{S})\boldsymbol{\varepsilon}$. This quantity is dominated by $\boldsymbol{\eta}^T\boldsymbol{\varepsilon}$. The other terms such as $\boldsymbol{\eta}^T\mathbf{S}^T\boldsymbol{\varepsilon}$ all have mean zero and variances of order $n^{1/5}$. This is because for a kernel type smoother there will be only $O(n\lambda)$ nonzero entries in any row or column of its smoother matrix and all such entries will be of order $(n\lambda)^{-1}$ (Exercise 30). Using the fact that $\boldsymbol{\eta}^T\boldsymbol{\varepsilon} = \sum_{i=1}^n \varepsilon_i\eta_i$ is a sum of independent random variables that are indentically distributed as $\varepsilon_1\eta_1$ it now follows that

$$\sqrt{n}(\widetilde{\gamma}-\gamma) \xrightarrow{d} \frac{\sigma}{\sigma_\eta} Z \tag{4.91}$$

for Z a standard normal random variable. In particular, this means that we can estimate γ with parametric efficiency using the estimator (4.83).

Due to (4.91) we can use normal theory type methods to conduct tests and create confidence intervals for the regression coefficient γ. Specifically, an asymptotic $100(1-\alpha)\%$ confidence interval for γ is

$$\widetilde{\gamma} \pm z_{\alpha/2}\sigma\sqrt{\mathbf{u}^T(\mathbf{I}-\mathbf{S})^T(\mathbf{I}-\mathbf{S})\mathbf{u}}. \tag{4.92}$$

To use this interval we must estimate σ^2. One approach to this could be based on averaging the squared residuals from the fit to the data as we did for the mildew control example in Section 3.7. Another approach is to use a version of the GSJS estimator that has been modified for the partially linear models setting.

One way to adjust the GSJS estimator (2.29)-(2.30) to accomodate the parametric part of (4.82) can be described as follows. First set

$$a_i = \frac{t_{i+1}-t_i}{t_{i+1}-t_{i-1}}, b_i = \frac{t_i - t_{i-1}}{t_{i+1}-t_{i-1}}, c_i = (1+a_i^2+b_i^2)^{-1/2},$$

for $i = 2, \ldots, n-1$, and take $\mathbf{A}$ to be the $(n-2)\times n$ matrix whose ith row has all zero entries except for its ith, $(i+1)$th and $(i+2)$th entries which are $a_ic_i, -c_i$ and b_ic_i, respectively. Then, the original GSJS estimator (2.29)-(2.30) is just $\mathbf{y}^T\mathbf{A}^T\mathbf{A}\mathbf{y}/(n-2)$. To account for the parametric part of the model we define a modified estimator as

$$\widehat{\sigma}^2 = \frac{\mathbf{y}^T\mathbf{A}^T(\mathbf{I}-\mathbf{P})\mathbf{A}\mathbf{y}}{\mathrm{tr}(\mathbf{A}^T(\mathbf{I}-\mathbf{P})\mathbf{A})} \tag{4.93}$$

with

$$\mathbf{P} = \frac{\mathbf{A}\mathbf{u}\mathbf{u}^T\mathbf{A}^T}{\mathbf{u}^T\mathbf{A}^T\mathbf{A}\mathbf{u}} . \tag{4.94}$$

It is straightforward to see that $\widehat{\sigma}^2$ is independent of $\mathbf{u}\gamma$. In addition, one can establish the following result whose proof is discussed in Exercise 31.

Proposition 4.1. *Assume that* $f \in C^2[0,1], \max|t_i - t_{i-1}| = O(n^{-1})$, $\max|t_i - t_{i-1}|/\min|t_i - t_{i-1}|$ *is bounded and that* $\varepsilon_1, \ldots, \varepsilon_n$ *are iid with* $\mathrm{E}\varepsilon_1 = \mathrm{E}\varepsilon_1^3 = 0$, $\mathrm{Var}(\varepsilon_1) = \sigma^2$ *and* $\mathrm{E}\varepsilon_1^4 < \infty$. *Then,* $\mathrm{E}\widehat{\sigma}^2 = \sigma^2 + O(n^{-2})$ *and* $\sqrt{n}(\widehat{\sigma}^2 - \sigma^2) = O_p(1)$.

The data in Figure 4.13 represents the results of a study concerning the effectiveness of a particular variety of mouthwash. It comes from Speckman (1988) and is listed in Table 3.8. The response is a measure of gum shrinkage called SBI. For each individual in the study an initial baseline SBI measurement was taken and then the subjects were randomized into either a control group, which used only a water rinse, or a treatment group that actually used the mouthwash. One way to ascertain if the mouthwash is effective would be to compute the difference, X, between baseline SBI and SBI at the end of the study and use the observed sample differences in a standard two-sample test comparing the means of X for the treatment and control groups. The resulting treatment and control group sample means are $\bar{X}_T = .19673$ and $\bar{X}_C = .410733$ with corresponding sample variances $s_T^2 = .0296$ and $s_C^2 = .1291$. This produces a test statistic value of $(\bar{X}_T - \bar{X}_C)/\sqrt{(s_T^2 + s_C^2)/15} = -2.08$. Thus, there is no indication that the mouthwash is more effective than a water rinse and the test statistic even suggests that the water rinse might be more effective in reducing SBI.

One might argue, however, that the previous analysis is not completely satisfactory because, as can be seen from Figure 4.13, the control group has received almost all of the subjects with the high baseline SBI values. An alternative analysis of this data that compensates for this disparity in the control and treatment groups' baseline SBI values can be obtained through the use of partially linear models methodology. With this approach the baseline SBI is treated as a covariate or independent variable and the u_i in (4.81) are taken to be indicator variables that are zero or one depending

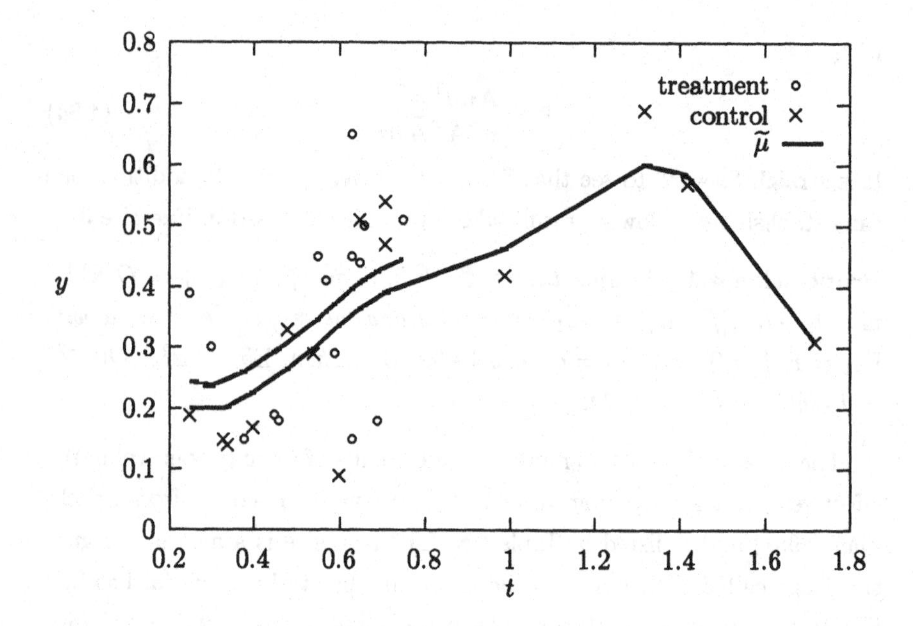

Figure 4.13: Partially Linear Model Fit to the Mouthwash Data.

on whether a subject is in the control or treatment group. The parameter γ in (4.81) represents the treatment effect. Such a model was fitted to the mouthwash data using (4.83)-(4.85) with a locally linear regression smoother matrix being used for **S**. The bandwidth for the smoother was selected by GCV and the resulting estimator of the mean function is shown in Figure 4.13.

The estimate of the treatment effect for this data was found to be $\tilde{\gamma} = .0425$ with $\hat{\sigma}^2 = .0139$ the value of the GSJS type variance estimator (4.93)-(4.94) suitably altered to allow repeated observations at "design" points (Exercise 32). The corresponding 95% confidence interval from (4.92) is then found to be $(-.0551, .1401)$. Thus, there is no evidence that the mouthwash is effective even if we adjust for the unusual allocation of subjects into the treatment and control groups.

There are other possibilities for estimators of γ beyond $\tilde{\gamma}$ in (4.83). One option is to directly generalize the series estimator formulae (3.135)-(3.137) and replace the projection smoother matrix in those equations with that of some other general smoother as suggested, e.g., by Green (1985) and Green, Jennison and Seheult (1985). For our case the resulting estimator of γ might look like

$$\widehat{\gamma} = (\mathbf{u}^T(\mathbf{I}-\mathbf{S})\mathbf{u})^{-1}\mathbf{u}^T(\mathbf{I}-\mathbf{S})\mathbf{y}. \tag{4.95}$$

This estimator has close connections to partial spline estimators discussed in Section 5.7 and can also be motivated from the work of Engle, Granger, Rice and Weiss (1986). Notice that $\widehat{\gamma}$ and $\widetilde{\gamma}$ are the same only when $\mathbf{S}$ is a projection operator which is not the case for the kernel smoothers being considered here.

Similar arguments to those used to establish (4.89) show that

$$n^{-1}\mathbf{u}^T(\mathbf{I}-\mathbf{S})\mathbf{u} \xrightarrow{P} \sigma_\eta^2. \tag{4.96}$$

However, one now finds that if the bandwidth for the kernel estimator behaves like $Cn^{-1/5}$ for some constant $C > 0$, then the numerator of (4.95) is dominated by

$$\mathbf{g}^T(\mathbf{I}-\mathbf{S})\mathbf{f} \sim \frac{1}{2}C^2M_2n^{3/5}\int_0^1 g(t)f''(t)w(t)dt. \tag{4.97}$$

In obtaining this last approximation we have assumed that the smoother admits a bias expansion of the form (4.22) in the interior and has the same order of bias in the interior and boundary regions.

Equation (4.97) has the consequence that the estimator $\widehat{\gamma}$ will have nonnegligible bias if the bandwidth is chosen to give an optimal level of smoothing for the nonparametric part of the model. Thus, one must undersmooth (i.e., choose the bandwidth too small) to effectively estimate γ with this estimator. From a theoretical viewpoint this may not matter if estimation of γ is the primary goal. However, data driven methods for choosing the bandwidth in this setting are likely to produce bandwidths decaying like $n^{-1/5}$ creating the practical problem of how much to shrink a data selected bandwidth in order to ensure that the bias of $\widehat{\gamma}$ is actually negligible. Problems of this nature make it difficult to recommend the use of $\widehat{\gamma}$. This is particularly true since the estimator $\widetilde{\gamma}$ has theoretically negligible bias at optimal levels of smoothing and is really no more difficult to compute than $\widehat{\gamma}$.

The work in this section is a special case of developments in Speckman (1988) who deals with more general random error structures and vector-valued u_i using kernel estimators. Similar results for local polynomial smoothers are given in Hamilton and Truong (1995). Heckman (1988) derives a minimax estimator for γ while Cuzick (1992) establishes a lower variance bound for estimating the regression coefficient and gives one possible efficient estimator.

4.10 Random t's

Let us now alter our original assumption about the probability structure of the (t_i, y_i) and allow the t_i to also be random. Specifically, assume that $(t_1, y_1), \ldots, (t_n, y_n)$ are *iid* copies of a continuous bivariate random variable (T, Y). In this case it is still possible to write down our model as

$$y_i = \mu(t_i) + \varepsilon_i,\ i = 1, \ldots, n.$$

However, now μ is the conditional mean,

$$\mu(t) = E[Y|T = t],$$

and the ε_i, while independent, will not be identically distributed. In particular, the assumption of a common variance will not typically be satisfied. Thus, a somewhat different approach is needed from that of previous sections to derive estimators of μ in this setting.

An understanding of how to estimate μ can be obtained by going back to the actual definition of the conditional mean. Let (T, Y) have an associated density $f(t, y)$ with

$$f_T(t) = \int_{-\infty}^{\infty} f(t, y)dy$$

the marginal density for T. Then, the conditional density for Y, given T, is

$$f_{Y|T}(y|t) = \frac{f(t, y)}{f_T(t)}$$

and μ is defined by

$$\mu(t) = \int_{-\infty}^{\infty} y f_{Y|T}(y|t)dy. \tag{4.98}$$

We can therefore estimate $f_{Y|T}$ and then plug this estimator into formula (4.98) to produce an estimator of the conditional mean function μ.

To estimate the conditional density we must first estimate the joint density f. For this purpose we can use the bivariate kernel density estimator (see, e.g., Silverman 1986, Scott 1992 and Exercise 35)

$$f_\lambda(t, y) = \frac{1}{n\lambda^2} \sum_{i=1}^{n} K_1\left(\lambda^{-1}(t - t_i)\right) K_2\left(\lambda^{-1}(y - y_i)\right),$$

where the kernel functions $K_r, r = 1, 2$ are nonnegative, supported on [-1,1] and satisfy (4.3)-(4.6). Substituting this into the formula for the marginal density, conditional density and conditional mean then gives

$$\begin{aligned} f_{\lambda T}(t) &= \int_{-\infty}^{\infty} f_\lambda(t, y)dy \\ &= \frac{1}{n\lambda} \sum_{i=1}^{n} K_1\left(\lambda^{-1}(t - t_i)\right) \end{aligned}$$

as our estimator of f_T,

$$f_\lambda(y|t) = f_\lambda(t, y)/f_{\lambda T}(t)$$

as our estimator of $f_{Y|T}$ and, finally, for λ sufficiently small,

$$\begin{aligned} \mu_\lambda(t) &= \frac{1}{f_{\lambda T}(t)} \int_{-\infty}^{\infty} y f_\lambda(t, y)dy \\ &= \frac{1}{f_{\lambda T}(t)} \frac{1}{n\lambda^2} \sum_{i=1}^{n} K_1\left(\lambda^{-1}(t - t_i)\right) \int_{-\infty}^{\infty} y K_2\left(\lambda^{-1}(y - y_i)\right) dy \\ &= \frac{1}{f_{\lambda T}(t)} \frac{1}{n\lambda} \sum_{i=1}^{n} K_1\left(\lambda^{-1}(t - t_i)\right) \int_{-1}^{1} (y_i + \lambda u) K_2(u)du \\ &= \sum_{i=1}^{n} K_1\left(\lambda^{-1}(t - t_i)\right) y_i \Big/ \sum_{i=1}^{n} K_1\left(\lambda^{-1}(t - t_i)\right) \end{aligned}$$

as our estimator of $\mu(t)$. But this is just the Nadaraya-Watson kernel estimator that was introduced in Section 4.2.

There is quite a bit known about the Nadaraya-Watson estimator in the case of random designs. Under certain restrictions it is known to have a variety of consistency and related properties. Some of these are explored by Nadaraya (1964, 1965, 1970), Noda (1976), Rosenblatt (1969), Schuster (1972), Schuster and Yakowitz (1979), Devroye (1978, 1981), Devroye and Wagner (1980), Johnston (1982), Mack and Silverman (1982), Greblicki, Krzyzak and Pawlak (1984) and Härdle (1989). For example, it follows from Mack and Müller (1989) that under certain restrictions

$$\mathrm{E}\,\mu_\lambda(t) \sim \mu(t) + \frac{M_2\lambda^2}{2}\frac{(\mu f_T)''(t)}{f_T(t)} \tag{4.99}$$

and

$$\mathrm{Var}(\mu_\lambda(t)) \sim \frac{\mathrm{Var}(Y|T=t)}{n\lambda f_T(t)} \tag{4.100}$$

if $n \to \infty, \lambda \to 0$ and $n\lambda \to \infty$. Thus, by choosing the bandwidth to decay at the rate $n^{-1/5}$ we can obtain an optimal $n^{-4/5}$ convergence rate for $\mathrm{E}\,(\mu_\lambda(t) - \mu(t))^2$ in the random design setting. The behavior of automated methods for selecting the bandwidth for μ_λ in the random design case has been studied by Härdle and Marron (1985), for example.

Since the Nadaraya-Watson estimator is but one possible version of a kernel estimator, it is also of interest to explore how the other kernel variants from Section 4.2 behave in the case of random t's. Of particular interest is the Gasser-Müller estimator that we studied in Section 4.3 for nonstochastic designs. In the case of random t_i a version of the Gasser-Müller estimator can be defined by

$$\mu_\lambda(t) = \sum_{i=1}^{n} y_{(i)} \int_{s_{(i-1)}}^{s_{(i)}} \lambda^{-1} K\left(\lambda^{-1}(t-s)\right) ds.$$

Here we take $t_{(1)} \le \cdots \le t_{(n)}$ to be the ordered, observed T values and $y_{(i)}$ is the ith concomitant, i.e., the observed Y value that corresponds to $t_{(i)}$. For the $s_{(i)}$ we might choose $s_{(0)} = -\infty, s_{(i)} = (t_{(i+1)} + t_{(i)})/2, i = 1, \ldots, n-1$, and $s_n = \infty$. Under this choice for the estimator it follows from Mack and Müller (1989) and Chu and Marron (1991a) that if $n \to \infty, \lambda \to 0$ and $n\lambda \to \infty$ then

$$\mathrm{E}\,\mu_\lambda(t) \sim \mu(t) + \frac{\lambda^2}{2}\mu''(t)M_2 \tag{4.101}$$

and

$$\mathrm{Var}(\mu_\lambda(t)) \sim 1.5\frac{\mathrm{Var}(Y|T=t)}{n\lambda f_T(t)}. \tag{4.102}$$

The factor of 1.5 in (4.102) can increase to 2 if we choose $s_{(i)} = t_{(i)}, i = 1, \ldots, n$.

Now it can be shown that the Nadaraya-Watson and Gasser-Müller estimators both have the same lead terms in their variance and bias in the case of a "fixed" design generated from a design density as in (4.37). Thus, comparing (4.99)-(4.100) with (4.101)-(4.102) we see that this is no longer the case when the t_i are random. To elaborate on this let us make the identification that $\mathrm{Var}(Y|T = t)$ and f_T occupy the roles previously held by σ^2 and w in the deterministic design scenario. From this we can conclude that the Gasser-Müller estimator has the same bias with random or fixed designs but its variance increases by a factor of 1.5 in the case of random t_i. While the Nadaraya-Watson estimator has the same variance with either fixed or random t's, its bias depends on the T density in the random design setting through the somewhat complicated factor $(\mu f_T)''/f_T$.

Since the Nadaraya-Watson and Gasser-Müller estimators no longer have similar asymptotic variances and biases, this raises question concerning which estimator is to be preferred for use with random t's. The increased variability of the Gasser-Müller estimator is certainly a drawback and one can devise situations where the bias of the Gasser-Müller estimator is also of larger magnitude than for the Nadaraya-Watson estimator (Exercise 36). However, there are reasons to prefer the less complicated bias of the Gasser-Müller estimator. In particular, one would expect a second order estimator to have near constant bias for a quadratic regression function and this will be true for the Nadaraya-Watson estimator only if the marginal T density is uniform.

In view of the above discussion one might conclude that an "ideal" estimator of μ in the random design case would have the bias of the Gasser-Müller estimator and variance of the Nadaraya-Watson estimator. It turns out that such an estimator exists and is provided by the local linear regression estimator of Section 4.7. Specifically, if $\widehat{\theta}_1$ and $\widehat{\theta}_2$ are the minimizers of

$$\sum_{i=1}^{n} K\left(\frac{t-t_i}{\lambda}\right)(y_i - \theta_1 - \theta_2(t_i - t))^2$$

with respect to θ_1 and θ_2, the local linear regression estimator of $\mu(t)$ is $\mu_\lambda(t) = \widehat{\theta}_1$. This estimator can be shown to satisfy

$$\mathrm{E}\,\mu_\lambda(t) \doteq \mu(t) + \frac{\lambda^2}{2}\mu''(t)M_2$$

and

$$\mathrm{Var}(\mu_\lambda(t)) \doteq \frac{\mathrm{Var}(Y|T=t)}{n\lambda f_T(t)}$$

when $n \to \infty, \lambda \to 0$ and $n\lambda \to \infty$ and certain other restrictions are met. See, e.g., Wand and Jones (1995, Sections 5.3-5.4) and Fan and Gijbels (1996, Section 3.2). Thus, in an asymptotic sense, the local linear smoother combines the best features of the Nadaraya-Watson and Gasser-Müller estimators for the case of random t's.

In terms of estimation efficiency, Fan and Gijbels (1996, Sections 3.4-3.5) show that local linear regression estimators are asymptotically fully minimax efficient relative to the set of linear estimators of μ in terms of their conditional pointwise squared error risk. In addition, we know from Section 4.7 that the locally linear regression estimator has the same essential asymptotic properties as the boundary corrected Gasser-Müller or Nadaraya-Watson estimators when the design is fixed. Thus, local linear smoothers appear to possess many of the ideal properties one might want from a kernel or any other type of smoother with either a fixed or random design.

4.11 Extensions

There are a number of ways to generalize kernel and local polynomial smoothers. A starting point in the case of kernel smoothing can be based on the fact that the Nadaraya-Watson estimator minimizes

$$\sum_{i=1}^{n} K\left(\lambda^{-1}(t-t_i)\right)(y_i - \theta)^2 \tag{4.103}$$

with respect to θ. Thus, one natural generalization is to replace the squared error measure of closeness in (4.103) and define a "kernel estimator" of $\mu(t)$ as the minimizer with respect to θ of the criterion

$$\sum_{i=1}^{n} K\left(\lambda^{-1}(t-t_i)\right) \rho(y_i-\theta) \tag{4.104}$$

with ρ some general distance measure. By choosing ρ appropriately this produces the robustified or M-type kernel estimators discussed in Chapter 6 of Härdle (1990). Parallel results for robustified local polynomial estimators are given in Wang and Scott (1994) and Fan and Gijbels (1996, Section 5.5).

One could also view (4.103) as a locally smoothed negative log-likelihood corresponding to the case of an assumed normal density for the random errors in a regression model. This suggests a way to extend kernel smoothing to fairly general problems. Suppose that our data consists of independent responses with the ith response having density function $f(\cdot;\mu(t_i))$ that depends on an unknown smooth function μ. Then, we could estimate μ by the minimizer of the local negative log-likelihood

$$-\sum_{i=1}^{n} K\left(\lambda^{-1}(t-t_i)\right) \ln f(y_i;\theta) \tag{4.105}$$

with respect to θ. Using variants of this approach kernel methods can be developed for estimation with exponential family models, censored survival data, etc. The large sample properties and applications of estimators derived from (4.105) have been studied by Staniswalis (1989a) and Severini and Staniswalis (1994). Similar results for local polynomial versions of these estimators are given in Fan, Heckman and Wand (1995) and Fan and Gijbels (1996, Chapter 5).

The case of kernel estimation with fixed designs but non-*iid* errors has been considered by several authors. In the case of uncorrelated, heteroskedastic errors the arguments in Section 4.3 can be modified fairly easily to obtain the asymptotic properties of kernel type estimators (see, e.g., Müller 1988 and Exercise 38). The more general problem of estimation and bandwidth selection with correlated errors has been studied by Hart and Wehrly (1986, 1993), Müller and Stadtmüller (1988), Altman (1990), Hart

(1991), Chu and Marron (1991b), Herrmann, Gasser and Kneip (1992) and Ray and Tsay (1997). The use of kernel estimators with stochastic processes is considered in Rosenblatt (1970), Roussas (1969), Bierens (1983), Yakowitz (1985), Roussas and Tran (1992) and Truong and Stone (1992).

All of the kernel estimators in Section 4.2 can be extended to include the case where t is vector valued. If $\mathbf{t} = (t_1, \ldots, t_p)^T$, a simple way this can be accomplished is to let $K(u_1, \ldots, u_p) = \prod_{j=1}^{p} K_j(u_j)$, with K_j satisfying (4.3)–(4.6) for $j = 1, \ldots, p$. Then, for example, a multivariate version of estimator (4.7) is

$$\mu_\lambda(\mathbf{t}) = \frac{1}{n} \sum_{i=1}^{n} y_i \prod_{r=1}^{p} \lambda_r^{-1} K_r\left(\lambda_r^{-1}\left(t_r - t_{ir}\right)\right),$$

where $\mathbf{t}_i = (t_{i1}, \ldots, t_{ip})^T$ is the vector of values for the p independent variables corresponding to y_i. Analogous results to those for the case of univariate t can be obtained for this estimator as well as estimators based on a more general p-variate kernel function or local polynomial fits. Discussions of these and related topics can be found in Müller (1988, Chapter 6), Härdle (1990, Chapter 10), Scott (1992, Chapter 8), Ruppert and Wand (1994), Wand and Jones (1995, Section 5.9), Simonoff (1995, Section 5.7), Fan and Gijbels (1996, Chapter 7) and Exercises 37.

4.12 Exercises

1. Suppose we observe data from the model

$$y_{ij} = \mu(t_i) + \varepsilon_{ij},\ j = 1, \ldots, n_i,\ i = 1, \ldots, n.$$

Discuss how one can construct kernel type estimators for this case. [Hint: Kernel estimators can be derived from criteria such as (4.63).]

2. Suppose that $\mu(t) = \theta_0 + \theta_1 t$. Show that, for example, estimator (4.10) is nearly unbiased in this case. Under what conditions is the estimator actually unbiased for a linear regression function with finite n? More generally, suppose that $\mu(t) = \sum_{j=1}^{m} \theta_j t^{j-1}$. What conditions are needed for (4.10) to be nearly unbiased in this case?

3. Consider the case where K is a symmetric kernel with support on the entire line. Show that, in general, there exists no $\lambda > 0$ such that, for estimator (4.10), $E\mu_\lambda(t) - \mu(t) = \lambda^2 C_t + o(\lambda^2)$, where C_t is a constant depending only on μ, t and K. What does this suggest about the difference between kernels with finite and nonfinite support?

4. Suppose that $\mu \in C^2[0,1], t_i = (2i-1)/2n, i = 1, \ldots, n$, and $\mu^{(j)}(0) = \mu^{(j)}(1), j = 0, 1, 2$. Give a form for the kernel estimator (4.7) in this case for which (4.22) holds uniformly in t. Derive an asymptotic approximation of the form (4.27) for the global risk of the estimator and compute the resulting asymptotically optimal bandwidth and risk. [Hint: Since μ is periodic we can define $\mu(-t) = \mu(1-t)$.]

5. Prove (4.39)–(4.42). [Hint: Observe that $s_i - s_{i-1} = (nw(\xi_i))^{-1} + O(n^{-2})$ for mean values $t_{i-1} \le \xi_i \le t_{i+1}, i = 2, \ldots, n-1$, when the design points are generated as in (4.37).]

6. Derive the choices of a_q, b_q that were used to produce the boundary kernel K_q in (4.45) and verify that $K_q = K$ when $q \ge 1$. Give the explicit form of this kernel for both the upper and lower boundary regions when K is the quadratic kernel in Table 4.1.

7. Let $\mathbf{Q} = \{q_j(t_i)\}_{i=1,n,j=0,1}$ for q_0, q_1 the polynomials in (3.117)-(3.118). Define the boundary corrected estimator $\widetilde{\boldsymbol{\mu}}_\lambda = \boldsymbol{\mu}_{\lambda c} + \mathbf{P}(\mathbf{y} - \boldsymbol{\mu}_{\lambda c})$ with $\boldsymbol{\mu}_{\lambda c}$ the vector of fitted values for the cut-and-normalize kernel estimator (4.26), $\mathbf{P} = \mathbf{G}(\mathbf{G}^T\mathbf{G})^-\mathbf{G}^T$ and $\mathbf{G} = (\mathbf{I} - \mathbf{S}_{\lambda c})\mathbf{Q}$ for $\mathbf{S}_{\lambda c}$ the smoother matrix for $\boldsymbol{\mu}_{\lambda c}$.

   (a) Derive an upper bound for the risk of this estimator using the Bias Reduction Lemma of Section 3.6 and use this to establish that the average risk $R_n(\lambda)$ for the estimator decays at an $n^{-4/5}$ rate if the bandwidth is chosen to be of order $n^{-1/5}$.

   (b) If the design is uniform and the bandwidth decay rate is $n^{-1/5}$, show that the asymptotic risk for $\widetilde{\mu}_\lambda$ will be smaller than that of a bias corrected Gasser-Müller estimator unless $\mu'(0) = \mu'(1)$.

[Hint: Compare the values for $J_2(\mu^*)$ and $J_2(\mu)$ with $\mu^* = \mu - \mu'(0)q_0 - \mu'(1)q_1$.]

8. Another approach to the removal of boundary bias can be obtained through the use of a generalized jackknife estimator for $\mu(t)$. This estimator will have the form

$$\widetilde{\mu}_\lambda(t) = \mu_{\lambda c}(t) + \beta[\mu_{\lambda c}(t) - \mu_{(\alpha\lambda)c}(t)]$$

with $\mu_{\lambda c}$ the cut-and-normalize estimator (4.26) and α, β selected appropriately.

   (a) If $t = q\lambda$, $0 < q < 1$, justify the approximation

$$\mathrm{E}\,\mu_{\lambda c}(t) - \mu(t) \doteq -\frac{\lambda\mu'(t)M_1(\lambda q)}{M_0(\lambda q)} + \frac{\lambda^2\mu''(t)M_2(\lambda q)}{2M_0(\lambda q)}$$

   with the $M_j(\cdot), j = 0, 1, 2$ defined in (4.21).

   (b) Now show that the choice

$$\beta = A(q)/[\alpha A(q/\alpha) - A(q)]$$

   with

$$A(q) = M_1(\lambda q)/M_0(\lambda q)$$

   results in the estimator $\widetilde{\mu}_\lambda(t)$ having bias whose leading term in λ is of the order λ^2.

   (c) Show that $\widetilde{\mu}_\lambda(t)$ is also a kernel estimator and derive the form of its kernel. A natural way to choose α is to select it so that the estimator is an average over an interval of width 2λ. This corresponds to taking $\alpha = 2 - q$. Plot the kernel for $\widetilde{\mu}_\lambda(t)$ under this choice of α and several choices of q.

   [Rice 1984b]

9. What is the value of λ_{opt} in (4.41) when the design is uniform, $\sigma = .15, n = 50$ and $\mu(t) = t + \exp\{-50(t - .5)^2\}$?

10. Verify (4.53) and discuss its implications.

11. Prove results (4.55)–(4.57).

12. Detail the properties that an mth order boundary kernel must satisfy and then use this to establish (4.58) and (4.60)–(4.61).

13. Suppose that the design is uniform and we are using estimator (4.7). The *twicing* approach to bias reduction can then be described as follows:

    i) Compute the residuals $e_j(\lambda) = y_j - \mu_\lambda(t_j)$.
    ii) Then compute $e_\lambda(t) = (n\lambda)^{-1}\sum_{i=1}^{n} K\left(\lambda^{-1}(t-t_i)\right) e_i(\lambda)$.
    iii) Now correct the original estimator using $e_\lambda(t)$ to obtain $\widetilde{\mu}_\lambda(t) = \mu_\lambda(t) + e_\lambda(t)$.

    Show that, apart from boundary effects, twicing amounts to use of the kernel

    $$K_2(u) = 2K(u) - \int_{-1}^{1} K(u-s)K(s)ds.$$

    Verify that if K is a kernel of order m then K_2 is a kernel of order $2m$. In light of this result discuss the utility of twicing compared to the use of $2m$th order kernels.

    [Stuetzle and Mittal 1979].

14. Consider the estimator $\mu_{\lambda d}(t)$ defined in (4.62) for the dth derivative of μ at t. If μ is in $C^m[0,1], n \to \infty, \lambda \to 0, n\lambda^{2d+1} \to \infty$, show that

    $$\mathrm{E}\left(\mu_{\lambda d}(t) - \mathrm{E}\,\mu_{\lambda d}(t)\right)^2 = \frac{\sigma^2}{n\lambda^{2d+1}} \int_{-1}^{1} K(u)^2 du(1+o(1))$$

    and

    $$\begin{aligned}\mathrm{E}\,\mu_{\lambda d}(t) - \mu^{(d)}(t) &= \frac{(-1)^m}{m!}\lambda^{m-d}\mu^{(m)}(t)\int_{-1}^{1} u^m K(u)du \\ &+ o(\lambda^{m-d}) + O(n^{-1}).\end{aligned}$$

    Use these expressions to obtain an asymptotic form for the pointwise risk $\mathrm{E}\left(\mu_{\lambda d}(t) - \mu^{(d)}(t)\right)^2$ of the estimator and find the corresponding asymptotically optimal bandwidth and risk.

15. Extend the results of the previous exercise to global assessment of the risk for derivative estimation. To accomplish this first determine the properties that are needed for boundary kernels in derivative estimation. Then approximate $R_{nd}(\lambda) = n^{-1}\sum_{i=1}^{n} \mathrm{E}\,(\mu_{\lambda d}(t_i) - \mu^{(d)}(t_i))^2$ for a suitably boundary modified estimator $\mu_{\lambda d}$. Use this approximation to derive parallels of (4.41)–(4.42) for kernel derivative estimators.

16. Suppose we estimate μ' using the boundary corrected derivative estimator from Exercise 15 and let $\widehat{\lambda}$ be an estimator of the asymptotically optimal global bandwidth for a second order kernel estimator of μ. Show how this bandwidth can be transformed to provide an estimator of the asymptotically optimal global bandwidth for estimating μ' using $\mu_{\lambda 1}$. [Hint: Note that the ratio of the asymptotically optimal global bandwidths for estimation of μ and μ' depends only on the kernels used for estimation of μ and μ'.]

17. Let K be a kernel for derivative estimation that satisfies

$$\int_{-1}^{1} u^j K(u)du = \begin{cases} 0, & j=0, \\ -1, & j=1, \\ M_2 \neq 0, & j=2. \end{cases}$$

Derive a boundary kernel of the form $K_q(u) = (a+bu)K(u)$ for estimation at a lower boundary point $t = \lambda q$ for $q < 1$ and give the explicit form of this kernel for when $K(u) = -3.75u(1-u^2)I_{[-1,1]}(u)$. Discuss how similar kernels can be obtained for estimating $\mu^{(d)}$ for any value of d.

18. Derive the minimizers of criterion (4.64) and thereby verify (4.66)–(4.67) as the local linear regression estimators of μ and μ'.

19. The data in Table 4.3 is the assay data used to produce Figure 4.1. Fit a local linear regression estimator to this data using an automated bandwidth selection method.

20. Fit a local linear regression estimator to the simulated data in Table 1.1 using GCV and unbiased risk bandwidth selection criteria. Compare the fit to the true regression function in this case and compare

Table 4.3: Assay Data

t	y	t	y	t	y	t	y
121	3.8999	123	4.0741	128	3.7278	129	4.0237
135	4.0182	136	3.7533	139	4.3692	140	3.5584
146	4.4066	150	4.2934	158	4.5164	159	5.0867
162	4.7741	164	3.1668	166	4.1855	168	4.6911
176	4.0410	178	4.9888	179	4.4330	181	5.3010
183	5.1303	188	5.3630	190	4.8382	192	4.5174
197	3.5303	198	5.4924	201	5.2837	206	4.5935
210	5.3836	211	5.8106	217	4.6011	218	5.8737
219	5.3590	220	5.5233	221	5.3141	223	5.4931
224	4.7092	227	4.9264	228	6.3602	229	4.7576
230	6.1446	231	4.5251	233	6.410	234	5.3718
235	5.4907	236	5.7140	237	6.0430	239	5.1485
240	5.2230	242	5.3271	243	5.5063	245	5.0223
246	5.5027	247	5.002	248	5.2350	249	4.9401
250	4.7401	251	5.2643	252	5.2714	253	4.8064
254	4.7242	255	5.6267	256	5.3220	257	4.8136
258	5.5088	259	6.3874	260	4.5459	261	4.5506
262	5.3783	263	4.8273	264	5.6071	265	4.8520
266	4.5395	267	5.2169	268	4.6378	271	5.2907
272	5.3923	274	4.7793	277	5.0956	283	5.8006
285	4.5902	288	5.2691	300	5.4131		

Source: Dr. Barbara Anderson, Department of Epidemiology, Texas A & M University.

your estimated bandwidths to the asymptotically optimal bandwidth (4.41) and the minimizer of the loss.

21. Fit a local linear regression estimator to the voltage drop data in Table 3.6 using GCV and unbiased risk bandwidth selection criteria. Compare your local linear regression fit to the polynomial regression

estimator obtained in Exercise 9 of Chapter 3.

22. Fit a local linear regression estimator to the windmill data in Table 3.7 using GCV and unbiased risk bandwidth selection criteria. Compare your local linear regression fit to the polynomial-cosine estimator obtained in Exercise 24 of Chapter 3.

23. Show that apart from quadrature error, the local linear derivative estimator (4.68) is of the form (4.62) with kernel $-uK(u)/M_2$ that satisfies the necessary moment conditions for derivative estimation.

24. Describe how to estimate $\mu^{(d)}$ using a local mth order polynomial regression estimator. What restrictions are needed on m and d?

25. The data set in Table 4.4 represents the growth data for the boy that was used to obtain Figure 4.7. The reported measurements (y) are standing heights in millimeters as a function of age in years (t).

Table 4.4: Growth Data

t	y	t	y	t	y
.083	525.	.25	608.	.5	665.
.75	717.	1.0	745.	1.5	803.
2.0	859.	3.0	940	4.0	1007
5.0	1065	6.0	1121	7.0	1183
8.0	1238	9.0	1298	10.0	1348
10.5	1369	11.0	1391	11.5	1422
12.0	1470	12.5	1525	13.0	1578
13.5	1638	14.0	1664	14.5	1692
15.0	1708	15.5	1723	16.0	1727
16.5	1727	17.0	1727	18.0	1729
19.0	1738	20.0	1738		

Source: Dr. Luciano Molinari, University of Zürich.

(a) Fit this data using a local cubic regression estimator with the quadratic kernel from Table 4.1 and an automated method for bandwidth selection.

(b) Obtain an estimator of the boy's acceleration curve μ'' from your local cubic estimator. How does the data driven bandwidth for estimating μ work in this case? Can you suggest another possible choice for the bandwidth that might work better?

(c) Which seems to provide the most information about the boy's growth: the growth or acceleration curve? Explain your choice.

26. Derive the limiting distribution of $(\mu_\lambda(t) - \mu(t))/\sqrt{\mathrm{Var}(\mu_\lambda(t))}$ for the Gasser-Müller estimator when $\lambda = \lambda_{opt}$ in (4.41).

27. Let K be a symmetric second order kernel for estimation of μ and let K^* be a kernel for second derivative estimation that satisfies

$$\int_{-1}^{1} u^j K^*(u)du = \begin{cases} 0, & j = 0, 1, 3, \\ 2, & j = 2, \\ M_4^* \neq 0, & j = 4. \end{cases}$$

Show that $K - M_2 K^*/2$ with $M_2 = \int_{-1}^{1} u^2 K(u)du$ is a fourth order kernel.

28. Compute the ordinary 95% normal theory intervals (4.73) and corresponding Bonferroni bands for your fit to the simulated data in Exercise 20 and compare how they perform in terms of width and containment of the true regression function.

29. Use the Bias Reduction Lemma of Section 3.6 to extend estimators (4.83)–(4.85) to the case where the u_i are vector valued in model (4.81). If $\widetilde{\boldsymbol{\mu}}$ is the resulting estimator, give an argument to the effect that

$$\mathrm{E}\,(\widetilde{\boldsymbol{\mu}} - \boldsymbol{\mu})^T(\widetilde{\boldsymbol{\mu}} - \boldsymbol{\mu}) = O(n^{-4/5})$$

if a second order, boundary corrected, kernel type smoother is used with a bandwidth of order $n^{-1/5}$.

30. Consider the partially linear model scenario of Section 4.9 with $\mathbf{S}$ corresponding to, e.g., a boundary corrected Gasser-Müller estimator with bandwidth of exact order $n^{-1/5}$. Show that the matrix $\mathbf{S}^T\mathbf{S}$ has only $O(n\lambda)$ nonzero entries in any row or column and that all nonzero entries are of order $n^{-4/5}$. Use this to show that $\boldsymbol{\eta}^T(\mathbf{I}-\mathbf{S})^T(\mathbf{I}-\mathbf{S})\boldsymbol{\varepsilon} = \boldsymbol{\eta}^T\boldsymbol{\varepsilon} + O_p(n^{1/10})$.

31. Establish Proposition 4.1. [Hint: Use Gershgorin's Theorem (Franklin 1968, Section 6.8) which has the consequence that every eigenvalue δ of an $n \times n$ matrix $\mathbf{B} = \{b_{ij}\}$ satisfies at least one of the inequalities $|\delta - b_{ii}| \leq \sum_{j \neq i} |b_{ij}|$.]

32. Give an alternative form of the GSJS estimator in (4.93)-(4.94) that can be used when there are repeated observations at one or more of the t_i. [Hint: Refer to Exercise 21 of Chapter 2.]

33. Compute estimators (4.83)-(4.85) for the mouthwash data in Table 3.8 using a locally linear smoother with the biweight kernel of Table 4.1 and a bandwidth selected by GCV. Compare your results to those for the quadratic kernel described in Section 4.9 and the polynomial-cosine estimator of Chapter 3.

34. Verify (4.96)–(4.97).

35. Let $x_1, \ldots, x_n$ be a random sample from a distribution with twice continuously differentiable density f. Define a kernel type density estimator by $f_\lambda(t) = (n\lambda)^{-1} \sum_{j=1}^{n} K\left(\lambda^{-1}(t - x_j)\right)$, where K is a symmetric probability density on [-1,1] which satisfies (4.3)-(4.6).

    (a) Give conditions under which $\mathrm{E}\, f_\lambda(t) \sim f(t) + \lambda^2 f''(t) M_2/2$ and $\mathrm{E}\,\left(f_\lambda(t) - \mathrm{E}\, f_\lambda(t)\right)^2 \sim \frac{f(t)}{n\lambda} \int_{-1}^{1} K(u)^2 du$.

    (b) Use the expressions from part (a) to obtain an asymptotic formula for the risk $R_n(t;\lambda) = \mathrm{E}\,\left(f_\lambda(t) - f(t)\right)^2$ and derive the resulting asymptotically optimal bandwidth and risk for the kernel density estimator.

(c) Give conditions under which the kernel density estimator f_λ will have an asymptotic normal distribution.

Table 4.5: ELISA Assay Data

t	y	t	y
.00	1.7, 1.66, 1.95, 2.07	.08	1.91, 2.27, 2.11, 2.39
.10	2.22, 2.25, 3.26, 2.92	.14	2.8, 2.94, 2.38, 2.7
.19	2.78, 2.64, 2.71, 2.85	.25	3.54, 2.86, 3.15, 3.32
.4	3.91, 3.83, 4.88, 4.21	.55	4.54, 4.47, 4.79, 5.68
.75	6.06, 5.07, 5. ,5.98	1.	5.84, 5.79, 6.1, 7.81
1.38	7.31, 7.08, 7.06, 6.87	1.85	9.88, 10.12, 9.22, 9.96
2.5	11.04, 10.46, 10.88, 11.65	3.25	13.51, 15.47, 14.21, 13.92
4.5	16.07, 14.67, 14.78, 15.21	6.0	17.34, 16.85, 16.74, 16.87
8.25	18.98, 19.85, 18.75, 18.51	11.25	21.67, 21.22, 19.79, 22.67
15.	23.21, 22.24, 22.44, 22.60	20.25	23.92, 24.87, 23.82, 24.87
27.5	25.75, 25.87, 24.91, 24.87	37.00	24.44, 25.87, 25.75, 27.27
50.	29.58, 26.70, 26.54, 27.18		

Source: Davidian, Carroll and Smith (1988).

36. Derive conditions for the random design setting where the lead term in the squared bias of the Gasser-Müller estimator will exceed that of the Nadaraya-Watson estimator. Provide an explicit example of a case where your conditions are met.

37. Suppose that we observe $(t_{i1}, t_{i2}, y_i), i = 1, \ldots, n$, for $\mathbf{t} = (t_1, t_2)$ a bivariate predictor variable. Let K satisfy (4.3)–(4.6) and define the estimator $\mu_\lambda(\mathbf{t}) = (n\lambda^2)^{-1} \sum_{j=1}^{n} y_j K\left(\lambda^{-1}(t_1 - t_{j1})\right) K\left(\lambda^{-1}(t_2 - t_{j2})\right)$ for the regression function. Give a formal asymptotic expression for the pointwise risk $\mathrm{E}\left(\mu(\mathbf{t}) - \mu_\lambda(\mathbf{t})\right)^2$. Find the resulting asymptotically optimal bandwidth and risk.

38. Suppose that model (4.1) holds except that $\mathrm{Cov}(\varepsilon_i, \varepsilon_j) = \sigma^2(t_i)\delta_{ij}$

with $\sigma(\cdot) \in C^1[0,1]$. Extend formulas (4.38)-(4.42) to include estimation with this type of heteroskedasticity.

39. The data in Table 4.5 represent the results of an ELISA assay. Fit this data using an appropriate kernel type estimator.

Chapter 5

Smoothing Splines

5.1 Introduction

In this chapter we will study smoothing spline estimators for the regression curve. To motivate the concept, suppose that we wish to fit a data set using a function that reflects the key features of the data but retains some degree of smoothness. A natural measure of smoothness associated with a function $f \in W_2^m[0,1]$ is $\int_0^1 f^{(m)}(t)^2 dt$ while a standard measure of goodness-of-fit to the data is the (average) residual sum-of-squares $n^{-1}\sum_{i=1}^n (y_i - f(t_i))^2$. Thus, an overall assessment of the quality of a candidate estimator f is provided by the convex sum

$$(1-q)n^{-1}\sum_{i=1}^n (y_i - f(t_i))^2 + q\int_0^1 f^{(m)}(t)^2 dt$$

for some $0 < q < 1$. An "optimal" estimator could then be obtained by minimizing this functional over $W_2^m[0,1]$. Upon setting $\lambda = q/(1-q)$ this becomes equivalent to estimating μ by the function μ_λ which minimizes

$$n^{-1}\sum_{i=1}^n (y_i - f(t_i))^2 + \lambda\int_0^1 f^{(m)}(t)^2 dt, \quad \lambda > 0, \tag{5.1}$$

over $f \in W_2^m[0,1]$. The result is the smoothing spline estimator of the regression function to be studied in this chapter.

The parameter λ in (5.1) governs the tradeoff between smoothness and goodness-of-fit and, for that reason, is usually referred to as the *smoothing parameter*. When λ is large (i.e., q is near 1) a premium is being placed on smoothness and potential estimators with large mth derivatives are penalized. The limiting case of $\lambda = \infty$ (or $q = 1$) produces an mth order (or $(m-1)$st degree) polynomial regression fit to the data. Conversely, a small value of λ corresponds to more emphasis on goodness-of-fit with $\lambda = q = 0$ giving an estimator that interpolates the data.

Another motivation for criterion (5.1) can be obtained from polynomial regression. Assume again the familiar regression model

$$y_i = \mu(t_i) + \varepsilon_i,\ i = 1, \ldots, n, \tag{5.2}$$

where $0 \le t_1 < \cdots < t_n \le 1$ are design points, $\mu \in W_2^m[0,1]$ and the $\varepsilon_i, i = 1, \ldots, n$, are zero mean, uncorrelated random errors with common variance σ^2. Using Taylor's Theorem (i.e., Theorem 3.4) we can rewrite this model as

$$y_i = \sum_{j=1}^{m} \theta_j t_i^{j-1} + \mathrm{Rem}(t_i) + \varepsilon_i,\ i = 1, \ldots, n,$$

for constants $\theta_1, \ldots, \theta_m$ and

$$\mathrm{Rem}(t) = [(m-1)!]^{-1} \int_0^1 \mu^{(m)}(\xi)(t-\xi)_+^{m-1} d\xi \,.$$

Now, from the Cauchy-Schwarz inequality we have

$$\max_{1 \le i \le n} \mathrm{Rem}(t_i)^2 \le \frac{J_m(\mu)}{(2m-1)[(m-1)!]^2},$$

where

$$J_m(\mu) = \int_0^1 \mu^{(m)}(t)^2 dt \tag{5.3}$$

is the smoothess measure from criterion (5.1). The value of $J_m(\mu)$ (or rather a constant multiple of $J_m(\mu)$) can therefore be viewed as providing a bound on how far (5.2) departs from a polynomial model.

If we knew, for example, that

$$J_m(\mu) \le \rho \tag{5.4}$$

for some $\rho \geq 0$, then this would provide us with information on how far the regression curve could depart from a polynomial and we could input this information into the estimation process. One way of accomplishing this would be to estimate μ by the minimizer of

$$RSS(f) = \sum_{i=1}^{n} (y_i - f(t_i))^2 \tag{5.5}$$

over all functions $f \in W_2^m[0,1]$ which satisfy (5.4). This is equivalent to minimizing $n^{-1}RSS(f) + \lambda(J_m(f) - \rho)$ with λ now the Lagrange multiplier for the constraint. But this is essentially criterion (5.1) and, not surprisingly, produces the same estimator of μ as (5.1). The relationship between the estimators obtained from (5.4)–(5.5) and (5.1) is made precise in the following theorem of Schoenberg (1964).

Theorem 5.1. *Assume that $n \geq m$ and let $\mu(\cdot;\rho)$ be the minimizer of (5.5) in $W_2^m[0,1]$ subject to the constraint (5.4). Let μ_λ denote the minimizer of (5.1) in $W_2^m[0,1]$. Then, there is a computable constant ρ_0 such that the sets $\{\mu(\cdot;\rho) : 0 \leq \rho \leq \rho_0\}$ and $\{\mu_\lambda(\cdot) : 0 \leq \lambda \leq \infty\}$ are identical in that for any value of λ there is a unique ρ such that $\mu_\lambda(\cdot) = \mu(\cdot;\rho)$ and conversely. If $\rho \leq \rho_0$, then $J(\mu(\cdot;\rho)) = \rho$.*

Theorem 5.1 has the consequence that the solution to the constrained minimization problem posed by (5.4)–(5.5) is a smoothing spline estimator of μ corresponding to some value of λ. On the other hand, the choice of a particular value for λ corresponds to the assumption that $J_m(\mu) \leq \rho_\lambda$ with $\rho_\lambda = J_m(\mu_\lambda)$ and, hence, reflects our beliefs about the magnitude of the remainder terms $\text{Rem}(t_i), i = 1, \ldots, n$. Since the choice $\rho = 0$ produces a polynomial regression estimator, it follows that smoothing splines give an extension of polynomial regression that attempts to guard against departures from an idealized polynomial regression model. Li (1982) takes this farther and shows that smoothng splines are minimax estimators that actually provide protection against certain "worst case" departures from polynomial models.

The origin of smoothing splines appears to lie in work on graduating data by Whittaker (1923). However, spline smoothing techniques were generally regarded as numerical analysis methods until extensive research by Grace Wahba demonstrated their utility for solving a host of statistical estimation problems. It has now become clear that smoothing splines, and their variants, provide extremely flexible data analysis tools. As a result, they have become quite popular and have found applications in such diverse areas as the analysis of growth data, medicine, remote sensing experiments and economics. Extensive developments of spline smoothing methods and related techniques can now be found in several books including Wahba (1990) and Green and Silverman (1994).

In the next section we derive the form of the minimizer of (5.1) and thereby justify the "spline" part of its name. Methods for selecting a value for the smoothing parameter are then described in Section 5.3 with computational techniques being the subject of Section 5.4. The large sample properties of spline smoothers are examined in Section 5.5. Smoothing splines can also be derived through a Bayesian extension of polynomial regression. This is established in Section 5.6 and utilized to motivate one approach to interval estimation and diagnostic analysis. Smoothing spline variants for estimation in other data smoothing problems are discussed in Section 5.7.

5.2 Form of the Estimator

In this section we will obtain an explicit expression for the smoothing spline estimator. To avoid clouding the issue it will be useful to first deal with the case where the t_i are distinct with $0 \leq t_1 < \cdots < t_n \leq 1$. Also we will assume that a value for λ has been selected so that the smoothing parameter value is fixed. All these restrictions will be relaxed presently.

To begin note that the smoothing spline estimator is necessarily a natural spline of order $2m$ with knots at $t_1, \ldots, t_n$. More precisely, μ_λ is a $2m$th order piecewise polynomial with $2m-2$ continuous derivatives that consists of different polynomial segments over each of the intervals $[t_i, t_{i+1}], i = 1, \ldots, n-1$, and is a polynomial of order m outside of $[t_1, t_n]$. This follows from Lemma 5.2 which ensures us that for any function $f \in W_2^m[0,1]$ criterion (5.1) can only become smaller if we replace f by the natural spline which agrees with f at the design points.

Since μ_λ is a natural spline, the problem of minimizing (5.1) over all functions in $W_2^m[0,1]$ reduces to the finite dimensional problem of minimization over the n dimensional set of natural splines. This results in the following theorem that gives a closed form for the estimator.

Theorem 5.2. *Let $x_1, \ldots, x_n$ be a basis for the set of natural splines of order $2m$ with knots at $t_1, \ldots, t_n$ and define $\mathbf{X} = \{x_j(t_i)\}_{i,j=1,n}$. Then, if $n \geq m$ the unique minimizer of (5.1) is $\mu_\lambda = \sum_{j=1}^n b_{\lambda j} x_j$, where $\mathbf{b}_\lambda = (b_{\lambda 1}, \ldots, b_{\lambda n})^T$ is the unique solution with respect to $\mathbf{c} = (c_1, \ldots, c_n)^T$ of the equation system*

$$(\mathbf{X}^T\mathbf{X} + n\lambda\boldsymbol{\Omega})\mathbf{c} = \mathbf{X}^T\mathbf{y}, \tag{5.6}$$

for $\mathbf{y} = (y_1, \ldots, y_n)^T$ the response vector and

$$\boldsymbol{\Omega} = \left\{ \int_0^1 x_i^{(m)}(t) x_j^{(m)}(t) dt \right\}_{i,j=1,n}. \tag{5.7}$$

Proof. Lemma 5.2 in Section 5.8 has the consequence that the residual sum-of-squares portion of criterion (5.1) is unaffected while $J_m(f)$ is reduced if we replace f in the criterion by the natural spline s satisfying $s(t_i) = f(t_i), i = 1, \ldots, n$. Thus, we may minimize (5.1) over functions of the form $s = \sum_{j=1}^n c_j x_j$ to obtain the normal equations (5.6). From Lemma 5.2 the matrix $\mathbf{X}$ has rank n and $J_m(s) = \mathbf{c}^T\boldsymbol{\Omega}\mathbf{c} \geq 0$ for all $\mathbf{c}$ entails that $\boldsymbol{\Omega}$ is positive semidefinite. Thus, the system (5.6) has a unique solution. •

Using (5.6) the vector of fitted values corresponding to the smoothing spline estimator is seen to be

$$\boldsymbol{\mu}_\lambda = (\mu_\lambda(t_1), \ldots, \mu_\lambda(t_n))^T = \mathbf{S}_\lambda \mathbf{y}, \tag{5.8}$$

where

$$\mathbf{S}_\lambda = \mathbf{X}(\mathbf{X}^T\mathbf{X} + n\lambda\boldsymbol{\Omega})^{-1}\mathbf{X}^T. \tag{5.9}$$

By analogy with linear regression terminology we will call $\mathbf{S}_\lambda$ the *hat matrix*. It is clear that $\mathbf{S}_\lambda$ is symmetric and positive definite and, hence, smoothing splines fall into the class of estimators discussed in Chapter 2. This has important consequences for the selection of λ that will be discussed in Section 5.3.

Equation (5.8) resembles the form one would have for fitted values from ordinary linear regression involving the "predictor variables" $x_1, \ldots, x_n$. In fact, by recalling our work in Section 2.1, we see that the smoothing spline estimator can be viewed as a type of generalized ridge regression estimator. This seeming connection with ridge regression is more a function of the shared Bayesian heritage of the two estimators (see Section 5.6) than any real relationship between the methodologies which actually stem from rather disparate estimation problems. In any case, the formal similarity between ridge regression and smoothing splines provides very little immediate help in understanding the properties of spline smoothers.

To gain some insight into how smoothing splines work, we will develop a canonical representation for the estimators which makes their properties more transparent. This is accomplished through a judicious choice of the basis elements $x_1, \ldots, x_n$ in Theorem 5.2 that simultaneously diagonalizes $\mathbf{X}^T\mathbf{X}$ and $\mathbf{\Omega}$ in (5.9). The resulting basis, due to Demmler and Reinsch (1975), is primarily a theoretical and motivational tool. More computationally expedient bases will be discussed in Section 5.4.

The Demmler-Reinsch basis functions appear to admit a simple closed form only in the special case of $m = 1$ and a uniform design. We will therefore deal with that case first and then use it to facilitate discussions of more general settings. Thus, suppose for the moment that we have data from model (5.2) at design points

$$t_i = (2i - 1)/2n, \ i = 1, \ldots, n, \tag{5.10}$$

and estimate $\mu \in W_2^1[0,1]$ by the minimizer μ_λ of

$$n^{-1} \sum_{i=1}^{n} (y_i - f(t_i))^2 + \lambda \int_0^1 f'(t)^2 dt \tag{5.11}$$

over $f \in W_2^1[0,1]$. This estimator is a linear smoothing spline and, from Theorem 5.2, $\mu_\lambda = \sum_{j=1}^{n} b_{\lambda j} x_j$, where $x_1, \ldots, x_n$ is a basis for the natural linear splines with knots at $t_1, \ldots, t_n$ and $\mathbf{b}_\lambda = (b_{\lambda 1}, \ldots, b_{\lambda n})^T$ is the solution of (5.6).

For our basis x_j, $j = 1, \ldots, n$, we will use the natural linear splines that interpolate the constant and the functions $\sqrt{2}\cos(j\pi t), j = 1, \ldots, n-1$, at the design points (5.10). These functions are given explicitly by $x_1(t) \equiv 1$ and

$$x_{j+1}(t) = \begin{cases} \sqrt{2}\cos(j\pi t_1), & 0 \le t < t_1, \\ \sqrt{2}\cos(j\pi t_i) + & \\ \sqrt{2}\frac{t-t_i}{t_{i+1}-t_i}[\cos(j\pi t_{i+1}) - \cos(j\pi t_i)], & t_i \le t < t_{i+1}, \\ & i = 1, \ldots, n-1, \\ \sqrt{2}\cos(j\pi t_n), & t_n \le t \le 1, \end{cases}$$

for $j = 1, \ldots, n-1$. Notice that the x_j are all natural linear splines in that they are all continuous, constant outside of $[t_1, t_n]$ and linear over each subinterval $[t_i, t_{i+1}], i = 1, \ldots, n-1$. In addition, they have the desired interpolation properties in that $x_j(t_i) = \sqrt{2}\cos((j-1)\pi t_i), i = 1, \ldots, n$, for $j = 2, \ldots, n$. The fact that the matrix

$$\mathbf{X} = [\mathbf{1}|\sqrt{2}\{\cos(j\pi t_i)\}_{i=1,n,j=1,n-1}]$$

satisfies $\mathbf{X}^T\mathbf{X} = \mathbf{X}\mathbf{X}^T = n\mathbf{I}$ ensures that $x_1, \ldots, x_n$ actually provide a basis.

While our particular choice for the x_j has a diagonal $\mathbf{X}^T\mathbf{X}$ matrix, it also turns out that the $\boldsymbol{\Omega}$ matrix in (5.7) is diagonal for this basis. To see this, differentiate the piecewise linear basis functions to obtain

$$\int_0^1 x_1'(t)x_j'(t)dt = 0,\ j = 1, \ldots, n,$$

and, using Lemma 3.5,

$$\begin{aligned} \int_0^1 x_{i+1}'(t)x_{j+1}'(t)dt &= 2n\sum_{r=1}^{n-1}[\cos(i\pi t_{r+1}) - \cos(i\pi t_r)] \\ &\quad \times [\cos(j\pi t_{r+1}) - \cos(j\pi t_r)] \\ &= 8n\sin\left(\frac{i\pi}{2n}\right)\sin\left(\frac{j\pi}{2n}\right)\sum_{r=1}^{n}\sin\left(\frac{r}{n}i\pi\right)\sin\left(\frac{r}{n}j\pi\right) \\ &= \delta_{ij}\gamma_j,\ i, j = 1, \ldots, n-1, \end{aligned}$$

with

$$\gamma_j = \left(2n\sin\left(\frac{j\pi}{2n}\right)\right)^2, \; j = 1,\ldots,n-1. \tag{5.12}$$

The γ_j are called the Demmler-Reinsch eigenvalues and the x_j are called the Demmler-Reinsch basis functions. Note that the γ_j in (5.12) are strictly increasing over $j \leq n$ and that

$$\gamma_j = (j\pi)^2(1 + o(1)) \tag{5.13}$$

as $n \to \infty$ for any fixed j.

Under the Demmler-Reinsch basis equation system (5.6) becomes

$$[n\mathbf{I} + n\lambda \text{diag}(0, \gamma_1, \ldots, \gamma_{n-1})]\mathbf{c} = n\mathbf{b},$$

where $\mathbf{b} = (b_1, \ldots, b_n)^T$ is the vector of sample cosine Fourier coefficients (3.42). That is, $b_1 = \bar{y}$, the response average, and

$$b_j = \frac{\sqrt{2}}{n}\sum_{i=1}^{n} y_i \cos((j-1)\pi t_i) \;, \quad j = 2, \ldots, n.$$

Thus, for any specific value of λ the linear smoothing spline is given by

$$\mu_\lambda = b_1 + \sum_{j=2}^{n} \frac{b_j}{1 + \lambda\gamma_{j-1}} x_j \;. \tag{5.14}$$

In particular, at the design points we have

$$\mu_\lambda(t_i) = b_1 + \sum_{j=2}^{n} \frac{b_j}{1 + \lambda\gamma_{j-1}} \sqrt{2}\cos((j-1)\pi t_i), \; i = 1, \ldots, n.$$

This expression makes it clear that the linear smoothing spline is essentially a type of weighted cosine series estimator. It smooths the data in a similar manner to that of kernel estimators by using information from all of the sample Fourier coefficients but weights them by a damping factor (i.e., $(1 + \lambda\gamma_j)^{-1}$) that decreases as the frequency increases for fixed λ. The smoothing parameter λ then controls the mix of high and low frequency information that is used in estimating μ. As $\lambda \to \infty$ damping becomes so severe that the estimator reduces to the average $\bar{y}$. The other extreme,

$\lambda = 0$, produces an estimator that interpolates the data in that $\mu_0(t_i) = y_i, i = 1, \ldots, n$. This corresponds to no smoothing of the data at all and an estimator that uses all the frequency information from the data without damping.

Figure 5.1 gives a plot of linear smoothing spline fits to a set of data simulated from model (5.2) using normal random errors with $\sigma = .05$ and regression function $\mu(t) = 16t^2(1-t)^2 + 32(t-.7)^3 I_{\{t \geq .7\}}(t)$ for $I_A(t)$ the indicator function for t falling in the set A. The fits correspond to the values $\lambda = .006, \lambda = .06$ and $\lambda = .6$ for the smoothing parameter. As expected from our discussion above, the smaller value $\lambda = .006$ produces an estimator that is more slavish to the data while the larger value $\lambda = .6$ gives a very smooth fit that ignores many of the data's features. The middle choice $\lambda = .06$ then represents a compromise that gives a better balance between the high and low frequency information being used by the estimator.

We have now seen that the linear smoothing spline is similar to a kernel estimator in that it can be represented as a damped cosine series estimator. One might therefore wonder if the connection is stronger in that there might be some kind of asymptotic equivalence between the two estimator types. By interchanging the order of summation in (5.14) we find that

$$\mu_\lambda(t) = n^{-1} \sum_{i=1}^{n} y_i K_n(t, t_i; \lambda)$$

with

$$K_n(t, s; \lambda) = 1 + \sqrt{2} \sum_{j=1}^{n-1} \frac{\cos(j\pi s) x_{j+1}(t)}{1 + \lambda \gamma_j} \tag{5.15}$$

for the x_j and γ_j the Demmler-Reinsch basis functions and eigenvalues. Plots of the linear smoothing spline weight function $K_n(.5, s; \lambda)$ are shown in Figure 5.2 for the three values of λ that were used with the estimators in Figure 5.1. These functions look very much like one would expect from a kernel estimator and we will see in Section 5.5 that $K_n(t, \cdot; \lambda)$ can be well approximated by $K((t - \cdot)/\sqrt{\lambda})/\sqrt{\lambda}$ with $K(u) = .5\exp\{-|u|\}$ the Laplace density on the line. This has the implication that for a uniform design the linear smoothing spline is essentially a kernel estimator that uses a Laplace

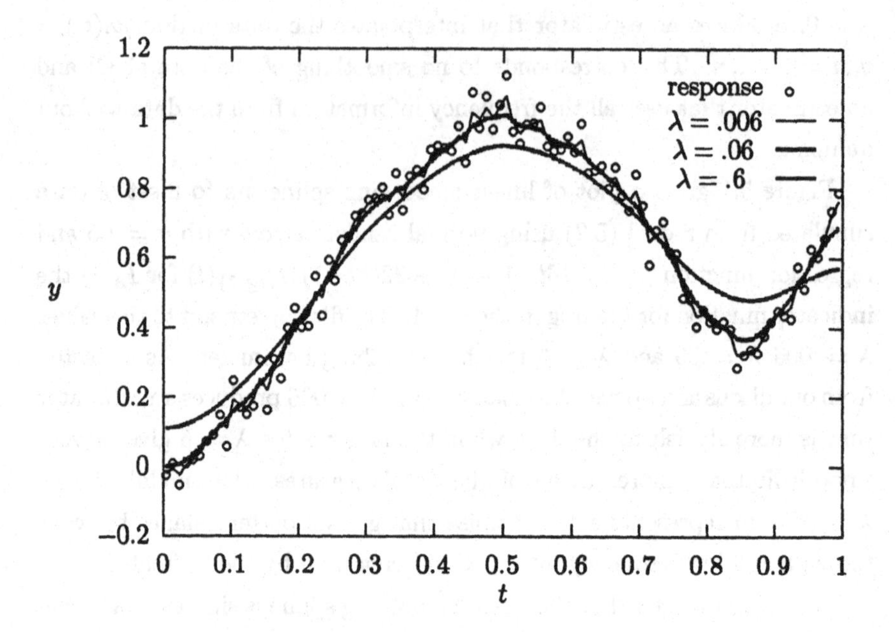

Figure 5.1: Linear Smoothing Spline Fits to Simulated Data.

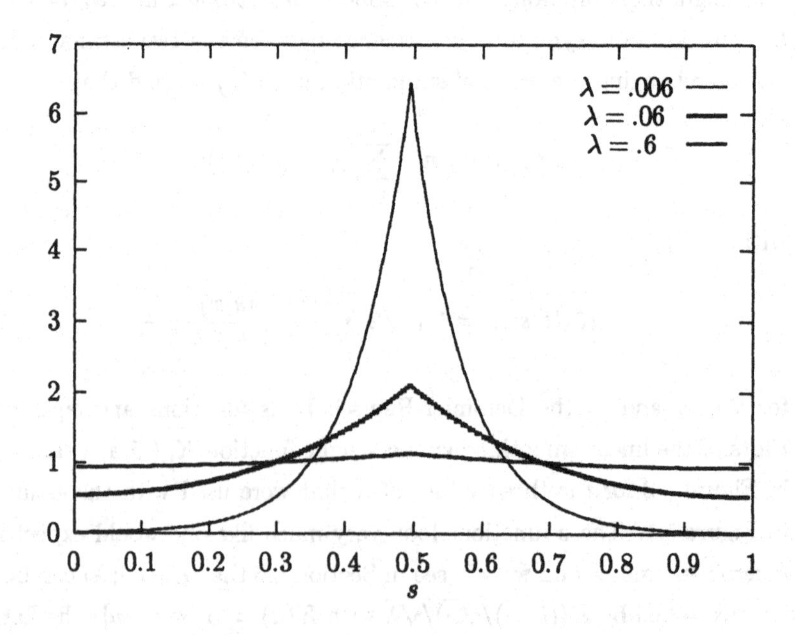

Figure 5.2: Linear Smoothing Spline Weight Functions.

density kernel and a bandwidth that is the square root of the smoothing parameter. Similar conclusions hold for other choices of m and general designs that will be discussed in Section 5.5.

For general m and/or nonuniform designs, there is no simple form for the Demmler-Reinsch basis functions and eigenvalues. Nevertheless, their properties are known to the extent that we can draw some useful parallels with the $m = 1$, uniform design scenario. For example, Demmler and Reinsch (1975) show that in general a natural spline basis $x_1, \ldots, x_n$ may be chosen so that

i) $x_1, \ldots, x_m$ span the space of polynomials of order m,

ii) the function x_j has at least $j-1$ sign changes in $(0, 1)$,

iii) $\mathbf{X}^T\mathbf{X} = n\mathbf{I} = \mathbf{X}\mathbf{X}^T$

and

iv) $\boldsymbol{\Omega} = \mathrm{diag}(\underbrace{0, \ldots, 0}_{m}, \gamma_1, \ldots, \gamma_{n-m})$ for eigenvalues $0 < \gamma_1 \leq \cdots \leq \gamma_{n-m}$.

It has also been shown (Nussbaum 1985, Speckman 1985 and Golubev and Nussbaum 1990) that as $n \to \infty$

v) $\gamma_j = C(j\pi)^{2m}(1 + o(1))$ for C a constant that depends only on m and the design.

Using properties i)-iv) the general Demmler-Reinsch representation for the mth order smoothing spline derived from (5.1) is

$$\mu_\lambda = \sum_{j=1}^{m} b_j x_j + \sum_{j=m+1}^{n} \frac{b_j}{1 + \lambda\gamma_{j-m}} x_j \,, \tag{5.16}$$

with $b_j = n^{-1} \sum_{i=1}^{n} y_i x_j(t_i), j = 1, \ldots, n$, the Demmler-Reinsch Fourier coefficients. Unlike the linear smoothing spline case, the b_j will not be cosine Fourier coefficients in general. However, the sign change property ii) for the basis functions parallels that of cosine functions. This allows us to continue to view the $b_j, j = 1, \ldots, n$, as providing a partitioning of the frequency content of the data with larger values of the coefficient index signifying higher frequencies. Combining this interpretation with property v) then gives us essentially the same conclusion as for the linear smoothing spline case: namely, a smoothing spline is a type of damped series estimator with λ controlling the relative amount of low and high frequency information that is used in estimating μ.

Notice that (5.16) reveals that smoothing splines consist of two distinct parts: a projection part $\sum_{j=1}^{m} b_j x_j$ that is the least-squares fit to the data by a polynomial of order m and a term $\sum_{j=m+1}^{n} b_j x_j/(1 + \lambda\gamma_{j-m})$ that estimates the departure from a polynomial model. Thus, the choice of $\lambda = \infty$ reduces the estimator to only its polynomial or projection part and gives the smoothest (as measured by J_m in (5.3)) possible fit to the data.

To conclude this section, we note that parallels of Theorem 5.2 can be established for various other versions of our original smoothing criterion. For example, (5.1) is most appropriate if each observation has the same variance. If this is not the case the criterion can be modified to take this into account. A more general smoothing criterion that arises from such considerations is

$$n^{-1}\sum_{i=1}^{n} w_i\,(y_i - f(t_i))^2 + \lambda\int_0^1 f^{(m)}(t)^2 dt, \tag{5.17}$$

with positive weights $w_i > 0, i = 1, \ldots, n$. Upon taking $w_i = [\mathrm{Var}(y_i)]^{-1}, i = 1, \ldots, n$, (5.17) becomes suitable for heteroskedastic observations.

Concerning the minimization of (5.17) we state the following result.

Theorem 5.3. *Under the notation and conditions of Theorem 5.2, the unique minimizer of (5.17) is* $\mu_\lambda = \sum_{j=1}^{n} b_{\lambda j} x_j$*, where* $\mathbf{b}_\lambda = (b_{\lambda 1}, \ldots, b_{\lambda n})^T$ *is the unique solution with respect to* $\mathbf{c}$ *of the equation system*

$$(\mathbf{X}^T\mathbf{W}\mathbf{X} + n\lambda\boldsymbol{\Omega})\mathbf{c} = \mathbf{X}^T\mathbf{y}$$

with $\mathbf{W} = \mathrm{diag}(w_1, \ldots, w_n)$.

Proof. Exercise 3.

Theorem 5.3 provides a key to determining how to handle repeated observations. Suppose that there are $n_i \geq 1$ responses at the design point

t_i and let $\bar{y}_i$ be their average. Then, minimization of criterion (5.1) is equivalent to minimizing

$$N^{-1}\sum_{i=1}^{n} n_i\left(\bar{y}_i - f(t_i)\right)^2 + \lambda\int_0^1 f^{(m)}(t)^2 dt,$$

where $N = \sum_{i=1}^{n} n_i$ (Exercise 4). Consequently, the smoothing spline estimator in this case is a natural spline with coefficients found by solving,

$$(\mathbf{X}^T\mathbf{W}\mathbf{X} + N\lambda\mathbf{\Omega})\mathbf{c} = \mathbf{X}^T\bar{\mathbf{y}}$$

with $\mathbf{W} = \text{diag}(n_1, \ldots, n_n)$ and $\bar{\mathbf{y}} = (\bar{y}_1, \ldots, \bar{y}_n)^T$.

5.3 Selection of λ

To this point we have dodged the issue of how to select a value for λ in (5.1). Thus, we now turn to the problem of selecting a suitable level of smoothing for a set of data.

There are a variety of ways to choose a value for the smoothing parameter including trial and error where arbitrary values of λ are tried until one is found which gives a visually satisfactory fit. This can be time consuming and, accordingly, it may be preferable to use some data driven method for selecting λ if for no other reason than to provide a starting value that can be used for further fine tuning.

The smoothing spline is a linear estimator with its corresponding hat matrix $\mathbf{S}_\lambda$ in (5.9) being symmetric and positive definite. Thus, our work in Chapter 2 can be used to suggest methods for selecting λ from the data. Specific selection methods would include minimization of the unbiased risk criterion

$$\widehat{P}(\lambda) = n^{-1}RSS(\mu_\lambda) + \frac{2\widehat{\sigma}^2}{n}\text{tr}\mathbf{S}_\lambda \tag{5.18}$$

with $\widehat{\sigma}^2$ being, for example, the GSJS estimator (2.29)-(2.30). Other possibilities are minimization of the cross-validation criterion

$$\text{CV}(\lambda) = n^{-1}\sum_{i=1}^{n}\left(y_i - \mu_{\lambda(i)}(t_i)\right)^2 \tag{5.19}$$

or generalized cross-validation (GCV) criterion

$$\mathrm{GCV}(\lambda) = \frac{nRSS(\mu_\lambda)}{\mathrm{tr}(\mathbf{I} - \mathbf{S}_\lambda)^2}, \tag{5.20}$$

where $\mu_{\lambda(i)}$ is the estimator obtained from (5.1) when the ith observation (t_i, y_i) is deleted from the data. As was true for ridge regression, $\mathrm{GCV}(\lambda)$ and $\mathrm{CV}(\lambda)$ are closely related with one being essentially a weighted version of the other (Exercise 6).

Efficient methods for computing (5.18)-(5.20) will be discussed in the next section. Given such methods, the corresponding criterion's minimizer can be found using algorithms for minimization of nonlinear functions. Methods which do not require the use of derivatives are to be preferred here and the golden section algorithm (see, e.g., Kennedy and Gentle 1980) is one such method that seems to have been successful in this context. It is sometimes feasible to conduct the minimization through a global search by evaluating a particular criterion over a grid of λ or, more typically, $\log \lambda$ values. This has the advantage of making it easy to produce plots of the criterion function that allow for the identification of local minima and other characteristics of possible interest.

The values of λ selected from (5.18)-(5.20) provide estimators of the smoothing level that minimizes the loss $L(\lambda) = n^{-1} \sum_{i=1}^{n} (\mu(t_i) - \mu_\lambda(t_i))^2$ and/or estimation risk $EL(\lambda)$ corresponding to μ_λ. In Section 5.5 we will see that spline smoothing is asymptotically equivalent to a variant of kernel smoothing with $\lambda^{1/2m}$ being essentially a bandwidth. Thus, our discussions in Section 4.5 would lead us to anticipate consistency properties for data driven choices of λ as estimators of the minimizers of the risk or loss. Various results of this type can be found in Craven and Wahba (1979), Cox (1983), Li (1986) and Nychka (1991). For example, Li (1986) shows that $L(\widehat{\lambda})/\inf_{\lambda>0} L(\lambda) \xrightarrow{P} 1$ as $n \to \infty$ with $\widehat{\lambda}$ either the unbiased risk or GCV estimator of the smoothing parameter. Nychka (1991) has established a parallel of the normal limiting distribution for kernel bandwidth estimators (see Section 4.5) for GCV estimators of λ with cubic smoothing splines.

Figures 5.3 and 5.4 show results from fitting the data in Figure 5.1 with a linear smoothing spline. Values of the GCV and unbiased risk (UBR)

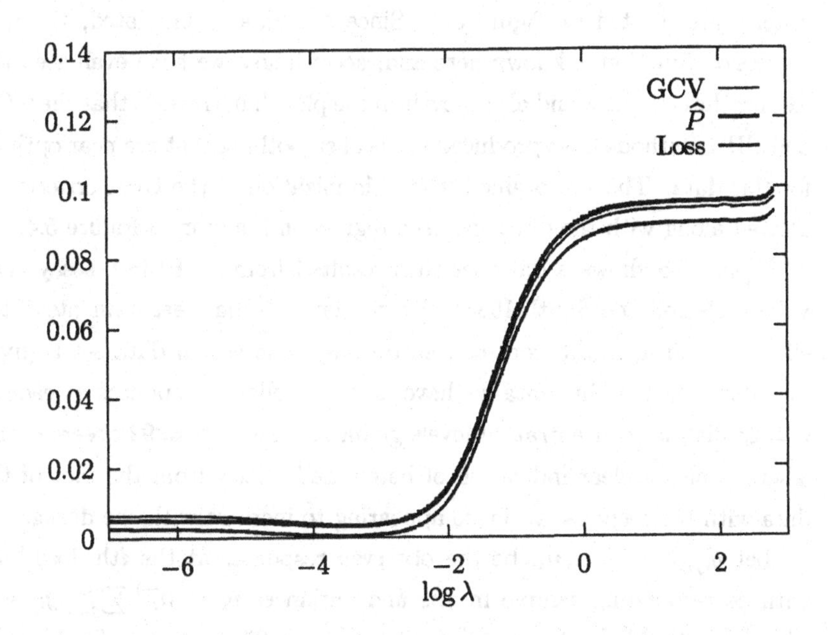

Figure 5.3: GCV and UBR Criteria for Simulated Data.

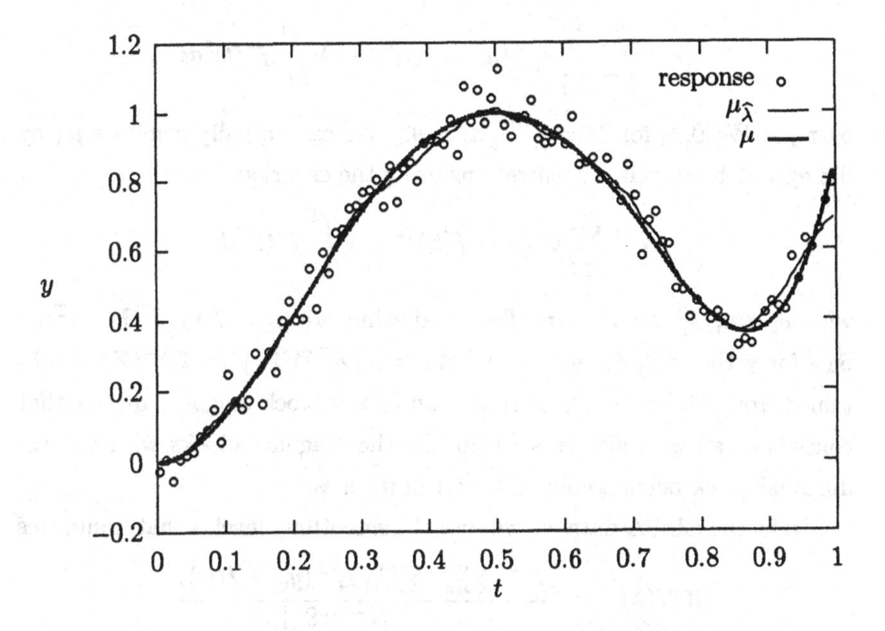

Figure 5.4: Linear Smoothing Spline Fit to Simulated Data.

criteria are plotted in Figure 5.3. Since the data is simulated, the true regression function is known here and, accordingly, we have evaluated the loss for the estimator and also show it in the plot. This reveals that the GCV and UBR methods have produced levels of smoothing that are near optimal for this data. The fit obtained from minimization of the UBR criterion is plotted along with the data and true regression function in Figure 5.4.

Figure 5.5 shows a data set that resulted from an ELISA assay (Davidian, Carroll and Smith 1988) with responses being measurements of the effect of a drug at different concentrations. The actual data set is given in Table 4.5. For this data we have $n_i = 4$ replicate responses at each of $n = 23$ distinct concentration levels giving a total of $N = 92$ observations. There is also a clear indication of heteroskedasticity from the plot of the data with the response variance appearing to increase with the dosage.

Let $y_{ij}, j = 1, \ldots, n_i$, be the observed responses at the ith dose level with corresponding sample means and variances $\bar{y}_i = n_i^{-1} \sum_{j=1}^{n_i} y_{ij}$ and $s_i^2 = (n_i - 1)^{-1} \sum_{j=1}^{n_i} (y_{ij} - \bar{y}_i)^2$ for $i = 1, \ldots, n$. Then, we will fit this data by the cubic smoothing spline μ_λ in Theorem 5.3 that minimizes

$$N^{-1} \sum_{i=1}^{n} \sum_{j=1}^{n_i} s_i^{-2} \left(y_{ij} - f(t_i)\right)^2 + \lambda \int_0^1 f''(t)^2 dt$$

over $f \in W_2^2[0,1]$ for $N = \sum_{i=1}^{n} n_i = 92$. We can actually compute μ_λ by fitting just the mean concentrations using the criterion

$$N^{-1} \sum_{i=1}^{n} w_i \left(\bar{y}_i - f(t_i)\right)^2 + \lambda \int_0^1 f''(t)^2 dt$$

with $w_i = n_i / s_i^2, i = 1, \ldots, n$. The fitted values are $\boldsymbol{\mu}_\lambda = \mathbf{J}\mathbf{A}_\lambda(\mathbf{J}^T\mathbf{J})^{-1}\mathbf{J}^T\mathbf{y} = \mathbf{S}_\lambda\mathbf{y}$ for $\mathbf{y}$ the response vector and $\mathbf{A}_\lambda = \mathbf{X}(\mathbf{X}^T\mathbf{W}\mathbf{X} + N\lambda\boldsymbol{\Omega})^{-1}\mathbf{X}^T\mathbf{W}$ obtained from Theorem 5.3. Here $\mathbf{J}$ is an $N \times n$ block diagonal matrix that consists of all zero elements except for the diagonal blocks with the ith diagonal block being a column vector of n_i ones.

To fit the ELISA data we will use the smoothing level $\hat{\lambda}$ that minimizes

$$\begin{aligned} \mathrm{GCV}(\lambda) &= \frac{N^{-1} \sum_{i=1}^{n} \sum_{j=1}^{n_i} s_i^{-2} \left(y_{ij} - f(t_i)\right)^2}{[1 - N^{-1}\mathrm{tr}\mathbf{S}_\lambda]^2} \\ &= \frac{N - n + \sum_{i=1}^{n} n_i s_i^{-2} \left(\bar{y}_i - \mu_\lambda(t_i)\right)^2}{N[1 - N^{-1}\mathrm{tr}\mathbf{A}_\lambda]^2} \end{aligned} \tag{5.21}$$

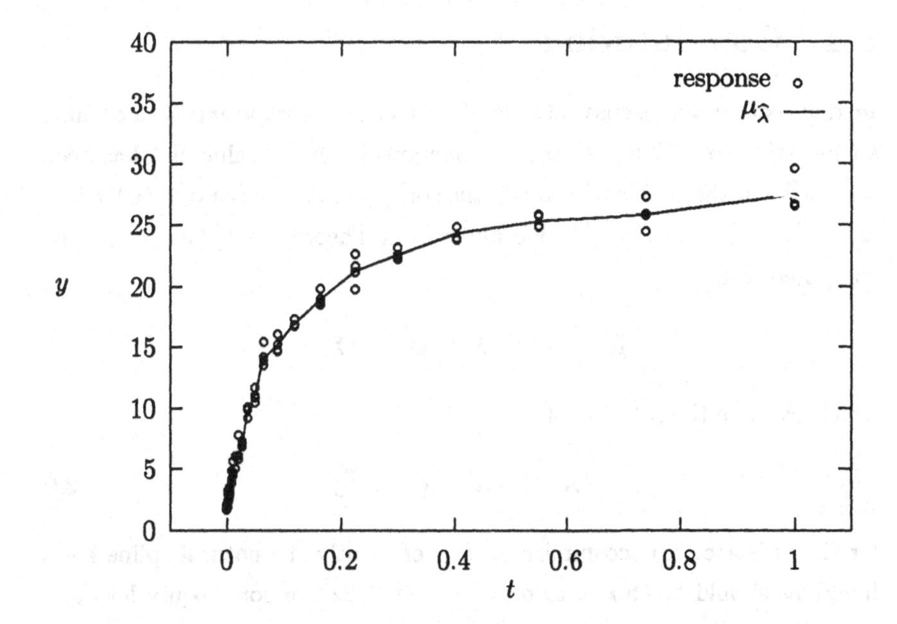

Figure 5.5: Smoothing Spline Fit to ELISA Data.

(Exercise 7). The resulting estimator of μ is shown in Figure 5.5. The small sample variances at low dose levels have produced large weights for their corresponding responses in the estimation criterion and this is reflected in the slavish nature of the fit at low concentrations.

Criterion (5.1) also involves the parameter m that must be chosen in some fashion. If certain smoothness properties are known or believed to hold for μ this can suggest a value for m. For example, the standard choice of $m = 2$ for a cubic smoothing spline corresponds to the assumption that μ is continuously differentiable with a square integrable second derivative. If a particular mth order polynomial model is being considered for μ this also provides a natural choice for m. A data selected value of $\lambda = \infty$ would then be viewed as an indication of the appropriateness of the candidate model. If one is unwilling to choose a value for m it may be included as an additional parameter for selection by the GCV or UBR criteria.

5.4 Computation

In this section we discuss methods for efficiently computing a smoothing spline estimator. Thus, assume for the moment that a value of λ has been specified and the goal is to evaluate the corresponding vector of fitted values $\boldsymbol{\mu}_\lambda = (\mu_\lambda(t_1), \ldots, \mu_\lambda(t_n))^T$. We know from Theorem 5.2 that $\boldsymbol{\mu}_\lambda$ can be computed from

$$\boldsymbol{\mu}_\lambda = \mathbf{X}(\mathbf{X}^T\mathbf{X} + n\lambda\boldsymbol{\Omega})^{-1}\mathbf{X}^T\mathbf{y}$$

which requires the solution of

$$(\mathbf{X}^T\mathbf{X} + n\lambda\boldsymbol{\Omega})\mathbf{c} = \mathbf{X}^T\mathbf{y} \tag{5.22}$$

for the vector $\mathbf{c}$. To accomplish all this efficiently the natural spline basis functions should be chosen to make $\mathbf{X}$ and (5.22), or some equivalent system, band limited and thereby allow the fitted values to be computed in $O(n)$ calculations. Specific ways to obtain such band limited structures are given in Anselone and Laurent (1968), Reinsch (1967, 1971) and Lyche and Schumaker (1973).

Perhaps the most important case is for $m = 2$ which gives the popular cubic smoothing spline. In that instance de Boor (1978) uses a piecewise polynomial representation for the estimator to show that

$$\boldsymbol{\mu}_\lambda = \mathbf{y} - n\lambda\mathbf{Q}(n\lambda\mathbf{Q}^T\mathbf{Q} + \mathbf{B})^{-1}\mathbf{Q}^T\mathbf{y}, \tag{5.23}$$

where $\mathbf{Q}^T$ is an $(n-2) \times n$ tridiagonal matrix with ith row

$$(\underbrace{0, \ldots, 0}_{i-1}, \frac{1}{t_{i+1} - t_i}, -\frac{1}{t_{i+2} - t_{i+1}} - \frac{1}{t_{i+1} - t_i}, \frac{1}{t_{i+2} - t_{i+1}}, \underbrace{0, \ldots, 0}_{n-i-2}) \tag{5.24}$$

and $6\mathbf{B}$ is a symmetric, $(n-2) \times (n-2)$, tridiagonal matrix having first and last rows $(t_2 - t_1, t_3 - t_2, \underbrace{0, \ldots, 0}_{n-4})$ and $(\underbrace{0, \ldots, 0}_{n-4}, t_{n-1} - t_{n-2}, t_n - t_{n-1})$, and with ith row

$$(\underbrace{0, \ldots, 0}_{i-2}, t_{i+1} - t_i, 2(t_{i+2} - t_i), t_{i+2} - t_{i+1}, \underbrace{0, \ldots, 0}_{n-i-3}) \tag{5.25}$$

for $i = 2, \ldots, n-3$. The fitted values for the cubic smoothing spline can therefore be obtained in $O(n)$ operations by first solving the 5-banded system

$$(n\lambda \mathbf{Q}^T\mathbf{Q} + \mathbf{B})\mathbf{c} = \mathbf{Q}^T\mathbf{y} \tag{5.26}$$

and then using $\boldsymbol{\mu}_\lambda = \mathbf{y} - n\lambda\mathbf{Q}\mathbf{c}$. FORTRAN code which implements this approach using a Cholesky decomposition to solve (5.26) is provided in de Boor (1978).

Given an $O(n)$ algorithm for computing $\boldsymbol{\mu}_\lambda$, the question then becomes how to efficiently compute the smoothing parameter selection criteria (5.18)-(5.20). If the GCV or unbiased risk criteria are to be used, then this will require the computation of both the residual sum-of-squares and trace of the smoothing spline hat matrix corresponding to any particular value of λ. While the residual sum-of-squares can be evaluated in $O(n)$ operations once $\boldsymbol{\mu}_\lambda$ is available, it is not so clear how to obtain the value of $\mathrm{tr}\mathbf{S}_\lambda$. Nevertheless, several $O(n)$ algorithms exist for evaluating this trace either exactly or approximately. See, e.g., Utreras (1981), Silverman (1984a), Hutchinson and de Hoog (1985, 1987), Ansley and Kohn (1987) and Girard (1989).

For band limited settings, one approach to computing $\mathrm{tr}\mathbf{S}_\lambda$ can be based on a result from Hutchinson and de Hoog (1985). They consider the situation where one has an $N \times N$, $2k+1$ banded, symmetric matrix $\mathbf{C}$ with Cholesky decomposed form $\mathbf{C} = \mathbf{A}^T\mathbf{D}^{-1}\mathbf{A}$ for $\mathbf{D} = \mathrm{diag}(d_1, \ldots, d_N)$ a diagonal matrix and $\mathbf{A} = \{a_{ij}\}$ an upper triangular, band limited matrix of bandwidth $k+1$. They show that the elements of $\mathbf{C}^{-1} = \{c^{ij}\}$ can be computed via a backward recursion algorithm with starting value $c^{NN} = d_n$ and general step

$$c^{i,i+l} = -\sum_{r=1}^{\min(k,N-i)} a_{i,i+r}c^{i+r,i+l} \;, l > 0, \tag{5.27}$$

and

$$c^{i,i} = d_i - \sum_{l=1}^{\min(k,N-i)} a_{i,i+l}c^{i,i+l}. \tag{5.28}$$

Notice that (5.27)-(5.28) have the important consequence that the number of calculations required to compute any element of $\mathbf{C}^{-1}$ depends only on the band size k and not on the dimension of the matrix.

To illustrate how (5.27)-(5.28) can be used for computing a smoothing spline trace, consider the case of a cubic smoothing spline. In that instance, (5.23) has the implication that

$$\mathrm{tr}\mathbf{S}_\lambda = n - n\lambda \mathrm{tr}(n\lambda \mathbf{Q}^T\mathbf{Q} + \mathbf{B})^{-1}\mathbf{Q}^T\mathbf{Q} \tag{5.29}$$

with $\mathbf{Q}$ and $\mathbf{B}$ defined in (5.24) and (5.25), respectively. Computation of $\mathrm{tr}\mathbf{S}_\lambda$ therefore reduces to computation of $\mathrm{tr}(n\lambda\mathbf{Q}^T\mathbf{Q} + \mathbf{B})^{-1}\mathbf{Q}^T\mathbf{Q}$. However, $\mathbf{Q}^T\mathbf{Q}$ is only a 5-banded matrix which means that the trace can be evaluated by computing only the diagonal elements and the elements in the two above-diagonal bands of the symmetric matrix $(n\lambda\mathbf{Q}^T\mathbf{Q}+\mathbf{B})^{-1}$. This can all be accomplished in $O(n)$ operations by, e.g., conducting a Cholesky decomposition of $\mathbf{C} = n\lambda\mathbf{Q}^T\mathbf{Q} + \mathbf{B}$ as $\mathbf{A}^T\mathbf{D}^{-1}\mathbf{A}$ and applying algorithm (5.27)-(5.28) with $N = n-2$ and $k = 2$. The actual leverage values or diagonal elements of $\mathbf{S}_\lambda$ can be obtained similarly (Exercise 9) and the algorithm for computing the trace (and/or leverage values) can be combined with the computation of the estimator so that only one Cholesky decomposition is required for any given λ.

5.5 Large Sample Properties

In this section we discuss some of what is known about the large sample behavior of smoothing splines. As in Section 5.2, we will initially focus on linear smoothing splines and then use this case to motivate more general developments.

Thus, for now, let us take μ_λ to represent the estimator of μ obtained from (5.11) and assume that the design is given by (5.10). The goal is to obtain large sample approximations for quantities such as the global risk

$$R_n(\lambda) = n^{-1}\sum_{i=1}^{n} \mathrm{E}\left(\mu(t_i) - \mu_\lambda(t_i)\right)^2 \tag{5.30}$$

under model (5.2). To accomplish this we will derive a global, kernel type approximation for the linear smoothing spline and then use this to aid in assessing the estimator's risk.

From our work in Section 5.2 we know that the linear smoothing spline can be expressed as $n^{-1}\sum_{i=1}^{n} y_i K_n(\cdot, t_i; \lambda)$ with K_n given by (5.15). We want to obtain a large sample approximation for K_n. To motivate the form of this approximation consider first the special case of $t = t_i$ a design point which gives

$$K_n(t, s; \lambda) = 1 + 2\sum_{j=1}^{n-1} \frac{\cos(j\pi s)\cos(j\pi t)}{1 + \lambda\gamma_j}.$$

Using this in combination with (5.13) leads us to expect that for large n the linear smoothing spline weight function should be given approximately by 1 $+2\sum_{j=1}^{\infty}\cos(j\pi s)\cos(j\pi t)/(1+\lambda(j\pi)^2)$ for large n. This can be simplified using the identities $2\cos(j\pi s)\cos(j\pi t) = \cos(j\pi(t-s)) + \cos(j\pi(t+s))$,

$$\sum_{k=0}^{\infty} \frac{\cos kx}{a^2+k^2} = \frac{\pi}{2a}\frac{e^{-a(\pi-|x|)} + e^{a(\pi-|x|)}}{e^{a\pi} - e^{-a\pi}} + \frac{1}{2a^2},\ |x| \le 2\pi, \tag{5.31}$$

and

$$\sum_{k=1}^{\infty} \frac{\cos kx}{a^2+k^2} = \frac{\pi}{2a}\frac{e^{-a(\pi-|x|)} + e^{a(\pi-|x|)}}{e^{a\pi} - e^{-a\pi}} - \frac{1}{2a^2},\ |x| \le 2\pi \tag{5.32}$$

(cf Gradshteyn and Ryzhik 1980, pg. 40). Specifically, by taking $a = (\sqrt{\lambda}\pi)^{-1}$ in (5.31)-(5.32) we see that $K_n(t, s; \lambda)$ should be approximately equal to

$$K^{+}(t, s; \lambda) = \frac{e^{-|t-s|/\sqrt{\lambda}} + e^{-2/\sqrt{\lambda}}e^{|t-s|/\sqrt{\lambda}} + e^{-(t+s)/\sqrt{\lambda}} + e^{(t+s-2)/\sqrt{\lambda}}}{2\sqrt{\lambda}(1 - e^{-2/\sqrt{\lambda}})}. \tag{5.33}$$

This approximation actually holds globally and not just at design points as described in the following theorem.

Theorem 5.4. *Assume that* $n \to \infty$, $\lambda \to 0$ *in such a way that* $n\lambda \to \infty$. *Then,*

$$K_n(t, s; \lambda) = K^{+}(t, s; \lambda) + O\left(\frac{1}{n\lambda}\right) \tag{5.34}$$

uniformly for $t, s \in [0, 1]$.

Proof. To prove (5.34), begin by observing that $|x_{j+1}(t) - \sqrt{2}\cos(j\pi t)| \leq \sqrt{2}(j\pi/n)^2$ for $t \in [0,1]$. From the Maclaurin series for $\sin(x)$ we then obtain

$$.5(j\pi) < (j\pi)\left(1 - \frac{1}{3!}\left(\frac{j\pi}{2n}\right)^2\right) \leq 2n\sin\left(\frac{j\pi}{2n}\right) \leq j\pi$$

and

$$|2n\sin(j\pi/(2n)) - j\pi| \leq (j\pi)^3/(24n^2)$$

for $1 \leq j \leq n-1$. Therefore,

$$\left|K_n(t,s;\lambda) - \left(1 + 2\sum_{j=1}^{n-1}\frac{\cos j\pi s \cos j\pi t}{1+\lambda\gamma_j}\right)\right| \leq \frac{8}{n\lambda}$$

and

$$\begin{aligned}
&\left|1 + 2\sum_{j=1}^{n-1}\frac{\cos(j\pi s)\cos(j\pi t)}{1+\lambda\gamma_j} - K^+(t,s;\lambda)\right| \\
&\leq \sum_{j=1}^{n-1}\frac{2\lambda|\gamma_j - (j\pi)^2|}{(1+\lambda\gamma_j)(1+\lambda(j\pi)^2)} + \sum_{j=n}^{\infty}\frac{2}{1+\lambda(j\pi)^2} \\
&\leq 2\lambda\sum_{j=1}^{n-1}\frac{|\sqrt{\gamma_j} - j\pi|(\sqrt{\gamma_j} + j\pi)}{(1+(.5)^2\lambda(j\pi)^2)(1+\lambda(j\pi)^2)} + \frac{2}{\lambda}\sum_{j=n}^{\infty}\frac{1}{(j\pi)^2} = O\left(\frac{1}{n\lambda}\right).
\end{aligned}$$

The proof is completed using (5.31)-(5.32). •

We will be most concerned with situations where λ decays to zero like n^{-a} for some $a \in (0,1)$. In that case $e^{-1/\sqrt{\lambda}} = o(n^{-q})$ for any finite $q > 0$ and (5.34) simplifies to

$$K_n(t,s;\lambda) = K(t,s;\lambda) + O\left(\frac{1}{n\lambda}\right) \tag{5.35}$$

uniformly for $t, s \in [0,1]$ with

$$K(t,s;\lambda) = \frac{1}{2\sqrt{\lambda}}\left\{e^{-|t-s|/\sqrt{\lambda}} + e^{-(t+s)/\sqrt{\lambda}} + e^{(t-1+s-1)/\sqrt{\lambda}}\right\}. \tag{5.36}$$

This latter approximation reveals that asymptotically K_n is the sum of an interior weight function $e^{-|t-s|/\sqrt{\lambda}}/(2\sqrt{\lambda})$ and terms $e^{-(t+s)/\sqrt{\lambda}}/(2\sqrt{\lambda})$ and

$e^{(t-1+s-1)/\sqrt{\lambda}}/(2\sqrt{\lambda})$ that give corrections at the lower and upper boundaries. In particular, for large n and fixed $t \in (0,1)$ the weight function behaves like that of a kernel estimator with Laplace kernel $K(u) = e^{-|u|}/2$ and bandwidth $\sqrt{\lambda}$. However, if t is a lower boundary point such as $t = \sqrt{\lambda}q$, for some $q > 0$, then

$$K_n(t,s;\lambda) = (2\sqrt{\lambda})^{-1}\left\{e^{-|t-s|/\sqrt{\lambda}} + e^{-2q}e^{(t-s)/\sqrt{\lambda}}\right\} + O\left(\frac{1}{n\lambda}\right).$$

The term $e^{-2q}e^{(t-s)/\sqrt{\lambda}}$ gives a first order boundary correction which makes the integral of $K(t,\cdot;\lambda)$ asymptotically the same as for t an interior point, i.e.,

$$\int_0^1 (2\sqrt{\lambda})^{-1}\left\{e^{-|t-s|/\sqrt{\lambda}} + e^{-2q}e^{(t-s)/\sqrt{\lambda}}\right\} ds = 1 + O\left(e^{-1/\sqrt{\lambda}}\right). \quad (5.37)$$

A similar result holds for the upper boundary (Exercise 10). Notice that in terms of its formulation the linear smoothing spline assumes only that $\mu \in W_2^1[0,1]$. It is therefore only expected to work as a first order estimator and (5.37) is precisely the type of moment condition that would be required for a first order kernel estimator.

Theorem 5.4 allows us to analyze the pointwise variance and bias of the linear smoothing spline using techniques similar to those employed for kernel estimators in Chapter 4. For this purpose continue to assume that λ is of exact order n^{-a} for $a \in (0,1)$. Then, use approximation (5.36) to see that if μ' is bounded

$$\begin{aligned} \mathrm{E}\mu_\lambda(t) &= n^{-1}\sum_{i=1}^{n}\mu(t_i)K_n(t,t_i;\lambda) \\ &= \int_0^1 \mu(s)K(t,s;\lambda)ds + O\left(\frac{1}{n\lambda}\right) \end{aligned} \quad (5.38)$$

and

$$\begin{aligned} \mathrm{Var}(\mu_\lambda(t)) &= \frac{\sigma^2}{n^2}\sum_{i=1}^{n}K_n^2(t,t_i;\lambda) \\ &= \frac{\sigma^2}{4n\sqrt{\lambda}}\left\{1 + e^{2(t-1)/\sqrt{\lambda}}\left(1 - \frac{2(t-1)}{\sqrt{\lambda}}\right)\right. \\ &+ \left. e^{-2t/\sqrt{\lambda}}\left(1 + \frac{2t}{\sqrt{\lambda}}\right) + o(1)\right\}. \end{aligned} \quad (5.39)$$

Results (5.38)-(5.39) are linear smoothing spline parallels of the variance and bias approximations (4.16) and (4.20) derived for kernel estimators in Chapter 4. Rigorous verification of these approximations is somewhat tedious and we therefore give the details only for the more difficult case of (5.39).

To establish (5.39) first write

$$n^{-1}\sum_{i=1}^{n} K_n^2(t,t_i;\lambda) = n^{-1}\sum_{i=1}^{n} K^2(t,t_i;\lambda)$$
$$+ \quad 2n^{-1}\sum_{i=1}^{n} K(t,t_i;\lambda)[K_n(t,t_i;\lambda) - K(t,t_i;\lambda)] + O\left(\frac{1}{(n\lambda)^2}\right).$$

Now combine $\max_i \max_t |K(t,t_i;\lambda)| = O(1/\sqrt{\lambda})$ with (5.35) to see that

$$|n^{-1}\sum_{i=1}^{n} K(t,t_i;\lambda)[K_n(t,t_i;\lambda) - K(t,t_i;\lambda)]| = O\left(\frac{1}{n\lambda^{3/2}}\right).$$

The next step is to approximate $n^{-1}\sum_{i=1}^{n} K^2(t,t_i;\lambda)$ by an integral. For this purpose set $s_0 = 0$, $s_i = i/n, i = 1,\ldots,n$, and observe that

$$n^{-1}K(t,t_i;\lambda)^2 - \int_{s_{i-1}}^{s_i} K(t,s;\lambda)^2 ds = n^{-1}[K(t,t_i;\lambda)^2 - K(t,\xi_i;\lambda)^2]$$

for mean values ξ_i with $t_i, \xi_i \in [s_{i-1}, s_i]$. Now if $t \notin [s_{i-1}, s_i]$, the Mean Value Theorem gives $|K(t,t_i;\lambda)^2 - K(t,\xi_i;\lambda)^2| \leq 9/(2\lambda^{3/2}n)$. The remaining case of $t \in [s_{i-1}, s_i]$ can occur for only one interval of our partition and this contributes at most $9/(2\lambda)$ to the quadrature error. Thus, we have

$$|\sum_{i=1}^{n}[n^{-1}K(t,t_i;\lambda)^2 - \int_{s_{i-1}}^{s_i} K(t,s;\lambda)^2 ds]| \leq \frac{9}{2}[\frac{1}{\sqrt{\lambda}n}\frac{1}{\lambda} + \frac{1}{n\lambda}] = O(\frac{1}{\sqrt{\lambda}n}\frac{1}{\lambda})$$

uniformly in $t \in [0,1]$. Result (5.39) now follows through direct integration.

One consequence of (5.39) is that for any fixed $t \in (0,1)$ $\mathrm{Var}(\mu_\lambda(t)) \sim \sigma^2/(4n\sqrt{\lambda})$ which is like the interior variance for a kernel estimator with bandwidth $\sqrt{\lambda}$. The estimator's variance remains of order $(n\sqrt{\lambda})^{-1}$ throughout all of $[0,1]$ but increases in the boundary region near 0 and 1 due to

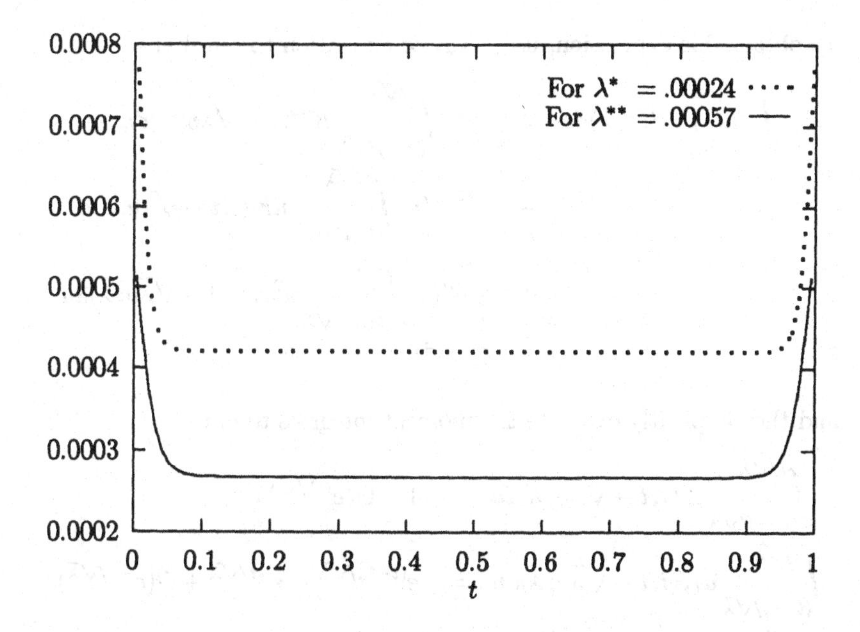

Figure 5.6: Linear Smoothing Spline Variance Functions.

the boundary terms $e^{-2t/\sqrt{\lambda}}\left(1+\frac{2t}{\sqrt{\lambda}}\right)$ and $e^{2(t-1)/\sqrt{\lambda}}\left(1-\frac{2(t-1)}{\sqrt{\lambda}}\right)$ in (5.39) that are largest at 0 and 1, respectively. A plot of linear smoothing spline variance functions for $\sigma = .05$ and $n = 100$ is shown in Figure 5.6 under two choices for λ: $\lambda^* = .00024$ and $\lambda^{**} = .00057$. Both functions show the expected increase in variance at the boundaries. The smaller choice for the smoothing parameter, which produces less averaging, is seen to result in uniformly higher variability across the design.

To derive a pointwise approximation to the bias of the linear smoothing spline we can use a Taylor expansion in (5.38) to see that if μ'' satisfies a Lipshitz condition of order 2η, then

$$\begin{aligned} \mathrm{E}\mu_\lambda(t) &= \mu(t) - \sqrt{\lambda}\mu'(t)\left\{e^{(t-1)/\sqrt{\lambda}} - e^{-t/\sqrt{\lambda}}\right\} \\ &+ \lambda\mu''(t)\left(1 + \frac{t-1}{\sqrt{\lambda}}e^{(t-1)/\sqrt{\lambda}} - \frac{t}{\sqrt{\lambda}}e^{-t/\sqrt{\lambda}}\right) \\ &+ O\left(\frac{1}{n\lambda} + \lambda^{1+\eta}\right). \end{aligned} \tag{5.40}$$

To obtain this expression, use a Taylor expansion to see that

$$\begin{aligned}\int_0^1 \mu(s)K(t,s;\lambda)ds &= \mu(t)\int_{(t-1)/\sqrt{\lambda}}^{t/\sqrt{\lambda}} K(t,t-\sqrt{\lambda}u;\lambda)du \\ &- \sqrt{\lambda}\mu'(t)\int_{(t-1)/\sqrt{\lambda}}^{t/\sqrt{\lambda}} uK(t,t-\sqrt{\lambda}u;\lambda)du \\ &+ \frac{\lambda}{2}\mu''(t)\int_{(t-1)/\sqrt{\lambda}}^{t/\sqrt{\lambda}} u^2K(t,t-\sqrt{\lambda}u;\lambda)du \\ &+ O(\lambda^{1+\eta})\end{aligned}$$

and then explicitly evaluate the moment integrals to obtain

$$\begin{aligned}\int_{(t-1)/\sqrt{\lambda}}^{t/\sqrt{\lambda}} K(t,t-\sqrt{\lambda}u;\lambda)du &= 1+O(e^{-1/\sqrt{\lambda}}), \\ \int_{(t-1)/\sqrt{\lambda}}^{t/\sqrt{\lambda}} uK(t,t-\sqrt{\lambda}u;\lambda)du &= e^{(t-1)/\sqrt{\lambda}} - e^{-t/\sqrt{\lambda}} + O(e^{-1/\sqrt{\lambda}}) \\ \int_{(t-1)/\sqrt{\lambda}}^{t/\sqrt{\lambda}} u^2K(t,t-\sqrt{\lambda}u;\lambda)du &= 2\left(1+\frac{t-1}{\sqrt{\lambda}}e^{(t-1)/\sqrt{\lambda}} - \frac{t}{\sqrt{\lambda}}e^{-t/\sqrt{\lambda}}\right) \\ &\quad + O(e^{-1/\sqrt{\lambda}})\end{aligned}$$

all uniformly in $t \in [0,1]$.

For a fixed $t \in (0,1)$ the exponential terms in (5.40) are negligible and the expected value of the estimator simplifies to

$$\mathrm{E}\mu_\lambda(t) = \mu(t) + \lambda\mu''(t) + O\left(\lambda^{1+\eta} + \frac{1}{n\lambda}\right).$$

Combining this with our approximation for the variance in this case reveals that μ_λ is a pointwise second order estimator since $\mathrm{E}(\mu_\lambda(t) - \mu(t))^2$ is of order $n^{-4/5}$ whenever λ is of order $n^{-2/5}$. However, things change if we are estimating at a boundary point such as $t = \sqrt{\lambda}q$ for some $q > 0$. In that case, while the variance remains of order $(n\sqrt{\lambda})^{-1}$, the bias becomes

$$\mathrm{E}\mu_\lambda(t) - \mu(t) = \sqrt{\lambda}\mu'(0)e^{-q} + O\left(\lambda + \frac{1}{n\lambda}\right)$$

because $\mu'(q\sqrt{\lambda}) = \mu'(0) + O(\sqrt{\lambda})$. Thus, if $n\lambda^{3/2} \to \infty$, μ_λ is only first order in boundary regions such as $[0, \sqrt{\lambda}]$ or $[1 - \sqrt{\lambda}, 1]$ unless $\mu'(0) = \mu'(1) = 0$.

To assess the global performance of the estimator we can now use our pointwise approximations (5.39) and (5.40) to approximate $\mathrm{E}(\mu_\lambda(t_i) - \mu(t_i))^2, i = 1,\ldots,n$, and then average across the design to approximate (5.30). This is permissable here because (5.39) and (5.40) are uniform in t and explicitly include boundary effects. Arguing in this manner we obtain that since $n\sqrt{\lambda} \to \infty$

$$\begin{aligned} n^{-1}\sum_{i=1}^{n} \mathrm{Var}(\mu_\lambda(t_i)) &= \frac{\sigma^2}{4n\sqrt{\lambda}}\left\{1 + \int_0^1 [e^{2(t-1)/\sqrt{\lambda}}\left(1 - \frac{2(t-1)}{\sqrt{\lambda}}\right)\right. \\ &+ \left. e^{-2t/\sqrt{\lambda}}\left(1 + \frac{2t}{\sqrt{\lambda}}\right)]dt + o(1)\right\} \\ &= \frac{\sigma^2}{4n\sqrt{\lambda}}(1 + o(1)) . \end{aligned} \tag{5.41}$$

If, in addition, $n\lambda^{3/2} \to \infty$ with one of $\mu'(0)$ or $\mu'(1)$ not being zero

$$\begin{aligned} n^{-1}\sum_{i=1}^{n}(\mathrm{E}\mu_\lambda(t_i) - \mu(t_i))^2 &= \lambda\int_0^1 \mu'(t)^2[e^{2(t-1)/\sqrt{\lambda}} + e^{-2t/\sqrt{\lambda}}]dt \\ &\quad + o(\lambda^{3/2}) \\ &= \frac{\lambda^{3/2}}{2}[\mu'(0)^2 + \mu'(1)^2] + o(\lambda^{3/2}). \end{aligned} \tag{5.42}$$

In the case that $\mu'(0) = \mu'(1) = 0$ and $n\lambda^2 \to \infty$, the last approximation improves to

$$n^{-1}\sum_{i=1}^{n}(\mathrm{E}\mu_\lambda(t_i) - \mu(t_i))^2 = \lambda^2\int_0^1 \mu''(t)^2dt + o(\lambda^2). \tag{5.43}$$

Approximations (5.41)-(5.43) can be combined to obtain asymptotic expressions for the global risk of the linear smoothing spline under different assumptions about the boundary properties of the regression function. For example, if μ' is bounded with at least one of $\mu'(0)$ or $\mu'(1)$ not being zero we have

$$R_n(\lambda) \sim \frac{\lambda^{3/2}}{2}[\mu'(0)^2 + \mu'(1)^2] + \frac{\sigma^2}{4n\sqrt{\lambda}}$$

if $n\lambda^{3/2} \to \infty$. This can be optimized with respect to λ to see that the optimal smoothing parameter in this case will decay at the exact rate $n^{-1/2}$ producing an $n^{-3/4}$ rate of decay for the risk of the estimator. Recall that this is the same rate one obtains from a second order kernel estimator without using a boundary correction. If instead μ'' is Lipshitz continuous, $\mu'(0) = \mu'(1) = 0$ with $n\lambda^2 \to \infty$

$$R_n(\lambda) \sim \lambda^2 \int_0^1 \mu''(t)^2 dt + \frac{\sigma^2}{4n\sqrt{\lambda}}.$$

So, in this latter case, the asymptotically optimal risk is of order $n^{-4/5}$ for a smoothing parameter value of order $n^{-2/5}$ and the estimator is globally of second order.

To give a specific illustration, consider the two regression functions

$$\mu^*(t) = 16t^2(1-t)^2 + 32(t-.7)^3 I_{\{t \geq .7\}}(t) \tag{5.44}$$

and

$$\mu^{**}(t) = 16t^2(1-t)^2 \tag{5.45}$$

shown in Figure 5.7. The derivative of the regression function μ^{**} vanishes both at 0 and 1 while that for μ^* vanishes only at zero. This means that a linear smoothing spline will provide a globally second order estimator for μ^{**} but will exhibit a boundary bias effect when estimating $\mu^*(t)$ near $t = 1$.

The asymptotically optimal smoothing parameter values for estimating μ^* and μ^{**} are $\lambda^* = [\sigma^2/(6(8.64)^2 n)]^{1/2}$ and $\lambda^{**} = [5\sigma^2/(16(32^2)n)]^{2/5}$, respectively (Exercise 11). In particular, if we take $n = 100$ and $\sigma = .05$, then $\lambda^* = .00024$, $\lambda^{**} = .00057$ and the corresponding variance functions for linear smoothing spline estimation are the ones in Figure 5.6. The biases for estimation in this case are plotted in Figure 5.8 where we see that the magnitude of the interior bias is smaller for estimating μ^* but increases substantially over that for estimating μ^{**} in the upper boundary.

Figure 5.9 shows a linear smoothing spline fit to a data set of size $n = 100$ that was simulated from model (5.2) using normal random errors with $\sigma = .05$ and regression function (5.45). The only difference between this data set and the one in Figure 5.4 is the regression function with Figure 5.4 having regression function μ^*. The linear smoothing spline fits shown in both plots were obtained using the UBR criterion and correspond to data

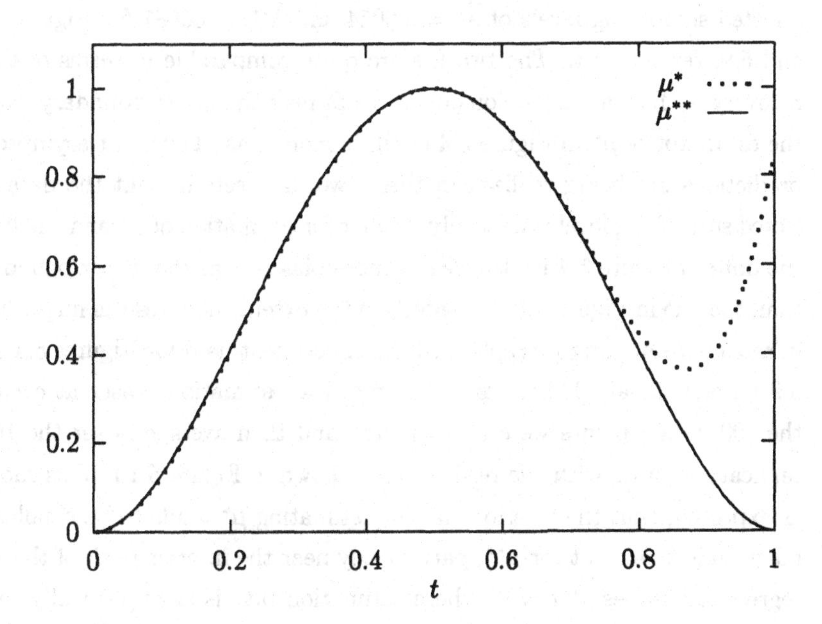

Figure 5.7: The Regression Functions μ^* and μ^{**}.

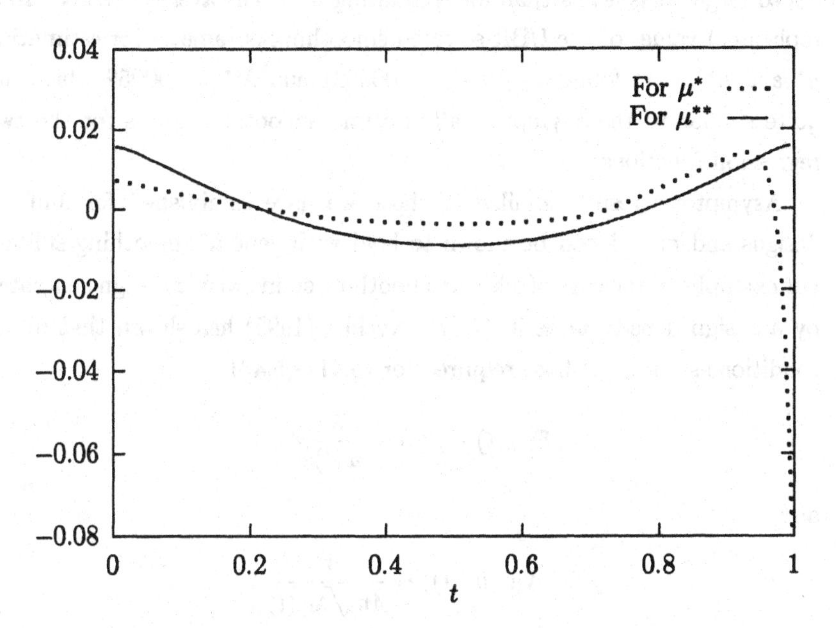

Figure 5.8: Linear Smoothing Spline Biases for μ^* and μ^{**}.

selected smoothing levels of $\widehat{\lambda}^* = .00054$ and $\widehat{\lambda}^{**} = .00087$ for Figures 5.4 and 5.9, respectively. The two fits are quite comparable in terms of their recovery of the true regression curve except near the upper boundary where the estimator of μ^* in Figure 5.4 exhibits more bias. Thus, our asymptotic predictions are being realized in these two data sets in that the data selected smoothing level is certainly smaller for estimation of μ^* and the fit to the data in Figure 5.4 is also clearly more biased near the upper boundary than the one in Figure 5.9. To ascertain the extent these results might hold in repeated sampling we replicated the experiment used for Figures 5.4 and 5.9 an additional 1000 times. The empirical estimation biases at each of the 100 design points were accumulated and then averaged over the 1000 replicate samples with the results being shown in Figure 5.10. This shows, as expected, that the interior bias for estimating μ^* tends to be smaller in magnitude than that for μ^{**}, particularly near the interior peak of the two regresssion curves at $t = .5$ where estimation bias is asymptotically most substantial. Near the upper boundary the fact that $(\mu^*)'(1) \neq 0$ has produced larger bias levels than for estimating μ^{**}. The average (across 1000 replicates) values of the UBR selected smoothing parameter for estimating μ^* and μ^{**} were found to be $\widehat{\lambda}^* = .000321$ and $\widehat{\lambda}^{**} = .00059$ which are quite similar to the asymptotically optimal smoothing levels for the two regression functions.

Asymptotic results similar to those we have established for uniform designs and $m = 1$ can be shown to hold with general smoothing splines. For example, in the case of a linear smoothing spline with a design generated by a design density w as in (4.37), Nychka (1995) has shown that under conditions similar to those required for (5.41)–(5.43),

$$\mathrm{E}\mu_\lambda(t) - \mu(t) \sim \frac{\lambda}{w(t)}\mu''(t)$$

and

$$\mathrm{Var}(\mu_\lambda(t)) \sim \frac{\sigma^2}{4n\sqrt{\lambda w(t)}}$$

for any fixed $t \in (0, 1)$. Comparing this with (4.38) indicates that the linear smoothing spline behaves like a second order kernel estimator at interior

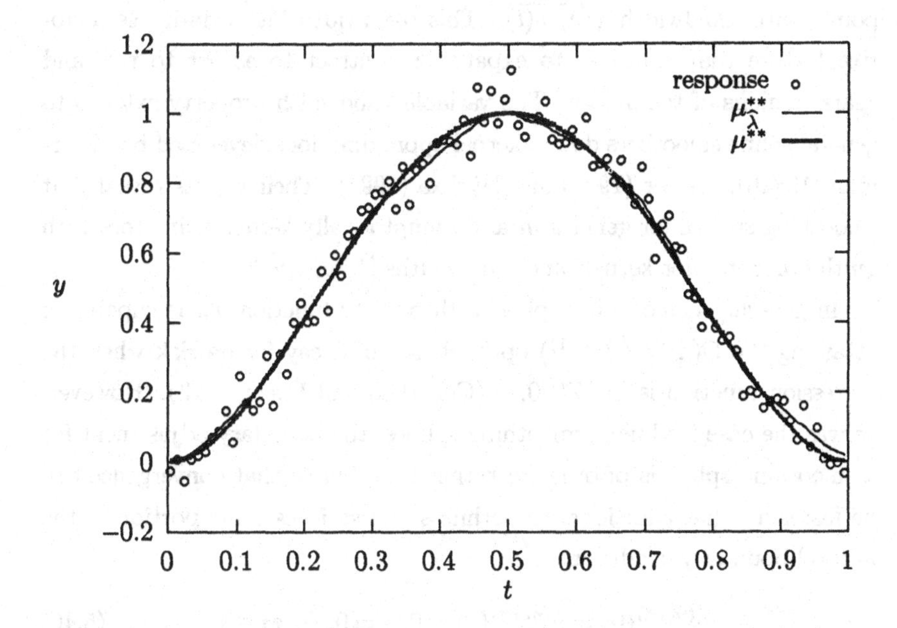

Figure 5.9: Linear Smoothing Spline Fit to Simulated Data.

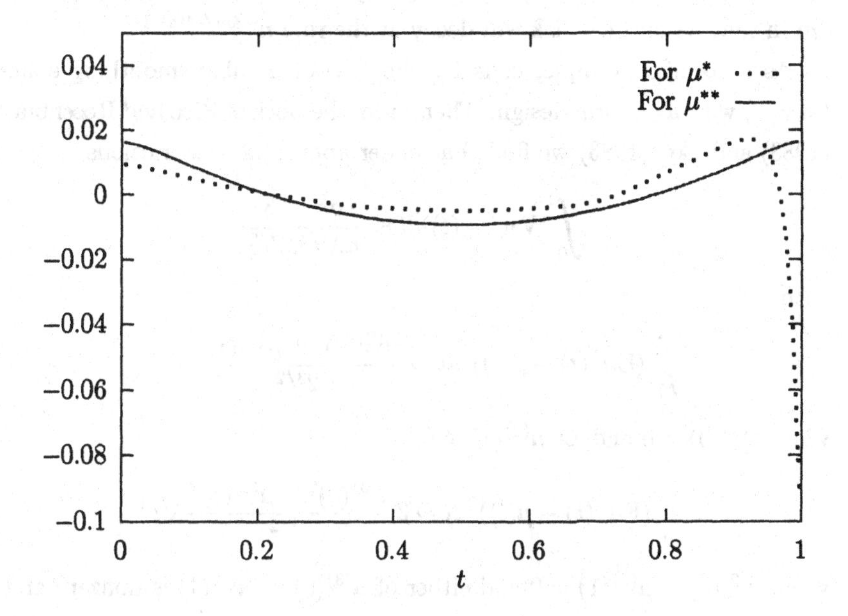

Figure 5.10: Average Empirical Biases for Estimating μ^* and μ^{**}.

points with bandwidth $\sqrt{\lambda/w(t)}$. This bandwidth has a desirable adaptive feature that allows it to expand or contract to adjust to rich and sparse regions of the design. The variable bandwidth property extends to general spline smoothers due to kernel approximations developed by Silverman (1984b), Messer (1991) and Nychka (1995). Their results reveal that smoothing splines for general m are asymptotically kernel estimators with $2m$th order interior kernels and bandwidths $[\lambda/w(\cdot)]^{1/2m}$.

In general, a smoothing spline with penalty function J_m is capable of attaining the $O(n^{-2m/(2m+1)})$ optimal rate of decay for its risk when the regression function is in $W_2^m[0,1]$ (Cox 1983 and Exercise 13). However, as was the case for linear smoothing splines, the boundary adjustment for a smoothing spline is of order m rather than $2m$ so that convergence can be faster if μ has additional smoothness and satisfies some portion of the natural boundary conditions

$$\mu^{(m+j)}(0) = \mu^{(m+j)}(1) = 0,\ j = 0, \ldots, m-1. \tag{5.46}$$

If, for example, $\mu \in W_2^{2m}[0,1]$ and (5.46) holds the estimator is of order $2m$ in that its optimal risk will decay at the rate $n^{-4m/(4m+1)}$.

As a specific example, consider the case of a cubic smoothing spline ($m = 2$) with a uniform design. Then, from the work of Rice and Rosenblatt (1983) and Cox (1983) we find that under appropriate restrictions

$$\int_0^1 \mathrm{Var}(\mu_\lambda(t))dt \sim \frac{\sigma^2}{n\lambda^{1/4}} \frac{3}{2^{7/2}}$$

with

$$\int_0^1 (\mathrm{E}\mu_\lambda(t) - \mu(t))^2 dt \sim \frac{\mu^{(2)}(0)^2 + \mu^{(2)}(1)^2}{2^{3/2}} \lambda^{5/4}$$

when $\mu^{(2)}(0) \neq 0$ and/or $\mu^{(2)}(1) \neq 0$,

$$\int_0^1 (\mathrm{E}\mu_\lambda(t) - \mu(t))^2 dt \sim 3\frac{\mu^{(3)}(0)^2 + \mu^{(3)}(1)^2}{2^{3/2}} \lambda^{7/4}$$

when $\mu^{(2)}(0) = \mu^{(2)}(1) = 0$ and either of $\mu^{(3)}(0)$ or $\mu^{(3)}(1)$ is nonzero and

$$\int_0^1 (\mathrm{E}\mu_\lambda(t) - \mu(t))^2 dt \sim \lambda^2 \int_0^1 \mu^{(4)}(t)^2 dt$$

when (5.46) holds for $m = 2$. Analogous results for more general designs and m can be deduced from the work of Cox (1983) and Utreras (1987).

The fact that the smoothing spline for a given m is essentially a $2m$th order kernel estimator with only an mth order boundary correction could be viewed as a type of "boundary bias" problem . It is actually fairly straightforward to remove the edge effects from the estimator using the Bias Reduction Lemma of Section 3.6 or other methods and thereby obtain a $2m$th order smoother. (See Oehlert 1992, Eubank and Speckman 1992 and Exercise 14). However, a word of caution is in order at this point. Due to the use of the penalty functional (5.3), a smoothing spline is only intended for estimation of functions in $W_2^m[0,1]$. If we believed that $\mu \in W_2^{2m}[0,1]$, then we should have used J_{2m} in our penalty and estimated μ with the corresponding higher order smoothing spline. From this viewpoint, the smoothing spline boundary behavior is more "boundary bonus" than "boundary bias" since it has the consequence that the estimator can produce faster than expected convergence rates for some situations where the regression function is smoother than anticipated.

Finally, one might wonder how the estimation risk for spline smoothing compares to that of other nonparametric estimators. Some results of this nature have been established by Carter, Eagleson and Silverman (1992) who compare the asymptotic risk behavior of smoothing splines with that of the minimax spline estimator of Speckman (1985). Speckman's estimator is the best possible in the sense of having the smallest average (across the design) risk over all regression functions $\mu \in W_2^m[0,1]$ for which $J_m(\mu) \leq \rho$. Carter, Eagleson and Silverman (1992) show that when $m = 2$ and under optimal levels of smoothing for both estimators, the cubic smoothing spline is asymptotically only 8.3% less efficient in terms of its average risk than the fully efficient minimax estimator. Consequently, the cubic smoothing spline is, in this respect, very nearly optimal as a second order estimator.

5.6 Smoothing Splines as Bayes Estimators

In previous discussions we have focused on the properties of spline smoothers as estimators of some unknown regression function in $W_2^m[0,1]$. However, as noted in Section 5.2, smoothing splines are formally generalized ridge regression estimators so, (cf Exercise 3 from Chapter 2) they also have a Bayesian interpretation. In this section we explore the Bayesian aspects of spline smoothing and demonstrate that smoothing splines can be derived from the perspective of a Bayesian approach to polynomial regression.

We begin with a simple formulation due essentially to Silverman (1985). Instead of model (5.2) let us now assume that we observe responses at distinct design points which, conditional on the value of a parameter vector $\boldsymbol{\beta} = (\beta_1, \ldots, \beta_n)^T$, satisfy

$$y_i = \sum_{j=1}^{n} \beta_j x_j(t_i) + \varepsilon_i,\ i = 1, \ldots, n, \tag{5.47}$$

for $\boldsymbol{\varepsilon} = (\varepsilon_1, \ldots, \varepsilon_n)^T$ a zero mean, normal random vector with covariance matrix $\sigma^2\mathbf{I}$ and $x_1, \ldots, x_n$ the Demmler-Reinsch basis for the natural splines of order $2m$ with knots at $t_1, \ldots, t_n$. We use the Demmler-Reinsch basis here to simplify some of the matrix algebra that follows. However, our final results will be stated in terms of quantities that are basis invariant thereby making it clear that this particular basis choice causes no loss of generality.

To complete the model specification $\boldsymbol{\beta}$ in (5.47) is taken to be n-variate normal with zero mean and covariance matrix

$$\mathrm{Var}(\boldsymbol{\beta}) = \frac{\sigma^2}{n\lambda}\mathbf{D}_\nu^{-1} = \frac{\sigma^2}{n\lambda}\mathrm{diag}(\underbrace{\nu, \ldots, \nu}_{m}, \gamma_1, \ldots, \gamma_{n-m})^{-1} \tag{5.48}$$

with $\gamma_1, \ldots, \gamma_{n-m}$ the Demmler-Reinsch eigenvalues. Note from our discussion of the Demmler-Reinsch basis in Section 5.2 that $\beta_1, \ldots, \beta_m$ are the coefficients for the polynomial portion of the spline basis and that (5.48) gives $\mathrm{Var}((\beta_1, \ldots, \beta_m)^T) = \sigma^2(n\lambda\nu)^{-1}\mathbf{I}$. We will eventually let $\nu \to 0$ to place a diffuse prior distribution on the polynomial part of the regression function.

Recall from Section 5.2 that under the Demmler-Reinsch basis the design matrix $\mathbf{X} = \{x_j(t_i)\}_{i,j=1,n}$ satisfies $\mathbf{X}\mathbf{X}^T = n\mathbf{I} = \mathbf{X}^T\mathbf{X}$ which also means that $n^{-1}\mathbf{X}^T = \mathbf{X}^{-1}$ and $n^{-1}\mathbf{X} = (\mathbf{X}^T)^{-1}$. A quick calculation then reveals that the joint density for the response vector $\mathbf{y}$ and coefficient vector $\boldsymbol{\beta}$ is proportional to

$$\begin{aligned} & \exp\{-\frac{1}{2\sigma^2}(\mathbf{y}-\mathbf{X}\boldsymbol{\beta})^T(\mathbf{y}-\mathbf{X}\boldsymbol{\beta}) - \frac{n\lambda}{2\sigma^2}\boldsymbol{\beta}^T\mathbf{D}_\nu\boldsymbol{\beta}\} \\ = & \exp\{-\frac{1}{2\sigma^2}\mathbf{y}^T(\mathbf{I}-\mathbf{S}_{\lambda,\nu})\mathbf{y} - \frac{1}{2\sigma^2}(\mathbf{X}\boldsymbol{\beta}-\mathbf{S}_{\lambda,\nu}\mathbf{y})^T\mathbf{S}_{\lambda,\nu}^{-1}(\mathbf{X}\boldsymbol{\beta}-\mathbf{S}_{\lambda,\nu}\mathbf{y})\}, \end{aligned}$$

where

$$\mathbf{S}_{\lambda,\nu} = \mathbf{X}(\mathbf{X}^T\mathbf{X}+n\lambda\mathbf{D}_\nu)^{-1}\mathbf{X}^T = n^{-1}\mathbf{X}(\mathbf{I}+\lambda\mathbf{D}_\nu)^{-1}\mathbf{X}^T. \tag{5.49}$$

This makes it clear that under (5.47)-(5.48) $\mathrm{E}[\boldsymbol{\mu}|\mathbf{y}] = \mathbf{S}_{\lambda,\nu}\mathbf{y}$ and $\mathrm{Var}(\boldsymbol{\mu}|\mathbf{y}) = \sigma^2\mathbf{S}_{\lambda,\nu}$ for $\boldsymbol{\mu} = \mathbf{X}\boldsymbol{\beta}$ the (conditional) mean vector for the responses in (5.47). Also, the unconditional $\mathbf{y}$ distribution is seen to be n-variate normal with mean zero and

$$\mathrm{Var}(\mathbf{y}) = \sigma^2(\mathbf{I}-\mathbf{S}_{\lambda,\nu})^{-1} = \frac{\sigma^2}{n}\mathbf{X}[\mathbf{I}-(\mathbf{I}+\lambda\mathbf{D}_\nu)^{-1}]^{-1}\mathbf{X}^T. \tag{5.50}$$

Now observe that $\mathbf{S}_{\lambda,\nu} \to \mathbf{S}_\lambda$ as $\nu \to 0$ with $\mathbf{S}_\lambda$ the smoothing spline hat matrix in (5.9). Thus,

$$\lim_{\nu\to 0}\mathrm{E}[\boldsymbol{\mu}|\mathbf{y}] = \boldsymbol{\mu}_\lambda \tag{5.51}$$

for $\boldsymbol{\mu}_\lambda$ the vector of smoothing spline fitted values (5.8),

$$\lim_{\nu\to 0}\mathrm{Var}(\boldsymbol{\mu}|\mathbf{y}) = \sigma^2\mathbf{S}_\lambda \tag{5.52}$$

and

$$\lim_{\nu\to 0}\mathrm{Var}((\mathbf{I}-\mathbf{S}_{\lambda,\nu})\mathbf{y}) = \sigma^2(\mathbf{I}-\mathbf{S}_\lambda). \tag{5.53}$$

Equation (5.51) reveals that the smoothing spline fitted values are the posterior mean of $\boldsymbol{\mu} = \mathbf{X}\boldsymbol{\beta}$ under (5.47)-(5.48) when we allow for a flat or diffuse prior over the polynomial part of the regression curve. Equations (5.52)-(5.53) give covariance matrices for $\boldsymbol{\mu}$ and the residual vector $\mathbf{e} = (\mathbf{I}-\mathbf{S}_\lambda)\mathbf{y}$ which bear a formal similarity to the covariance matrices for fitted values and residuals from ordinary least-squares linear regression. We will exploit this property subsequently to motivate one possible approach to developing confidence intervals and diagnostic methods that can be used to accompany the smoothing spline estimator.

There are two additional unknowns, λ and σ^2, in (5.47)-(5.48) that must be dealt with in some manner. One approach is to estimate the parameters by maximum likelihood type methods. To produce such estimators we will follow arguments in Wahba (1990, Section 4.8). First partition the matrix $\mathbf{X}$ as $[\mathbf{X}_1|\mathbf{X}_2]$ with $\mathbf{X}_1 = \{x_j(t_i)\}_{i=1,n,j=1,m}$ corresponding to only the polynomial part of the basis. Now set $\mathbf{u} = \mathbf{X}_2^T\mathbf{y}$ and observe that $\mathbf{u}$ and $\mathbf{X}_1^T\mathbf{y}$ are uncorrelated with $\text{Var}(\mathbf{X}_1^T\mathbf{y}) = O(\nu^{-1})$. Thus, as $\nu \to 0$ only the distribution of $\mathbf{u}$ will contain information about λ and we will restrict our attention to $\mathbf{u}$ for developing an estimator for this parameter.

Using the orthogonality of the Demmler-Reinsch basis $\mathbf{u}$ is found to have an $(n-m)$-dimensional normal distribution with mean zero and

$$\begin{aligned}\text{Var}(\mathbf{u}) &= \sigma^2\mathbf{X}_2^T(\mathbf{I}-\mathbf{S}_{\lambda,\nu})^{-1}\mathbf{X}_2 \\ &= \frac{\sigma^2}{n}\mathbf{X}_2^T\mathbf{X}[\mathbf{I}-(\mathbf{I}+\lambda\mathbf{D}_\nu)^{-1}]^{-1}\mathbf{X}^T\mathbf{X}_2 \\ &= n\sigma^2\text{diag}(\frac{1+\lambda\gamma_1}{\lambda\gamma_1},\ldots,\frac{1+\lambda\gamma_{n-m}}{\lambda\gamma_{n-m}}).\end{aligned}$$

Consequently,

$$|\text{Var}(\mathbf{u})| = (n\sigma^2)^{(n-m)}|\mathbf{I}-\mathbf{S}_\lambda|_+^{-1}$$

with

$$|\mathbf{I}-\mathbf{S}_\lambda|_+ = \Pi_{j=1}^{n-m}\frac{\lambda\gamma_j}{1+\lambda\gamma_j}$$

the product of the nonzero eigenvalues of $\mathbf{I}-\mathbf{S}_\lambda$. The natural log of the $\mathbf{u}$ likelihood is therefore

$$-\frac{n-m}{2}\log\sigma^2 + \frac{1}{2}\log|\mathbf{I}-\mathbf{S}_\lambda|_+ - \frac{1}{2}\mathbf{u}^T\text{Var}(\mathbf{u})^{-1}\mathbf{u} \tag{5.54}$$

apart from constants that do no involve either λ or σ^2. We can maximize (5.54) first with respect to σ^2 to see that, for fixed or known λ, a maximum likelihood variance estimator is

$$\tilde{\sigma}^2 = \frac{\mathbf{y}^T(\mathbf{I}-\mathbf{S}_\lambda)\mathbf{y}}{n-m}. \tag{5.55}$$

Substituting (5.55) into (5.54) gives the so-called generalized maximum likelihood (GML) estimator of λ : namely,

$$\widetilde{\lambda} = \text{argmin}_{\lambda} \frac{\mathbf{y}^T(\mathbf{I} - \mathbf{S}_{\lambda})\mathbf{y}}{|\mathbf{I} - \mathbf{S}_{\lambda}|_+^{1/(n-m)}}. \tag{5.56}$$

Notice that for a fixed value of λ, $\widetilde{\sigma}^2$ in (5.55) is an unbiased estimator of σ^2 under our Bayesian model (Exercise 16). Other unbiased estimators included the linear regression type variance estimator

$$\widetilde{\sigma}^2_W = \frac{\mathbf{y}^T(\mathbf{I} - \mathbf{S}_{\lambda})^2\mathbf{y}}{\text{tr}(\mathbf{I} - \mathbf{S}_{\lambda})} \tag{5.57}$$

suggested by Wahba (1990, Section 4.7). When the Bayesian model does not hold and the true regression function is deterministic, both $\widetilde{\sigma}^2$ and $\widetilde{\sigma}^2_W$ will be biased with their biases being a function of the level of smoothing.

In spite of the somewhat complicated form for the GML criterion in (5.56), its value can be computed efficiently in $O(n)$ operations for any given value of λ using the diffuse Kalman filter as in Kohn and Ansley (1987). Results in Wahba (1990, Section 4.9) suggest that both the GML and GCV criteria should perform similarly for estimating λ in the case of a stochastic regression function like that of model (5.47)-(5.48) but that the GML method may tend to undersmooth when selecting the level of smoothing for data from the frequentist model (5.2). Extensive simulation experiments conducted by Kohn, Ansley and Tharm (1991) have instead found GML to be competitive with, and often superior to, GCV for selecting the level of smoothing even when μ is nonstochastic.

The developments to this point show that smoothing splines can be derived from a Bayesian regression model wherein the regression function is a random natural spline whose distribution is diffuse over polynomials of order m. By now letting $\lambda \to \infty$ in the Bayesian model we recover the usual least-squares fit to the data using an mth order polynomial. One could therefore view this approach to estimation as being an extension of polynomial regression with a specific choice of λ reflecting one's prior belief about the extent that the regression function might depart from a polynomial.

Although the Bayesian framework (5.47)-(5.48) simplifies mathematical developments, the use of a model like this in practice could be rather hard to justify. Even though the assumed regression function is high dimensional, the model is not truly nonparametric and the choice of μ as a natural spline looks to be motivated more by convenience than any actual belief in this form for the regression curve. It turns out, however, that there are other, more satisfactory, Bayesian formulations that produce identical results to those described above for (5.47)-(5.48).

Suppose, for example, that $(t_1, y_1), \ldots, (t_n, y_n)$ are obtained by sampling from a stochastic process $\{Y(t) : t \in [0,1]\}$ at $t_1, \ldots, t_n$. The Y process is assumed to have the form

$$Y(t) = \mu(t) + \varepsilon(t),\ t \in [0,1], \tag{5.58}$$

for $\{\varepsilon(t) : t \in [0,1]\}$ a zero mean normal process with

$$\mathrm{Cov}(\varepsilon(s), \varepsilon(v)) = \begin{cases} 0, & v \neq s, \\ \sigma^2, & v = s, \end{cases}$$

and

$$\mu(t) = \sum_{j=0}^{m-1} \theta_j t^j + \frac{\sigma}{\sqrt{n\lambda}} Z(t), \tag{5.59}$$

for $\boldsymbol{\theta} = (\theta_0, \ldots, \theta_{m-1})^T$ a zero mean, normal random vector with covariance matrix $\nu^{-1}\mathbf{I}$ and $\{Z(t) : t \in [0,1]\}$ an $(m-1)$-fold integrated Brownian motion process that is uncorrelated with $\boldsymbol{\theta}$. More specifically, $Z(\cdot)$ is a zero mean normal process with covariance kernel

$$\mathrm{Cov}(Z(s), Z(v)) = \int_0^1 \frac{(s-u)_+^{m-1}(v-u)_+^{m-1}}{[(m-1)!]^2} du. \tag{5.60}$$

Model (5.58)-(5.60) treats μ as being a polynomial plus a special type of random, rather than deterministic, error function. The use of models of this nature has been proposed by Blight and Ott (1975) and others as a strategy for handling possible departures from polynomial models. The basic premise behind such formulations is to model μ as far as possible by a polynomial and then treat the remainder as some type of zero mean

Gaussian process such as $Z(\cdot)$ in (5.59). This produces a model that is nonparametric in the sense that, unlike parametric models, no shapes are ruled out for the regression function. However, it is not equivalent to model (5.2). The difficulty lies in the fact that if μ were to have an mth derivative it would be the derivative of a Wiener process and the Wiener process is well known to be nowhere differentiable with probability one. So, the sample paths of (5.58)-(5.60) actually have zero probability of being in $W_2^m[0,1]$.

The basic properties of the smoothing spline estimator that we established under model (5.47)-(5.48) continue to hold under model (5.58)-(5.60) (Wahba 1978, 1990). In particular, μ_λ is again found to be the posterior mean of μ in (5.59) when the distribution for the polynomial coefficients becomes diffuse (i.e., $\nu \to 0$) and identities (5.51)-(5.53) also continue to hold. Disparities between the two approaches arise only when we attempt to compute posterior covariances at non-design points and we will not be concerned with that here. The connection between models (5.47)-(5.48) and (5.58)-(5.60) has been explored by van der Linde (1993) who shows that these two seemingly different formulations can be reconciled and actually derive from a common prior distribution.

It is possible to express model (5.58)-(5.60) in a state space formulation that allows for efficient, order n, computation of smoothing splines via the Kalman filter. Details can be found in Kohn and Ansley (1987). This can also be combined with Monte Carlo Markov chain methodology to obtain estimators of all the model's parameters as in Carter and Kohn (1996).

One of the most useful applications of the Bayesian interpretation of smoothing splines is to the development of interval estimators for μ. As a result of identity (5.52) we see that under the Bayesian model with known λ and σ^2

$$P\left(|\mu(t_i) - \mu_\lambda(t_i)| \leq z_{\alpha/2}\sigma\sqrt{s_i}\right) = 1 - \alpha,$$

for any particular design point t_i with s_i the ith diagonal element of the smoothing spline hat matrix $\mathbf{S}_\lambda$ in (5.9). From this we obtain the pointwise $100(1-\alpha)\%$ "confidence intervals"

$$\mu_\lambda(t_i) \pm z_{\alpha/2}\sigma\sqrt{s_i}\,, i = 1, \ldots, n, \tag{5.61}$$

for $\mu(t_i), i = 1, \ldots, n$. These intervals have the pleasing intuitive property that they directly parallel the form of interval estimators for the mean function in parametric linear regression analysis. However, strictly speaking they are not confidence intervals but rather prediction intervals since the $\mu(t_i)$ are random variables under the Bayesian model.

Figures 5.11 and 5.12 show the 95% Bayesian confidence intervals (5.61) (using the GSJS variance estimator for σ^2) that were obtained from the linear smoothing spline fits to the data sets in Figures 5.4 and 5.9. The true regression functions are also shown in the plots from which we see that, with the exception of a few points in the upper boundary region of Figure 5.11, the intervals perform remakably well at capturing the true regression curve. However, we should point out that there is no obvious justification for the use of the Bayesian intervals with these data sets since we know for a fact that they come from model (5.2) and not either of our Bayesian models. Nevertheless, the intervals seem to be effective, and this raises the question of whether or not intervals such as (5.61) could also work from a frequentist perspective.

One frequentist justification for use of the Bayesian confidence intervals (5.61) has been provided by Wahba (1983) and Nychka (1988, 1990). Under certain restrictions and with an optimally chosen level of smoothing, Nychka (1988, 1990) shows that

$$n^{-1}\sum_{i=1}^{n}\text{Prob}(|\mu_\lambda(t_i) - \mu(t_i)| \leq z_{\alpha/2}\sigma\sqrt{s_i}) \doteq 1 - \alpha. \tag{5.62}$$

This states that under model (5.2) the average coverage for the Bayesian intervals across the design should be approximately the same as their nominal, pointwise coverage. Consequently, the Bayesian intervals give a curvewise (rather than pointwise) coverage level of roughly $1 - \alpha$. While results such as (5.62) are encouraging, they tells us very little about the pointwise coverage properties of any particular interval since these can be unacceptably lower or higher than the nominal level and still satisfy (5.62).

Another intuitive notion of how the Bayesian intervals can be expected to work from a pointwise frequentist perspective can be obtained by referring back to the discussions in Sections 5.5. There we observed that

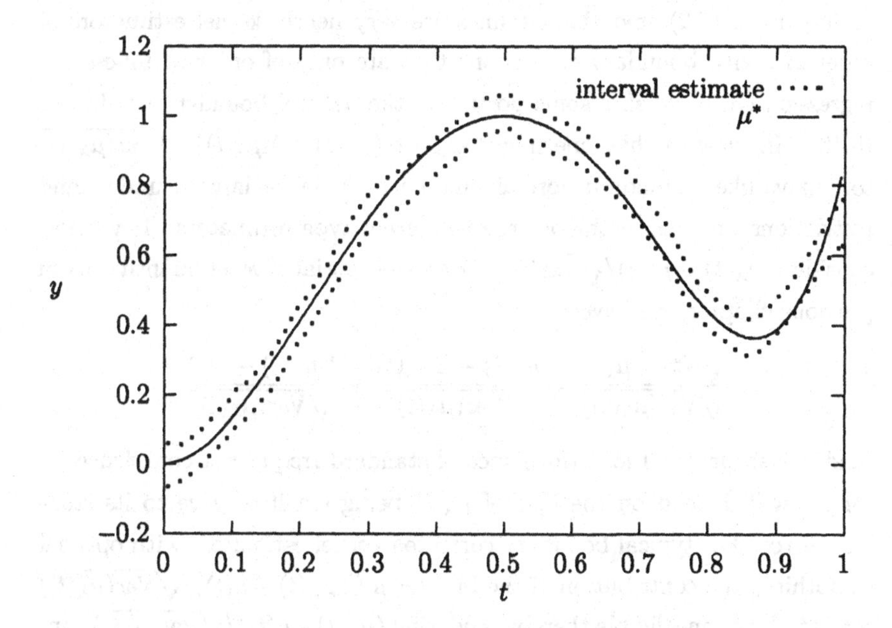

Figure 5.11: Bayesian Intervals for the Data in Figure 5.4.

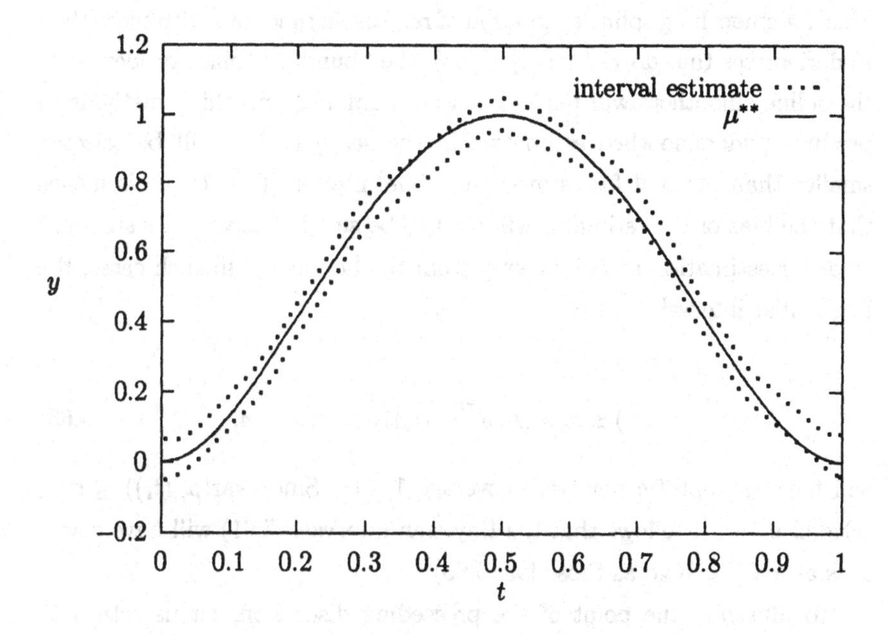

Figure 5.12: Bayesian Intervals for the Data in Figure 5.9.

under model (5.2) smoothing splines are very nearly kernel estimators of order $2m$ with boundary corrections that are only of order m unless the regression curve satisfies some portion of the natural boundary conditions (5.46). In view of this, one might expect $(\mu_\lambda(t) - \mathrm{E}\mu_\lambda(t))/\sqrt{\mathrm{Var}(\mu_\lambda(t))}$ to behave like a standard normal random variable for large n under mild restrictions on λ since this is true for kernel type estimators. If we now consider $(\mu_\lambda(t) - \mu(t))/\sqrt{\mathrm{Var}(\mu_\lambda(t))}$ as a potential pivotal quantity, as in previous chapters we have

$$\frac{\mu_\lambda(t) - \mu(t)}{\sqrt{\mathrm{Var}(\mu_\lambda(t))}} = \frac{\mu_\lambda(t) - \mathrm{E}\mu_\lambda(t)}{\sqrt{\mathrm{Var}(\mu_\lambda(t))}} + \frac{\mathrm{E}\mu_\lambda(t) - \mu(t)}{\sqrt{\mathrm{Var}(\mu_\lambda(t))}}$$

and, consequently, the performance of standard frequentist confidence intervals will depend on the bias of $\mu_\lambda(t)$ being small relative to its standard error. For typical boundary corrected kernel estimators with optimal smoothing, the contribution of the bias term $(\mathrm{E}\mu_\lambda(t) - \mu(t))/\sqrt{\mathrm{Var}(\mu_\lambda(t))}$ tends to be nonnegligible thereby rendering $(\mu_\lambda(t) - \mu(t))/\sqrt{\mathrm{Var}(\mu_\lambda(t))}$ unsuitable for a pivotal quantity without some type of bias correction. However, for smoothing spline estimation of regression functions with more than m derivatives that do *not* satisfy (5.46), the boundary "bias" properties of the spline smoothers will tend to cause automated smoothing methods to produce undersmoothed estimators in the sense that λ will be selected smaller than optimal for estimation in the interior of [0, 1]. This means that the bias of the estimator will tend to be small relative to its standard error for estimation at points away from the boundary. In such cases, the frequentist intervals

$$\mu_\lambda(t_i) \pm z_{\alpha/2}\sqrt{\mathrm{Var}(\mu_\lambda(t_i))}\ , i = 1, \ldots, n, \tag{5.63}$$

will have asymptotic pointwise coverage $1 - \alpha$. Since $\mathrm{Var}(\mu_\lambda(t_i)) \leq \sigma^2 s_i$ (Exercise 18), it follows that the Bayesian intervals (5.61) will have coverages at least as high as those for (5.63).

To illustrate the point of the preceeding discussion, let us return to the two regression functions μ^* and μ^{**} in (5.44)-(5.45). We discussed the results of a simulation using these two regression functions in Section 5.5

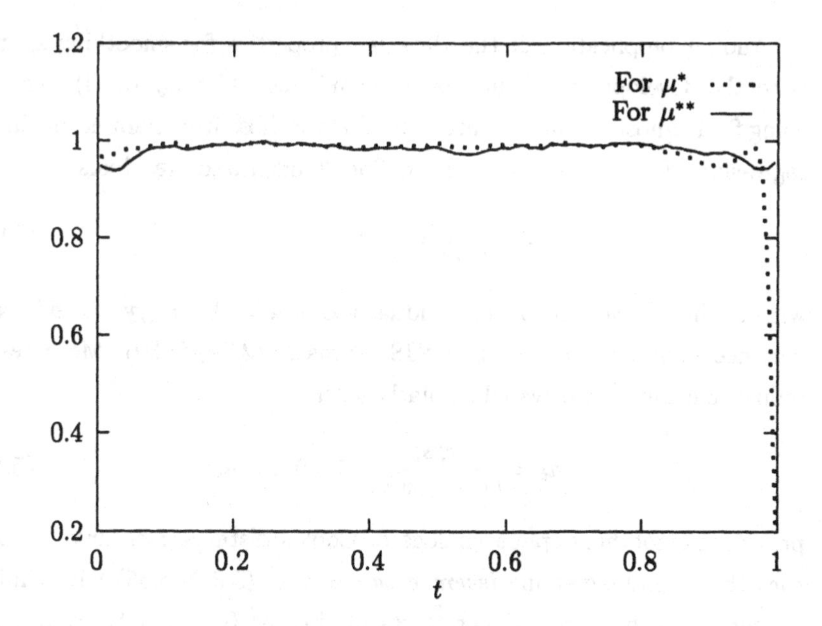

Figure 5.13: Empirical Coverages for μ^* and μ^{**}.

wherein 1000 replicate data sets of size 100 were generated from model (5.2) with normal errors having $\sigma = .05$ and a uniform design. The data sets were all fit using linear smoothing splines with λ selected by UBR and empirical biases were accumulated for estimation of μ^* and μ^{**}. For this same simulation the empirical average (over 1000 replicates) coverage proportions for 95% Bayesian confidence intervals (using (5.61) with the GSJS variance estimator) were recorded at each of the design points and the results are summarized in Figure 5.13. The coverage levels are quite good for both regression curves away from the upper boundary region. In the upper boundary the fact that μ^* does not satisfy the natural boundary conditions for $m = 1$ was seen in Figure 5.10 to cause a boundary bias effect for estimation of μ^* and this, in turn, has produced very significant undercoverage. However, away from the upper boundary the coverages associated with estimation of μ^* are at least as high as those for estimating μ^{**} and there is some increase in coverage for μ^* over μ^{**} in the regions in Figure 5.13 where estimation of μ^* produces smaller empirical bias.

Another application of the Bayesian properties for smoothing splines is to the development of diagnostic techniques. Identity (5.53) suggests using fit diagnostics that are direct parallels of those from parametric linear regression. For example, we might define "studentized" residuals by

$$r_i = \frac{e_i}{\hat{\sigma}\sqrt{1 - s_i}},\ i = 1, \ldots, n, \tag{5.64}$$

with e_i the ith element of the residual vector $\mathbf{e} = (\mathbf{I} - \mathbf{S}_\lambda)\mathbf{y}$ and $\hat{\sigma}^2$ some variance estimator such as the GSJS estimator (2.29)-(2.30). Measures of influence could be motivated similarly with

$$d_i = \frac{r_i^2 s_i}{(1 - s_i)\mathrm{tr}\mathbf{S}_\lambda},\ i = 1, \ldots, n, \tag{5.65}$$

providing smoothing spline analogs of Cook's distance measures. Notice that the smoothing spline *leverage values* s_i in (5.64)-(5.65) have similar properties to those from linear regression in that $0 \le s_i \le 1, i = 1, \ldots, n$, and s_i gives a measure of distance between the ith row of the design matrix $\mathbf{X}$ and its centroid $\bar{\mathbf{x}} = n^{-1}\mathbf{X}^T\mathbf{1}$ with $\mathbf{1}$ an n-vector of all unit elements (Exercise 18).

As an illustration of the use of the diagnostics discussed above, consider the German hyperinflation data shown in Figure 5.14. This data consists of values for the logarithm of the money supply as a function of the logarithm for the premium, or discount, on a forward contract for foreign exchange (renormalized onto the interval [0, 1]) during the German hyperinflation (Ansley and Wecker 1981). A cubic smoothing spline was fit to this data with the smoothing parameter value $\widehat{\lambda}$ selected by GCV. The resulting curve estimate is also depicted in the plot.

Figures 5.15-5.17 give plots of the leverage values, studentized residuals and Cook's distance measures that were obtained from the fit in Figure 5.14. The marked observations, numbers 19, 24, 28, 29 and 31, in the figures represent the most influential observations in terms of Cook's distance measures. Their influence on the fit stems from different reasons which can be understood by examination of our diagnostic measures.

From Figure 5.15 we see that the influence of observations 19 and 31 is primarily due to high leverage. The large leverage values for these two

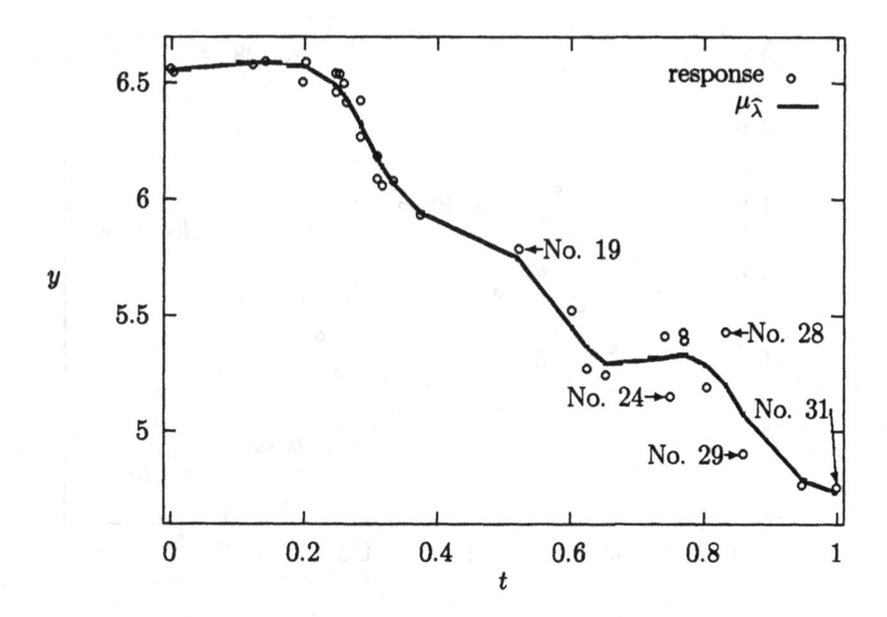

Figure 5.14: Smoothing Spline Fit to Hyperinflation Data.

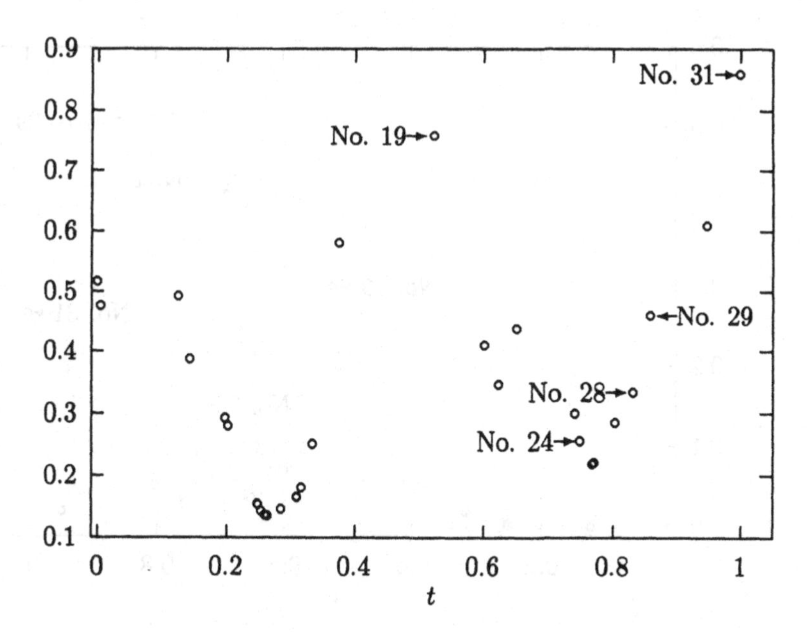

Figure 5.15: Leverage Values for Hyperinflation Data.

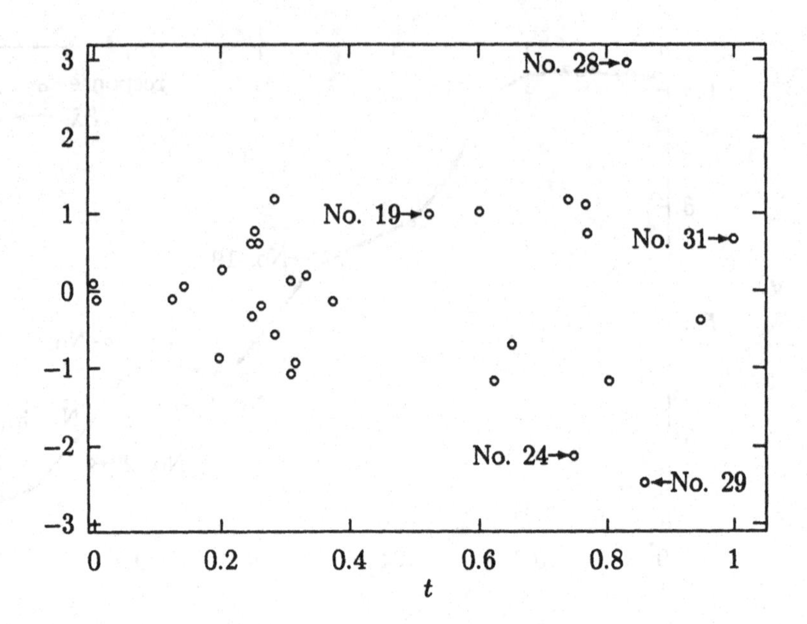

Figure 5.16: Studentized Residuals for Hyperinflation Data.

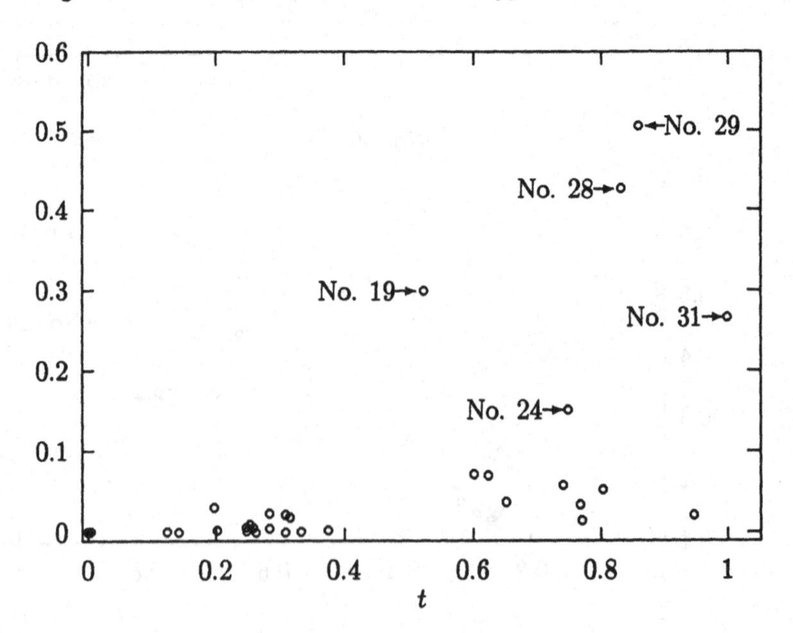

Figure 5.17: Cook's Distance Measures for Hyperinflation Data.

observations have resulted from the location of their t ordinates in a sparse region of the "design" and the upper boundary.

The influence of observations 28 and 29 derives from the fact that they lie significantly far from the fitted curve as measured by their values for studentized residuals in Figure 5.16. Observation 24 also has a "significantly" large studentized residual. However, it is not a high leverage point and, as indicated by the lower value for its corresponding Cook's distance measure, is less influential on the fit.

Our diagnostic analysis of the fit to the hyperinflation data has therefore pointed out several possible sources of difficulty. Specific remedial measures might include consideration of alternative transformations for the data, down-weighting certain points in the smoothing criterion (5.1) or the use of robust smoothing methodology which takes leverage into account.

5.7 Extensions

One of the advantages of the penalized least-squares approach to data smoothing is its flexibility for handling other problems. In this section we discuss ways that criterion (5.1) can be adapted for use in other data smoothing environments.

Throughout the majority of this book we have concentrated on a model of the form

$$y_i = \mu_i + \varepsilon_i,\ i = 1, \ldots, n,$$

where the unknown mean values $\mu_1, \ldots, \mu_n$ were the objects of interest. For example, in model (5.2) we have $\mu_i = \mu(t_i)$ for some regression function μ and our emphasis was on the recovery of the function μ from the data. A more general formulation can be obtained by assuming that the μ_i represent transformations of an unknown function $\eta \in W_2^m[0, 1]$. For example, suppose that $\mu_i = L_i\eta, i = 1, \ldots, n$, for some known set of functionals $L_1, \ldots, L_n$. Then, model (5.2) can be re-expressed as

$$y_i = L_i\eta + \varepsilon_i,\ i = 1, \ldots, n. \tag{5.66}$$

The objective is then to estimate η rather than the $\mu_i = L_i\eta, i = 1, \ldots, n$.

One specific example of transformations in (5.66) is point evaluation where $L_i\eta = \eta(t_i), i = 1,\ldots,n$, which allows us to recover our original model (5.2). A fairly general type of functional that is often useful arises from integration against a bivariate kernel K to obtain

$$L_i\eta = \mu(t_i) = \int_0^1 K(t_i, s)\eta(s)ds,\ i = 1,\ldots,n.$$

In particular, the kernel

$$K(t, s) = [(m-1)!]^{-1}(t-s)_+^{m-1}$$

gives $\eta = \mu^{(m)}$ so that derivative estimation falls into this framework. When $K(t, s)$ depends on s and t only through their difference it is possible to write

$$\mu(t) = \int_0^1 K(t-s)\eta(s)ds$$

and the regression function is obtained by convolving η with a univariate kernel K. The problem of estimating η is then called *deconvolution*.

An obvious way to extend the concept of smoothing splines for estimation under (5.66) is to estimate η by the minimizer η_λ of

$$n^{-1}\sum_{i=1}^{n}(y_i - L_if)^2 + \lambda\int_0^1 f^{(m)}(t)^2dt \tag{5.67}$$

over all $f \in W_2^m[0,1]$. The resulting estimator is called a *method of regularization* (MOR) estimator of η.

It turns out that η_λ is a linear estimator which can be computed by solving an equation system similar to (5.6). To select λ for MOR estimators one might again use cross-validation type methods. If we let $\widetilde{\mu}_{\lambda i} = L_i\eta_\lambda, i = 1,\ldots,n$, then the fact that η_λ is a linear estimator means that $(\widetilde{\mu}_{\lambda 1},\ldots,\widetilde{\mu}_{\lambda n})^T = \widetilde{\mathbf{S}}_\lambda\mathbf{y}$ for some "hat" matrix $\widetilde{\mathbf{S}}_\lambda$. Thus, a generalized cross-validation criterion could be defined as

$$\mathrm{GCV}(\lambda) = n^{-1}\sum_{i=1}^{n}(y_i - L_i\eta_\lambda)^2/[1 - n^{-1}\mathrm{tr}\widetilde{\mathbf{S}}_\lambda]^2$$

and this could be minimized to effect a choice for λ. The problem is that such a level of smoothing is aimed at estimating the minimizer of the *range risk* $n^{-1}\sum_{i=1}^{n} \mathrm{E}(\mu_i - L_i\eta_\lambda)^2$. However, when η is of interest it makes more sense to estimate the value of λ which minimizes a ***domain risk*** such as $\int_0^1 \mathrm{E}(\eta(t) - \eta_\lambda(t))^2\, dt$. This is complicated by the fact that there is no direct data on η and, hence, no obvious parallels of the GCV and UBR criteria that could be used for estimation of this risk function. Wahba (1990, Section 8.4) shows that in certain cases minimizing the range risk can provide an effective estimate of η. However, Rice (1986b) notes this will not always be true and that in some cases it will be preferable to attempt to estimate the minimizer of the domain risk. (See Rice 1986b and O'Sullivan 1986 for discussion of how this might be done.) The question of whether or not GCV/UBR levels of smoothing for estimating the regression function will also work for estimation of η hinges, to a large extent, on whether the optimal rates of decay for the λ's minimizing the range and domain risks are the same. There are instances such as derivative estimation where the two rates will coincide (see, e.g., Rice and Rosenblatt 1983 and Exercise 16 of Chapter 4). There are also cases where they will not.

There are various ways to generalize the concept of MOR estimators. One useful formulation allows for the presence of nonlinear as well as linear functionals in model (5.66). (Wahba 1990, Section 8.3). It is also possible to incorporate convexity/concavity type constraints into the estimation problem. Minimization of (5.66) subject to f in some convex subset of $W_2^m[0,1]$ is typically a difficult problem. However, it is possible to show that there are unique solutions in some cases that can at least be partially characterized. See, e.g., Laurent (1969), Wright and Wegman (1980), and Utreras (1985). Things become somewhat simpler if the contraints are of a linear inequality nature since, in that case, computation of the estimator is a quadratic programming problem (Villalobos and Wahba 1987 and Wahba 1990, Section 9.4).

One of the assumptions in model (5.2) was that the design points were scalar valued. More generally, we might let $\mathbf{t}_i = (t_{i1}, \ldots, t_{ip})^T, i = 1, \ldots, n$, be known p-vectors. In that case model (5.2) becomes

$$y_i = \mu(\mathbf{t}_i) + \varepsilon_i,\ i = 1, \ldots, n,$$

with μ now being a function of p variables. There are a number of ways to extend criterion (5.1) to allow for estimation in this case depending on how one defines a parallel of J_m in (5.3) for the multivariate case. One possibility is to use

$$J_{m,p}(f) = \sum_{k_1+\cdots+k_p=m} \frac{m!}{k_1!\cdots k_p!} \int_{-\infty}^{\infty} \cdots \int_{-\infty}^{\infty} \left(\frac{\partial^m f(\mathbf{t})}{\partial t_1^{k_1} \cdots \partial t_p^{k_p}} \right)^2 dt_1 \cdots dt_p$$

which produces what are known as ***Laplacian*** or ***thin plate smoothing splines*** (Wahba 1990, Section 2.4). These are linear estimators that can be computed by solving a parallel of equation system (5.6) with λ then being selected by cross-validation or related methods. Estimation in the multivariate setting can be extended to multivariate MOR problems and/or constraints as in Wahba and Wendelberger (1980) and Villalobos and Wahba (1987).

Another approach to modeling multivariate data is to use an ANOVA type decomposition for μ into a sum of functions involving only single variables, pairs of variables, triples of variables, etc. For example, if $p = 2$ we would write $\mu(t_1, t_2) = \mu_1(t_1) + \mu_2(t_2) + \mu_{12}(t_1, t_2)$, where the functions μ_1, μ_2 and μ_{12} are required to satisfy conditions which ensure that all three functions are identifiable. The functions μ_1 and μ_2 are called the main effect functions and μ_{12} is termed the interaction function. Similar decompositions can be developed for general p which look like

$$\mu(\mathbf{t}) = \sum_{i=1}^{p} \mu_i(t_i) + \sum_{i<j} \mu_{ij}(t_i, t_j) + \cdots + \mu_{12\cdots p}(t_1, \ldots, t_p).$$

General developments and theory for this type of modeling can be found in Stone (1985, 1986). In some cases it may be reasonable to assume that all the interaction functions are negligible in which case estimation of μ reduces to estimating the main effects $\mu_i(\cdot), i = 1, \ldots, p$. Estimation techniques that can be used in this and related settings are discussed in Hastie and Tibshirani (1990). The spline smoothing paradigm can be extend to additive modeling through use of a criterion such as

$$n^{-1} \sum_{i=1}^{n} (y_i - \sum_{j=1}^{p} f_j(t_{ij}))^2 + \sum_{j=1}^{p} \lambda_j J(f_j)$$

with J a suitable penalty function and smoothing parameters $\lambda_1, \ldots, \lambda_p$. This basic idea can be extended to include estimation of interactions as well and the resulting estimators are known as ***interaction splines.*** Discussions of these estimators can be found in Wahba (1990, Chapter 10) and Wahba, et al (1995).

Another practically important variation of model (5.2) allows the random errors to be correlated with positive definite variance-covariance matrix $\boldsymbol{\Sigma}$. A smoothing spline estimator of μ in this case can be obtained from the criterion

$$n^{-1} \sum_{i=1}^{n} \sum_{j=1}^{n} \sigma^{ij} (y_i - f(t_i))(y_j - f(t_j)) + \lambda J_m(f), \tag{5.68}$$

where σ^{ij} is the ijth element of $\boldsymbol{\Sigma}^{-1}$. The resulting estimator is again a natural spline whose basis function coefficients can be found by solving a parallel of system (5.6) (Exercise 29). However, the elements of $\boldsymbol{\Sigma}$ will generally be unknown which causes difficulties in actually using this type of smoother. Methods for estimating the covariance parameters in (5.68) and selection of the level of smoothing for the associated smoothing spline are considered in Diggle and Hutchinson (1989), Kohn, Ansley and Wong (1992) and Wang (1998).

A general approach to smoothing that is suitable for a myriad of problems would have us optimize

$$n^{-1} \sum_{i=1}^{n} \rho_i(y_i, L_i f) + \lambda J(f) \tag{5.69}$$

over $f \in C \subset W$, where C is some convex subset of a general function space W with associated smoothness measure J, and the ρ_i and $L_i, i = 1, \ldots, n$, are known functionals. If, for example, $L_i f = f(t_i), \rho_i(y_i, f) = (y_i - f(t_i))^2, i = 1, \ldots, n$, $C = W = W_2^m[0,1]$ and $J = J_m$, we obtain our original smoothing spline. However, if we alter this slightly and instead use $\rho_i(y_i, f) = \rho(y_i - f(t_i)), i = 1, \ldots, n$, for some general distance measure ρ, this allows us to produce robustifed smoothing splines that parallel the robust kernel estimators discussed in Section 4.11 (Cox 1983). Alternatively, if we choose $\rho_i(y_i, f), i = 1, \ldots, n$, to be the negative of a log likelihood function, we get a smoothing criterion that can be used with likelihood based models such as those arising in the generalized linear models environment. See, e.g., O'Sullivan, Yandell and Raynor (1986), Wahba (1990, Section 9.2) and Wood and Kohn (1998).

Finally, we note that it is straightforward to generalize spline smoothing to a partially linear model setting. Suppose, for example, that the responses satisfy

$$y_i = \mathbf{u}_i^T \gamma + f(t_i) + \varepsilon_i,\ i = 1, \ldots, n,$$

for $\varepsilon_1, \ldots, \varepsilon_n$ zero mean uncorrelated random errors, $\mathbf{u}_1, \ldots, \mathbf{u}_n$ known p-vectors, γ an unknown parameter vector and f an unknown function in $W_2^m[0,1]$. Then, to estimate γ and f we can minimize

$$n^{-1} \sum_{i=1}^{n} \left(y_i - \mathbf{u}_i^T \mathbf{c} - f(t_i) \right)^2 + \lambda J_m(f)$$

over all $\mathbf{c} \in \mathbb{R}^p$ and $f \in W_2^m[0,1]$. Setting $\mathbf{U}^T = [\mathbf{u}_1, \ldots, \mathbf{u}_n]$ we obtain the estimators

$$\hat{\mathbf{f}} = (\hat{f}(t_1), \ldots, \hat{f}(t_n))^T = \mathbf{S}_\lambda(\mathbf{y} - \mathbf{U}\hat{\gamma}) \tag{5.70}$$

and

$$\hat{\gamma} = [\mathbf{U}^T(\mathbf{I} - \mathbf{S}_\lambda)\mathbf{U}]^{-1}\mathbf{U}^T(\mathbf{I} - \mathbf{S}_\lambda)\mathbf{y}, \tag{5.71}$$

of $\mathbf{f} = (f(t_1), \ldots, f(t_n))^T$ and γ with $\mathbf{y}$ the response vector and $\mathbf{S}_\lambda$ the smoothing spline hat matrix in (5.9). The estimator of the mean function that results from all this is known as a ***partial spline*** and originated from work by Rice (1982), Engle et al (1986) and Wahba (1984).

The smoothing parameter for the partial spline estimator can, of course, be chosen via cross-validation type methods. The extreme case of $\lambda = \infty$ corresponds to linear regression of the responses on $\mathbf{U}$ and a polynomial of order m in t (Exercise 24). Thus, when $m = 2$ partial splines provide a generalization of the classical approach to analysis of covariance.

Notice that an application of the Bias Reduction Lemma as in Section 3.6 leads us to instead estimate the partially linear model parameters by

$$\tilde{\mathbf{f}} = \mathbf{S}_\lambda(\mathbf{y} - \mathbf{U}\tilde{\gamma}_\lambda) \tag{5.72}$$

and

$$\widetilde{\gamma} = [\mathbf{U}^T(\mathbf{I}-\mathbf{S}_\lambda)^2\mathbf{U}]^{-1}\mathbf{U}^T(\mathbf{I}-\mathbf{S}_\lambda)^2\mathbf{y}\,. \tag{5.73}$$

Our discussions in Section 4.9 suggest that (5.73) might be preferred over the partial spline coefficient estimator from a bias perspective. Results in Heckman (1986), Rice (1986c), Speckman (1988) and Shiau and Wahba (1988) tend to support this conclusion.

The computation of (5.72)-(5.73) is actually not much more difficult than computing an ordinary smoothing spline. One first transforms $\mathbf{y}$ and $\mathbf{U}$ to $\widetilde{\mathbf{y}} = (\mathbf{I}-\mathbf{S}_\lambda)\mathbf{y}$ and $\widetilde{\mathbf{U}} = (\mathbf{I}-\mathbf{S}_\lambda)\mathbf{U}$. Then, $\widetilde{\gamma}$ can be obtained by ordinary linear regression of $\widetilde{\mathbf{y}}$ on $\widetilde{\mathbf{U}}$: i.e., by solving the normal equations $\widetilde{\mathbf{U}}^T\widetilde{\mathbf{U}}\mathbf{c} = \widetilde{\mathbf{U}}^T\widetilde{\mathbf{y}}$. It follows that (5.72)-(5.73) can be calculated in order n operations using one of the efficient methods from Section 5.4 to conduct the transformations of $\mathbf{y}$ and $\mathbf{U}$.

For fixed λ the covariance matrix for $\widetilde{\gamma}$ is

$$\begin{aligned}\mathrm{Var}(\widetilde{\gamma}) &= \sigma^2(\widetilde{\mathbf{U}}^T\widetilde{\mathbf{U}})^{-1}\widetilde{\mathbf{U}}^T(\mathbf{I}-\mathbf{S}_\lambda)^2\widetilde{\mathbf{U}}(\widetilde{\mathbf{U}}^T\widetilde{\mathbf{U}})^{-1}\\ &= \sigma^2(\widetilde{\mathbf{U}}^T\widetilde{\mathbf{U}})^{-1}\check{\mathbf{U}}^T\check{\mathbf{U}}(\widetilde{\mathbf{U}}^T\widetilde{\mathbf{U}})^{-1} \end{aligned}\tag{5.74}$$

with $\check{\mathbf{U}}^T = (\mathbf{I}-\mathbf{S}_\lambda)\widetilde{\mathbf{U}}$. The value of σ^2 in (5.74) can be estimated using, for example, the GSJS type estimator $\widehat{\sigma}^2$ (4.93)-(4.94). The covariances can then be estimated by solving a $p\times p$ linear equation system (Exercise 25).

The estimator of the mean vector $\boldsymbol{\mu} = \mathbf{U}\gamma + \mathbf{f}$ that arises from (5.72)-(5.73) is

$$\widetilde{\boldsymbol{\mu}} = \mathbf{U}\widetilde{\gamma} + \widetilde{\mathbf{f}} = [\mathbf{S}_\lambda + \widetilde{\mathbf{U}}(\widetilde{\mathbf{U}}^T\widetilde{\mathbf{U}})^{-1}\widetilde{\mathbf{U}}^T(\mathbf{I}-\mathbf{S}_\lambda)]\mathbf{y}. \tag{5.75}$$

To use either the GCV or UBR criteria for selecting the level of smoothing we will therefore need to compute $\mathrm{tr}\mathbf{S}_\lambda$ and $\mathrm{tr}(\widetilde{\mathbf{U}}^T\widetilde{\mathbf{U}})^{-1}\widetilde{\mathbf{U}}^T\check{\mathbf{U}}$. We have already discussed efficient methods for computing the trace of $\mathbf{S}_\lambda$ in Section 5.4. To compute the second trace, we first transform $\widetilde{\mathbf{U}}$ to $\check{\mathbf{U}}$, an $O(n)$ computation, and then solve the "normal" equation system $\widetilde{\mathbf{U}}^T\widetilde{\mathbf{U}}\mathbf{C} = \widetilde{\mathbf{U}}^T\check{\mathbf{U}}$ for the $p\times p$ matrix $\mathbf{C}$. If the solution is $\widetilde{\mathbf{C}} = \{\widetilde{c}_{ij}\}_{i,j=1,p}$, then the desired trace is just $\widetilde{c}_{11} + \cdots + \widetilde{c}_{pp}$.

The basic algorithm we have just described allows for computation of (5.72)-(5.73) and the corresponding GCV or UBR criteria efficiently in $O(n)$

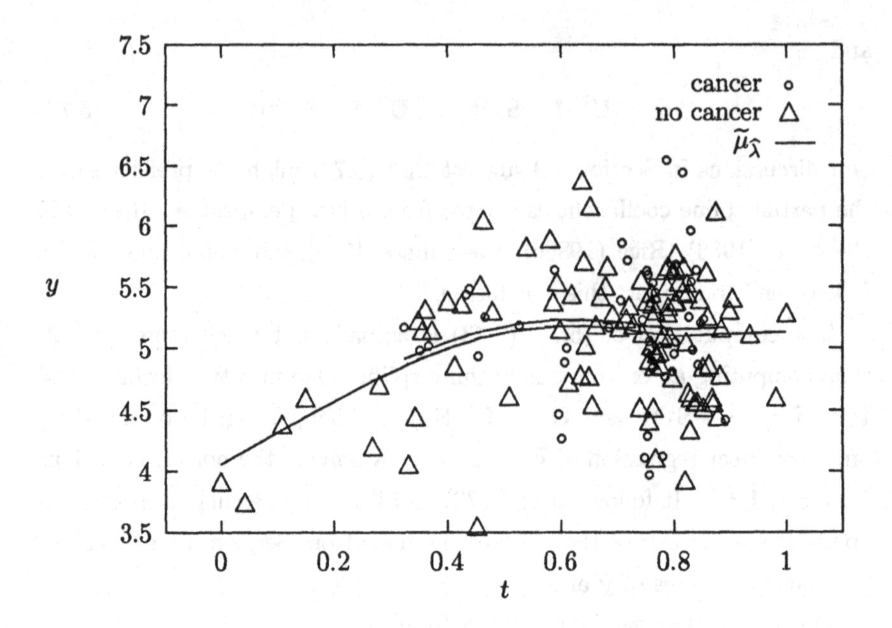

Figure 5.18: Spline Fit to Assay Data.

calculations provided only that an order n method is used for computing univariate spline smooths of $\mathbf{y}$, $\mathbf{U}$ and $\widetilde{\mathbf{U}}$. A similar $O(n)$ algorithm can be developed for use with the partial spline estimators (5.70)-(5.71) (Exercise 26).

To illustrate the use of estimators (5.72)-(5.73), consider the data in Figure 5.18. This data represents the levels of a chemical in blood taken from pregnant women at day t of pregnancy (normalized onto [0,1]) with the u_i being scalar valued indicator variables that are either 1 or 0 depending on whether or not cancer occurred in the subject. A plot of the data is shown in Figure 5.18 along with the mean estimator (5.75) with λ selected using the UBR criterion. The estimator of γ was found to be $\widetilde{\gamma} = .054$ with estimated standard error .088. Thus, the data does not appear to support a relationship between cancer and levels of this chemical.

The discussions in this section have done little more than touch the surface of the range of applications for spline and penalized smoothing methodology. Further reading on the topics treated here as well as many others can be found in Wahba (1990), Green and Silverman (1994) and references therein.

5.8 Appendix

In this section we detail some of the properties of natural splines that are used in this chapter. First let us define a *spline* of order r with *knots* at $\xi_1, \ldots, \xi_k$ to be any function s of the form

$$s(t) = \sum_{j=0}^{r-1} \theta_j t^j + \sum_{j=1}^{k} \eta_j (t - \xi_j)_+^{r-1} \tag{5.76}$$

for some set of coefficients $\theta_0, \ldots, \theta_{r-1}, \eta_1, \ldots, \eta_k$. This definition is equivalent to saying that

i) s is a piecewise polynomial of order r on any subinterval $[\xi_i, \xi_{i+1})$,

ii) s has $r-2$ continuous derivatives and

iii) s has a discontinuous $(r-1)$st derivative with jumps at $\xi_1, \ldots, \xi_k$.

Thus, a spline is a piecewise polynomial whose different polynomial segments have been joined (tied) together at the knots $\xi_1, \ldots, \xi_k$ in a fashion which insures certain continuity properties. Notice, in particular, that a spline is the smoothest possible piecewise polynomial which still retains a segmented nature.

Let $S^r(\xi_1, \ldots, \xi_k)$ denote the set of all functions of the form (5.76). Then, $S^r(\xi_1, \ldots, \xi_k)$ is a vector space in the sense that sums of functions in $S^r(\xi_1, \ldots, \xi_k)$ remain in the set, etc. Since the functions $1, t, \ldots, t^{r-1}$, $(t - \xi_1)_+^{r-1}, \ldots, (t-\xi_k)_+^{r-1}$ are linearly independent, it follows that $S^r(\xi_1, \ldots, \xi_k)$ has dimension $k+r$.

Of particular importance for this chapter is the set of natural splines of order $r = 2m$ with $k = n$ knots at the design points. A spline function is a *natural spline* of order $2m$ with knots at $t_1, \ldots, t_n$ if, in addition to properties i)-iii), it satisfies

iv) s is a polynomial of order m outside of $[t_1, t_n]$.

The name natural spline stems from the fact that, as a result of iv), s satisfies the natural boundary conditions (5.46).

Let $NS^{2m}(t_1,\ldots,t_n)$ denote the collection of all natural splines of order $2m$ with knots at $t_1,\ldots,t_n$. Then, $NS^{2m}(t_1,\ldots,t_n)$ is a subspace of $S^{2m}(t_1,\ldots,t_n)$ obtained by placing $2m$ linear restrictions arising from property iv) on the coefficients in (5.76). In particular, we see that in order for s to be a natural spline we must have

$$\theta_m = \cdots = \theta_{2m-1} = 0 \tag{5.77}$$

in (5.76) since s must be a polynomial of order m for $t < t_1$. One may verify that $NS^{2m}(t_1,\ldots,t_n)$ has dimension n.

A useful quadrature property of natural splines is detailed in the following lemma from Lyche and Schumaker (1973).

Lemma 5.1. *Let $x_1,\ldots,x_n$ be a basis for $NS^{2m}(t_1,\ldots,t_n)$. Then, there are coefficients $\theta_{0j},\ldots,\theta_{(m-1)j},\eta_{1j},\ldots,\eta_{nj}$ such that*

$$x_j(t) = \sum_{i=0}^{m-1}\theta_{ij}t^i + \sum_{i=1}^{n}\eta_{ij}(t-t_i)_+^{2m-1}. \tag{5.78}$$

If $s(t) = \sum_{j=1}^{n} b_j x_j(t)$ and $f \in W_2^m[0,1]$ then

$$\int_0^1 f^{(m)}(t)s^{(m)}(t)dt = (-1)^m(2m-1)!\sum_{i=1}^{n} f(t_i)\sum_{j=1}^{n} b_j\eta_{ij}. \tag{5.79}$$

Proof. The first part of the lemma is merely a restatement of the fact that $NS^{2m}(t_1,\ldots,t_n) \subset S^{2m}(t_1,\ldots,t_n)$ and that (5.77) holds. Thus, we need only verify (5.79). To accomplish this, first observe that $s^{(m+j)}$, $j = 0,\ldots,m-1$, vanishes outside of $[t_1,t_n]$ as a result of property iv). Then, integrate by parts to obtain

$$\begin{aligned}
\int_0^1 f^{(m)}(t)s^{(m)}(t)dt &= (-1)^{m-1}\int_0^1 f'(t)s^{(2m-1)}(t)dt \\
&= (-1)^{m-1}\sum_{i=1}^{n-1}\int_{t_i}^{t_{i+1}} f'(t)s^{(2m-1)}(t)dt \\
&= (-1)^{m-1}(2m-1)!\sum_{i=1}^{n-1}\Bigg\{[f(t_{i+1})-f(t_i)] \\
&\quad \times\left(\sum_{j=1}^{n} b_j\sum_{r=1}^{i}\eta_{rj}\right)\Bigg\} \\
&= (-1)^{m-1}(2m-1)!\sum_{j=1}^{n}\Bigg[b_j\Bigg\{\sum_{i=2}^{n} f(t_i)\sum_{r=1}^{i-1}\eta_{rj} \\
&\quad -\sum_{i=1}^{n-1} f(t_i)\sum_{r=1}^{i}\eta_{rj}\Bigg\}\Bigg] \\
&= (-1)^{m}(2m-1)!\sum_{j=1}^{n} b_j\sum_{i=1}^{n} f(t_i)\eta_{ij}.
\end{aligned}$$

In the last step we used the fact that $x_j^{(2m-1)}(t) = (2m-1)!\sum_{i=1}^{n}\eta_{ij} = 0$ for $t > t_n$ as a result of property iv). •

Lemma 5.1 is the key to establishing the following optimal interpolation property for natural splines.

Lemma 5.2. *Let $x_1, \ldots, x_n$ be a basis for $NS^{2m}(t_1, \ldots, t_n)$ with associated design matrix $\mathbf{X} = \{x_j(t_i)\}_{i,j=1,n}$ and let $\mathbf{a} = (a_1, \ldots, a_n)^T$ be a specified vector of constants. Then, if $n \geq m$, the unique minimizer of $J_m(f)$ over all $f \in W_2^m[0,1]$ that satisfy $f(t_i) = a_i, i = 1, \ldots, n$, is $s = \sum_{j=1}^{n} b_j x_j$, where $\mathbf{b} = (b_1, \ldots, b_n)^T$ is the unique solution to $\mathbf{Xc} = \mathbf{a}$. In particular, the matrix $\mathbf{X}$ has full rank n.*

Proof. Let us first show that $\mathbf{X}$ has full rank. For this purpose it suffices to show that $\mathbf{Xc} = \mathbf{0}$ if and only if $\mathbf{c} = \mathbf{0}$.

Now for any $\mathbf{c}$ with $\mathbf{Xc} = \mathbf{0}$ the associated natural spline function $s = \sum_{j=1}^{n} c_j x_j$ satisfies $(s(t_1), \ldots, s(t_n))^T = \mathbf{Xc} = \mathbf{0}$. Consequently, by Lemma 5.1, $\int_0^1 s^{(m)}(t)^2 dt = 0$ which entails that s is a polynomial of order m that interpolates the zero function at $n \geq m$ points. Hence, $s = \sum_{j=1}^{n} c_j x_j \equiv 0$ which means that $\mathbf{c} = \mathbf{0}$ because the x_j are a basis.

To see that $s = \sum_{j=1}^{n} b_j x_j$ is the minimizer of $J_m(f)$, let $g \in W_2^m[0,1]$ be any other function that satisfies $g(t_i) = a_i, i = 1, \ldots, n$. Then,

$$J_m(g) = J_m(s) + 2\int_0^1 s^{(m)}(t)(g^{(m)}(t) - s^{(m)}(t))dt + J_m(g - s).$$

However, $\int_0^1 s^{(m)}(t)(g^{(m)}(t) - s^{(m)}(t))dt = 0$ due to Lemma 5.1. So, $J_m(g) > J_m(s)$ unless $J_m(g - s) = 0$. But this latter event can occur only if $g - s$ is a polynomial of order m that agrees with the zero function at $n \geq m$ points, i.e., only if $g - s \equiv 0$. •

5.9 Exercises

1. Obtain plots of the linear smoothing spline weight function $K_n(t, \cdot; \lambda)$ in (5.15) for various values of λ and t. Use these plots to discuss the behavior of the weight function for large and small λ and for estimation in the interior and boundaries of [0, 1].

2. Show that $\sum_{i=1}^{n} K_n(t, t_i; \lambda) = n$ for all $t \in [0,1]$ with $K_n(t, \cdot; \lambda)$ defined in (5.15) and a uniform design (5.10). More generally, show that $\mathbf{S}_\lambda \mathbf{t}^j = \mathbf{t}^j, j = 0, \ldots, m-1$, for $\mathbf{S}_\lambda$ the smoothing spline hat matrix for general m and $\mathbf{t}^j = (t_1^j, \ldots, t_n^j)^T$.

3. Prove Theorem 5.3.

4. Suppose that our data follows the model $y_{ij} = \mu(t_i) + \varepsilon_{ij}, j = 1, \ldots, n_i$, $i = 1, \ldots, n$. Show that the smoothing spline fit in this case is obtained by minimizing

$$N^{-1}\sum_{i=1}^{n} n_i \left(\bar{y}_i - f(t_i)\right)^2 + \lambda \int_0^1 f^{(m)}(t)^2 dt,$$

where $N = \sum_{i=1}^{n} n_i$ and $\bar{y}_i = n_i^{-1} \sum_{j=1}^{n_i} y_{ij}, i = 1, \ldots, n$.

5. Schoenberg (1964) shows that $RSS(\mu(\cdot; \rho))$ is a differentiable function of ρ. Use this fact to show that $\lambda = -n^{-1} \frac{d}{d\rho} RSS(\mu(\cdot; \rho))$ and interpret the meaning of this result.

6. Let $\mu_{\lambda(j)}$ be the smoothing spline fit to the data excluding the jth observation (t_j, y_j). Show that $\mu_{\lambda(j)}(t_j) = (\mu_\lambda(t_j) - s_j y_j)/(1 - s_j)$, where s_j is the jth diagonal element of $\mathbf{S}_\lambda$ in (5.9). Use this to conclude that $\mathrm{CV}(\lambda) = n^{-1}\sum_{i=1}^n (y_i - \mu_\lambda(t_i))^2/(1-s_i)^2$ and, hence, that the GCV criterion is a weighted version of the CV criterion.

7. Suppose we fit a smoothing spline using criterion (5.17) and choose λ by minimizing $\mathrm{GCV}(\lambda) = n^{-1}\sum_{i=1}^n w_i(y_i - \mu_\lambda(t_i))^2/(1 - n^{-1}\mathrm{tr}\mathbf{S}_\lambda)^2$ with $\mathbf{S}_\lambda$ the smoothing spline hat matrix from Theorem 5.3. Derive the form of this GCV criterion in the case of n_i replicate observations at each design point t_i all of which are to be weighted by a common factor w_i. In particular, derive criterion (5.21) that was used to select the level of smoothing for the ELISA data in Section 5.3. Comment on the computational implications these criteria have for smoothing with replicate data.

8. Define a natural linear spline basis by

$$B_1(t) = \begin{cases} 1, & t < t_1, \\ (t_2 - t)/(t_2 - t_1), & t_1 \le t < t_2, \\ 0, & t \ge t_2, \end{cases}$$

$$B_r(t) = \begin{cases} 0, & t \le t_{r-1}, \\ (t - t_{r-1})/(t_r - t_{r-1}), & t_{r-1} < t < t_r, \\ (t_{r+1} - t)/(t_{r+1} - t_r), & t_r \le t < t_{r+1}, \\ 0, & t \ge t_{r+1}, \end{cases}$$

for $r = 2, \ldots, n-1$, and

$$B_n(t) = \begin{cases} 0, & t \le t_{n-1}, \\ (t - t_{n-1})/(t_n - t_{n-1}), & t_{n-1} < t < t_n, \\ 1, & t \ge t_n. \end{cases}$$

   (a) Use this basis to obtain a band limited form for equation system (5.6).

   (b) Use your results from part (a) to develop an efficient $O(n)$ algorithm for computing the linear smoothing spline.

(c) Develop an $O(n)$ algorithm for computing the diagonal elements and trace of the hat matrix for a linear smoothing spline.

(d) Implement your results from (a)-(c) in a computer program that includes automated selection of the smoothing parameter using the GCV and unbiased risk criteria.

9. Develop an $O(n)$ algorithm for computing the diagonal elements and trace of the hat matrix for cubic smoothing splines. Combine this with the algorithm for computing the cubic spline fitted values in Section 5.4 to develop computer code for cubic spline smoothing that includes automated selection of the smoothing parameter.

10. Derive the asymptotic form of the linear smoothing spline weight function under the uniform design (5.10) when estimation is being conducted at a point $t = 1 - \sqrt{\lambda}q$ for q a positive constant and verify that the moment condition (5.37) continues to hold in this case.

11. Derive the asymptotically optimal smoothing parameters for estimation of regression functions (5.44)-(5.45) using a linear smoothing spline in the case of a uniform design (5.10). Discuss the reasons for the difference in the resulting levels of smoothing for the two regression curves.

12. Show that for general m we have $n^{-1}\sum_{i=1}^{n}(\mathrm{E}\mu_\lambda(t)-\mu(t))^2 \le \lambda J_m(\mu)$.

13. Cox (1983) and Speckman (1985) show that under certain conditions $n^{-1}\sum_{i=1}^{n}\mathrm{Var}(\mu_\lambda(t_i)) \sim \sigma^2 C/(n\lambda^{1/2m})$ for C a constant that depends only on m and the design. Use this with your bound from Exercise 12 to give an argument that $n^{-1}\sum_{i=1}^{n}\mathrm{E}(\mu(t_i) - \mu_\lambda(t_i))^2 = O(n^{-2m/(2m+1)})$ whenever λ is of order $n^{-2m/(2m+1)}$. Discuss the implications of this result.

14. Detail how to boundary correct the linear smoothing spline using the Bias Reduction Lemma of Section 3.6.

(a) Give an explicit expression for your boundary corrected linear smoothing spline and combine this with your algorithm from

Exercise 8 to obtain an $O(n)$ algorithm along with computer code for automated fitting using this new estimator.

(b) Show that your boundary corrected linear smoothing spline is guaranteed to be a second order estimator when the design is uniform and μ'' is Lipshitz continuous.

15. Verify the form of the GML estimator of the smoothing parameter in (5.56).

16. Show that for fixed λ, $E\tilde{\sigma}^2 = E\tilde{\sigma}_W^2 = \sigma^2$ under model (5.47)-(5.48).

17. Consider the case of estimation using a linear smoothing spline when the design satisfies (5.10). Show that with asymptotically optimal global smoothing levels and fixed $t \in (0,1)$, $(\mu_\lambda(t)-\mu(t))/\sqrt{\text{Var}(\mu_\lambda(t))}$ will have a limiting standard normal distribution when μ is given by (5.44) but not when μ is (5.45). Discuss why this occurs and the implications it has for both frequentist and Bayesian confidence interval estimation of these two functions.

18. Show that the diagonal elements $s_1, \ldots, s_n$ of the smoothing spline hat matrix $\mathbf{S}_\lambda$ in (5.9) are at least as large as those for $\mathbf{S}_\lambda^2$. In addition, show that $0 \le s_i \le 1$ and $s_i = n^{-1} + (\mathbf{x}_i - \bar{\mathbf{x}})^T(\mathbf{X}^T\mathbf{X} + n\lambda\mathbf{\Omega})^{-1}(\mathbf{x}_i - \bar{\mathbf{x}}), i = 1, \ldots, n$, for $\mathbf{x}_i^T$ the ith row of $\mathbf{X}$ and $\bar{\mathbf{x}} = n^{-1}\sum_{i=1}^n \mathbf{x}_i$.

19. Fit the simulated data in Table 1.1 using a smoothing spline and automated selection of the smoothing parameter. Compute interval estimators for the values of the regression function at the design points and conduct diagnostic analyses using the Bayesian methods of Section 5.6. Discuss the performance of the Bayesian interval estimators in terms of their coverages of the true regression function values.

20. Fit the voltage drop data in Table 3.6 using a smoothing spline and automated selection of the smoothing parameter. Compute interval estimators for the values of the regression function at the design points and conduct diagnostic analyses using the Bayesian methods of Section 5.6.

21. Fit the windmill data in Table 3.7 using a smoothing spline and automated selection of the smoothing parameter. Compute interval estimators for the values of the regression function at the design points and conduct diagnostic analyses using the Bayesian methods of Section 5.6.

22. Fit the assay data in Table 4.3 using a smoothing spline and automated selection of the smoothing parameter. Compute interval estimators for the values of the regression function at the design points and conduct diagnostic analyses using the Bayesian methods of Section 5.6.

23. Discuss how to use a cubic smoothing spline for estimating the derivative of a regression function. How should λ be chosen in this case? Can you give an argument that cross-validation type methods for estimating μ might also be effective for estimating μ'? If so, then use this to obtain an estimator of the velocity curve for the growth data in Table 4.4.

24. Let $\widehat{\boldsymbol{\mu}} = \mathbf{U}\widehat{\boldsymbol{\gamma}} + \widehat{\mathbf{f}}$ be the partial spline estimator of the mean function from (5.70)-(5.71). If $\mathbf{A} = [\mathbf{U}|\mathbf{T}]$, for $\mathbf{T} = \{t_i^{j-1}\}_{i=1,n,j=1,m}$, has full rank, show that $\lim_{\lambda\to\infty} \widehat{\boldsymbol{\mu}} = \mathbf{A}(\mathbf{A}^T\mathbf{A})^{-1}\mathbf{A}^T\mathbf{y}$ for $\mathbf{y}$ the response vector.

25. Derive a formula that shows how to compute the elements of the covariance matrix (5.74) using the solution $\widetilde{\mathbf{C}}$ of the linear equation system $\widetilde{\mathbf{U}}^T\widetilde{\mathbf{U}}\mathbf{C} = \check{\mathbf{U}}^T$ with $\check{\mathbf{U}} = (\mathbf{I} - \mathbf{S}_\lambda)\widetilde{\mathbf{U}}$, $\widetilde{\mathbf{U}} = (\mathbf{I} - \mathbf{S}_\lambda)\mathbf{U}$ and $\mathbf{S}_\lambda$ the smoothing spline hat matrix from (5.9).

26. Develop an $O(n)$ algorithm for computing the partial spline estimator from (5.70)-(5.71) that includes estimation of the covariance matrix for the coefficient estimators and automated selection of λ using the GCV or unbiased risk criteria.

27. Adapt the code you developed in Exercises 9 and 14 to produce data driven versions of the estimators (5.70)-(5.73).

28. Fit a partial spline estimator to the mouthwash data in Table 3.8 using a data driven choice for the level of smoothing. Is the mouthwash treatment effective at the 5% level?

29. Show that the minimizer of criterion (5.68) over functions in $W_2^m[0,1]$ is a natural spline of order $2m$ with knots at the design points whose basis function coefficients are obtained by solving the linear system $(\mathbf{X}^T\boldsymbol{\Sigma}^{-1}\mathbf{X} + n\lambda\boldsymbol{\Omega})\mathbf{c} = \mathbf{X}^T\boldsymbol{\Sigma}^{-1}\mathbf{y}$.

Chapter 6

Least-Squares Splines

6.1 Introduction

In Chapters 3-5 we have studied series, kernel and smoothing spline estimators which represent what are arguably the most popular approaches to nonparametric regression. In this final chapter we study another way to use spline functions for data smoothing: namely, least-squares splines. These estimators have local fitting qualities similar to those for kernel and smoothing spline estimators. However, least-squares splines do not admit kernel or series representations, even asymptotically, which distinguishes them from the smoothing methods of previous chapters.

6.2 Form of the Estimator

To motivate the estimator's development let us again assume the basic nonparametric regression model where observations $(t_i, y_i), i = 1, \ldots, n$, satisfy

$$y_i = \mu(t_i) + \varepsilon_i,\ i = 1, \ldots, n, \tag{6.1}$$

with the ε_i representing zero mean, uncorrelated random errors having common variance σ^2 and $0 \le t_1 \le \cdots \le t_n \le 1$. Then, if $\mu \in W_2^m[0,1]$ a

Taylor expansion allows us to write (6.1) as

$$y_i = \sum_{j=1}^{m} \theta_j t_i^{j-1} + \text{Rem}(t_i) + \varepsilon_i,\ i = 1, \ldots, n, \tag{6.2}$$

where

$$\text{Rem}(t) = [(m-1)!]^{-1} \int_0^1 \mu^{(m)}(t)(t-\xi)_+^{m-1} d\xi. \tag{6.3}$$

If $\text{Rem}(t_1), \ldots, \text{Rem}(t_n)$ are uniformly small in magnitude, polynomial regression provides a reasonable method of analyzing the data. When this is not the case, some type of protection against non-polynomial structure needs to be built into any scheme for estimating μ.

One method of estimation which attempts to guard against departures from polynomial models is smoothing splines which were studied in Chapter 5. When the smoothing parameter is selected by data driven methods, they accomplish this by, loosely speaking, estimating a bound on the magnitude of $\text{Rem}(\cdot)$ in (6.3) and then incorporating information about this bound into the estimator.

Least-squares splines represent another generalization of polynomial regression estimators which also utilize information in the data about $\text{Rem}(\cdot)$ to compensate for possible inadequacies of a polynomial model. The basic premise is that the integral in (6.3) can be approximated using the quadrature formula

$$\sum_{j=1}^{k} \delta_j (t-\xi_j)_+^{m-1}$$

for coefficients $\delta_1, \ldots, \delta_k$ and points $0 < \xi_1 < \cdots < \xi_k < 1$. Combining this with the original polynomial approximation leads to an overall approximation of the regression function by

$$s(t) = \sum_{j=1}^{m} \theta_j t^{j-1} + \sum_{j=1}^{k} \delta_j (t-\xi_j)_+^{m-1}. \tag{6.4}$$

Functions of this type are discussed in Section 5.8 and are called splines. In particular, s in (6.4) is a spline of order m with knots $\xi_1, \ldots, \xi_k$. The set of all such functions, $S^m(\xi_1, \ldots, \xi_k)$, is an $m + k$ dimensional vector space consisting of mth order piecewise polynomials having $k+1$ segments that are tied together at the knots in a way that provides the maximum amount of smoothness for a piecewise polynomial without losing its segmented nature.

Given a choice for $\lambda = \{\xi_1, \ldots, \xi_k\}$ we can then estimate μ by estimating the coefficients in the approximation (6.4). One method of accomplishing this is to use least squares. Let

$$\begin{aligned} x_j(t) &= t^{j-1}, j = 1, \ldots, m, & (6.5) \\ x_{m+j}(t) &= (t - \xi_j)_+^{m-1}, j = 1, \ldots, k, & (6.6) \end{aligned}$$

and set

$$\boldsymbol{\beta} = (\theta_1, \ldots, \theta_m, \delta_1, \ldots, \delta_k)^T. \tag{6.7}$$

The *least-squares spline estimator* of μ is then given by

$$\mu_\lambda = \sum_{j=1}^{m+k} b_{\lambda j} x_j, \tag{6.8}$$

where $\mathbf{b}_\lambda = (b_{\lambda 1}, \ldots, b_{\lambda(m+k)})^T$ is the estimator of $\boldsymbol{\beta}$ obtained by minimizing the residual sum-of-squares

$$RSS(\mathbf{c}; \lambda) = \sum_{i=1}^{n} \left(y_i - \sum_{j=1}^{m+k} c_j x_j(t_i) \right)^2 \tag{6.9}$$

with respect to $\mathbf{c} = (c_1, \ldots, c_{m+k})^T$. More specifically, if we define

$$\mathbf{X}_\lambda = \{x_j(t_i)\}_{i=1,n,j=1,m+k}, \tag{6.10}$$

then $\mathbf{b}_\lambda$ is found as any solution to the normal equations

$$\mathbf{X}_\lambda^T \mathbf{X}_\lambda \mathbf{c} = \mathbf{X}_\lambda^T \mathbf{y}, \tag{6.11}$$

where, as usual, $\mathbf{y}$ is the response vector. If $\mathbf{X}_\lambda$ has rank $m+k$, which we will assume to be the case,

$$\mathbf{b}_\lambda = \left(\mathbf{X}_\lambda^T \mathbf{X}_\lambda\right)^{-1} \mathbf{X}_\lambda^T \mathbf{y}. \tag{6.12}$$

It follows from (6.8) and (6.12) that, given $\lambda = \{\xi_1, \ldots, \xi_k\}$, μ_λ is a linear estimator of μ. The set λ is actually a smoothing parameter. Its choice requires the selection of both the number and positions of the knots for the spline. As the number of knots, k, is increased, this produces increasingly flexible and potentially more wiggly estimators since the placement of a knot in an area allows the estimator to adapt more to the data in that region. When only a few knots are used, the estimator tends to be smoother in the respect that it becomes more like a polynomial of order m.

6.3 Selecting λ

To use a least-squares spline we must first choose the number and locations of the knots for the estimator. Methods for making these choices are the subject of this section.

Perhaps the simplest approach to knot selection is through visual inspection of the data. This is the analog of the trial and error method for selecting bandwidths for kernel estimators or the smoothing parameter for a smoothing spline. While this approach is heuristic, it often tends to work quite well in terms of giving a visually pleasing fit to a set of data and has a definite computational advantage over the other methods that will be discussed subsequently.

The idea to keep in mind when making knot placements through visual inspection is that more knots will be needed where μ seems to change more rapidly. This is because the estimator will need more flexibility in such regions. By combining this premise with knowledge of the pliancy of the individual polynomial segments for a spline, guidelines can be advanced for knot placement. A set of *ad hoc* rules for locating knots which arise from such consideration is the following:

1. For linear splines ($m = 2$), place knots at points where the data exhibits a change in slope.

2. For quadratic splines ($m = 3$), locate knots near local maxima, minima or inflection points in the data.

3. For cubic splines ($m = 4$), arrange the knots so that they are close to inflection points in the data and not more than one extreme point

(maximum or minimum) and one inflection point occurs between any two knots.

The motivation for guideline 1 is intuitively clear while guideline 2 derives from the fact that a quadratic function has constant curvature and must behave symmetrically about its extreme point. Recognition that a cubic polynomial can accommodate different types of behavior on each side of its maximum or minimum but must behave asymmetrically about its inflection point provides the reasoning behind guideline 3 (Wold 1974).

It will also usually be necessary to decide on a choice of m, the order of the spline. If this is done by examining a scatter plot of the data, instances where linear or quadratic splines are suitable will usually be apparent. The determination of m in other cases may not be so simple. Fortunately, when $m = 2$ or 3 is not appropriate, there appear to be few cases where cubic splines ($m = 4$) will not suffice if enough knots are used.

To illustrate the use of visual inspection for selecting λ consider the data on the ratio of weight to height, as a function of age (in months), for boys shown in Figure 6.1 (Eppright, et al 1972). Examination of the scatter plot suggests that we might fit this data with a quadratic spline with one knot located around what appears to be an inflection point at roughly 12 months. The resulting fit obtained from (6.8) and (6.11) with $\lambda = \{12\}$ is

$$\mu_{\{12\}}(t) = .423 + .055t - .002t^2 + .002(t - 12)_+^2.$$

A simple calculation shows us that $\mu_{\{12\}}$ can be expressed in piecewise polynomial form as

$$\mu_{\{12\}} = \begin{cases} .423 + .055t - .002t^2, & t < 12, \\ .731 + .004t, & t \geq 12. \end{cases}$$

Consequently the estimator is linear for $t \geq 12$ and, therefore, is a piecewise polynomial consisting of two different order segments. This provides an illustration of a situation where it would be useful to depart from the basis functions (6.5)-(6.6) and allow for the possibility of different order polynomial segments. We will discuss this further in Section 6.6.

Another illustration is provided by the data in Figure 6.2. The response values in this case are the natural logarithms of specific retention volumes

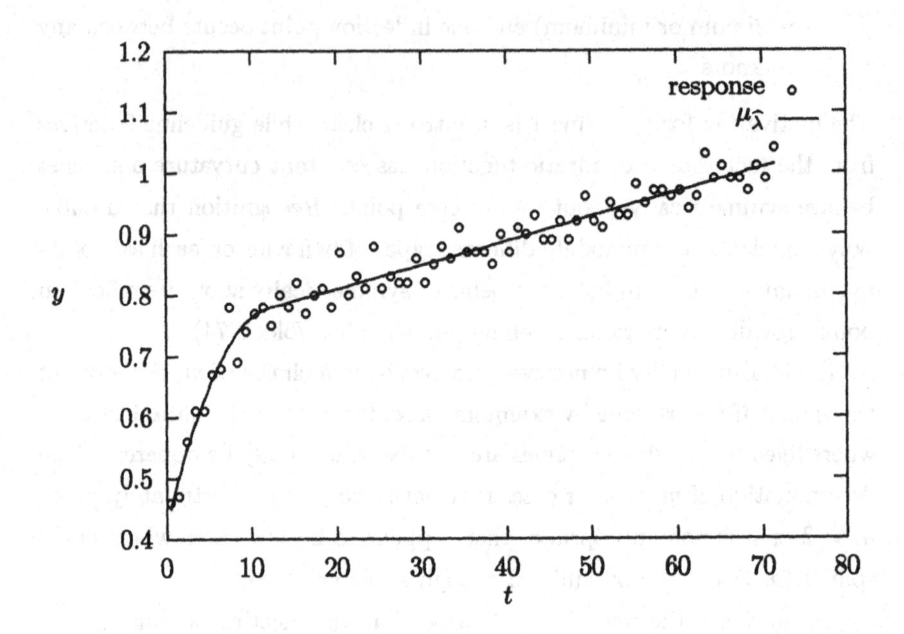

Figure 6.1: Least-Squares Spline Fit to Weight-Height Ratio Data.

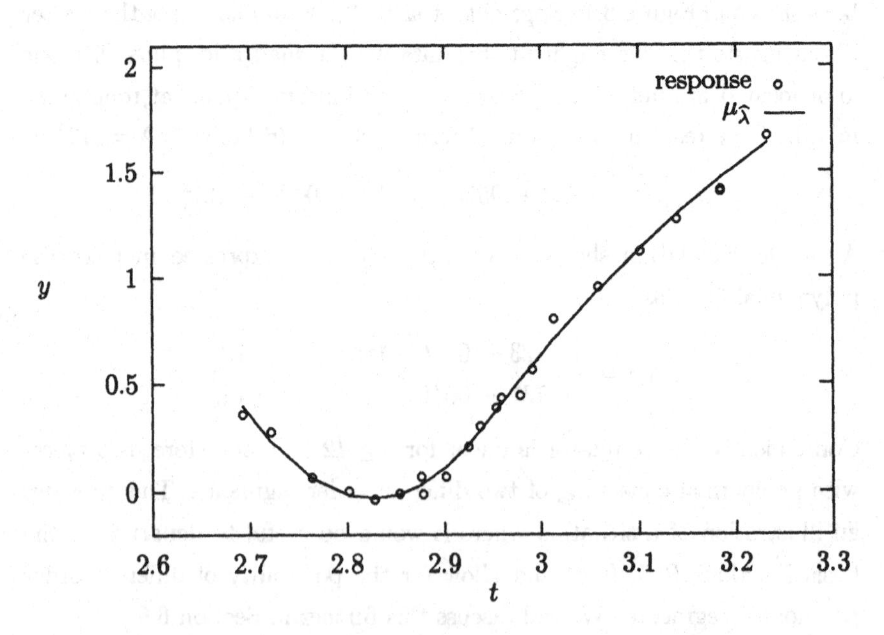

Figure 6.2: Least-Squares Spline Fit to Cycloheptene Data.

of cycloheptene in polyethylene terephthalate which have been plotted as a function of t, the reciprocal times 10^3 of temperature in degrees Kelvin (Gallant 1974). One might attempt to fit this data also using a quadratic spline. An application of guideline 2 would suggest the use of a single knot at about 2.9 or 3.

Given a choice of k and m there is an alternative, more objective, method of selecting $\lambda = \{\xi_1, \ldots, \xi_k\}$ than by looking at scatter plots. One can simply minimize $RSS(\mathbf{c}; \lambda)$ in (6.9) with respect to both $\mathbf{c}$ and the knot locations. Since a spline is nonlinear in the knot locations, this is a nonlinear problem which can be solved using nonlinear least-squares methodology such as the modified Gauss-Newton algorithm (Gallant and Fuller 1973).

To illustrate the type of results which can be obtained from nonlinear least squares, consider again the data in Figure 6.2. A quadratic spline with one knot was fit to this data using the modified Gauss-Newton algorithm with an initial or starting value of 3 being used for the knot location. The starting values for the coefficients of the basis functions in (6.5)-(6.6) were obtained by first fitting a quadratic spline to the data with a knot at 3. The resulting optimal value for the knot was found to be $\widehat{\lambda} = \{\widehat{\xi}_1\} = \{2.94\}$ and the fitted function is the one plotted in Figure 6.2.

In the previous example we selected a starting value for the knot and then used the least-squares estimates of the θ's and δ's to initialize the nonlinear least-squares algorithm. This suggests an alternative to the full-blown nonlinear least-squares algorithm used for the example.

Let $\lambda = \{\xi_1, \ldots, \xi_k\}$ be an arbitrary choice for the knot set and let $\mathbf{b}_\lambda$ be any corresponding solution to (6.11). Then

$$RSS(\lambda) = \sum_{i=1}^{n} \left(y_i - \sum_{j=1}^{m+k} b_{\lambda j} x_j(t_i) \right)^2 \tag{6.13}$$

is a function of λ alone that can be minimized to obtain an estimate of μ. The latter approach exploits the partially linear character of the spline function in a more effective fashion than minimizing the residual sum-of-squares simultaneously with respect to both the basis function coefficients and knot locations. It has been studied in some detail by Jupp (1972, 1978).

A natural question is when will minimization of $RSS(\lambda)$ and $RSS(\mathbf{c};\lambda)$ coincide. The relationship between problems of this type has been investigated by Golub and Pereyra (1973). Applying their work to our problem gives the conclusion that one can expect a local minimum of $RSS(\lambda)$ to correspond to a local minimum of $RSS(\mathbf{c};\lambda)$. It will therefore suffice in many cases to search among the critical points of $RSS(\lambda)$ in computing the estimator.

The minimization of functionals such as (6.13) is called the *variable projection method.* This technique has also been studied by Barham and Drane (1972) and Guttman, Pereyra and Scolnik (1973). The consensus seems to be that, although each iteration may be more computationally intensive, the variable projection approach tends to significantly reduce the number of iterations and thereby results in decreased run times.

Unfortunately, selection of knots by nonlinear least squares is a rather nasty nonlinear optimization problem. Neither $RSS(\lambda)$ nor $RSS(\mathbf{c};\lambda)$ will be convex functionals in general (Jupp 1972), and both will have many solutions on the boundary of the feasible region (Jupp 1978). Thus, the search for a (feasible) solution can be quite tedious and time consuming. In addition, Jupp (1975, 1978) notes that the variable knot problem exhibits a lethargy property which stems from the behavior of $RSS(\lambda)$ and $RSS(\mathbf{c};\lambda)$ on the boundary of the solution space where some of the ξ_j can coincide. Algorithms attempting to find optimal knots will have a tendency to become trapped on the boundary and, in many cases, obtain numerical convergence quite far from the boundary solution. A solution on the boundary will correspond to one or more coincident knots which can be shown to imply a reduction in the number of continuous derivatives for the estimator. If a spline is really considered appropriate such solutions would usually not be acceptable. Jupp (1978) has demonstrated that some of these difficulties can be overcome by transforming from the ξ_j to a new parameter set which reduces the tendency of the knots to coalesce. Another approach that employs an estimation criterion which penalizes coalescing knots has been developed by Lindstrom (1998).

Still other methods for selecting λ can be devised from the techniques in Chapter 2. It is even possible to effect an objective choice for k and m through this approach. To illustrate the idea, suppose it is desired to obtain a data driven choice for both the location and number of knots. Adopting the notation $\lambda_k = \{\xi_1, \ldots, \xi_k\}$ to specifically indicate dependence on k, a GCV criterion corresponding to λ_k is

$$\mathrm{GCV}(\lambda_k) = \frac{n^{-1}RSS(\lambda_k)}{(1-(m+k)/n)^2}, \tag{6.14}$$

where $RSS(\lambda_k)$ is defined as in (6.13). Let $\widehat{\lambda}_k$ be the minimizer of $\mathrm{GCV}(\lambda_k)$ for any fixed k. Then an optimal knot set is $\widehat{\lambda}_{\widehat{k}}$, where $\widehat{k}$ is the minimizer of $\mathrm{GCV}(\widehat{\lambda}_k)$. Note that, since the denominator of (6.14) is independent of the knot location, finding $\widehat{\lambda}_k$ for any given k is just the variable projection method for estimating the ξ_j that was discussed above. For other discussions of the use of automated selection methods with least-squares splines see, e.g., Chen (1987) and Shi and Tsai (1997).

Extensive developments of nonlinear least squares and related methods for estimation with spline functions in a variety of practical settings can be found in Ramsay (1988) and the references therein. Most of this work assumes that the function being estimated is actually a spline which means that interval estimation techniques developed in this context will not generally be effective for nonparametric problems. However, the point estimation methodology is equally applicable to parametric and nonparametric settings.

6.4 Computational Considerations

The set of functions defined in (6.5)-(6.6) forms a basis for $S^m(\xi_1, \ldots, \xi_k)$ that is usually called the ***truncated power basis***. As noted by Smith (1979), this basis has some advantages from a statistical viewpoint. However, the basis tends to be ill-conditioned which can be reflected in a nearly singular $\mathbf{X}_\lambda$ matrix that makes the normal equations (6.11) difficult to solve. Thus, it is desirable to explore the use of other basis functions which are better conditioned. One set of functions with this property is the so-called B-splines.

To develop the B-spline basis for $S^m(\xi_1,\ldots,\xi_k)$ first define $2m$ additional knots $\xi_{-(m-1)},\ldots,\xi_{-1},\xi_0,\xi_{k+1},\ldots,\xi_{k+m}$ by

$$\xi_{-(m-1)} = \cdots = \xi_0 = 0$$

and

$$\xi_{k+1} = \cdots = \xi_{k+m} = 1 .$$

The B-splines of order m with knots at $\xi_1,\ldots,\xi_k$ can then be defined recursively by

$$N_{i,m}(t) = \frac{t-\xi_i}{\xi_{i+m-1}-\xi_i}N_{i,m-1}(t) + \frac{\xi_{i+m}-t}{\xi_{i+m}-\xi_{i+1}}N_{i+1,m-1}(t), \qquad (6.15)$$

for $i = -(m-1),\ldots,k$, with

$$N_{i,1}(t) = \begin{cases} 1, & t \in [\xi_i, \xi_{i+1}), \\ 0, & \text{otherwise}, \end{cases}$$

being used to initialize the recursion. In using this definition we must employ the property that a B-spline of order d corresponding to $d+1$ coincident knots is the zero function. A proof that B-splines provide a basis for $S^m(\xi_1,\ldots,\xi_k)$ along with a direct derivation of B-splines from the truncated power basis can be found in de Boor (1978).

To illustrate the use of (6.15) consider the case of linear ($m = 2$) B-splines. In this instance $2m = 4$ additional knots are defined by $\xi_{-1} = \xi_0 = 0$ and $\xi_{k+1} = \xi_{k+2} = 1$. Since $N_{-1,1}$ has coincident knots ξ_{-1} and ξ_0 (both equal to 0), $N_{-1,1} \equiv 0$ and, similarly, $N_{k+1,1} \equiv 0$. Thus, (6.15) gives

$$\begin{aligned}
N_{-1,2}(t) &= \frac{\xi_1 - t}{\xi_1 - \xi_0}N_{0,1}(t), \\
N_{0,2}(t) &= \frac{t-\xi_0}{\xi_1-\xi_0}N_{0,1}(t) + \frac{\xi_2 - t}{\xi_2 - \xi_1}N_{1,1}(t), \\
&\;\vdots \\
N_{k,2}(t) &= \frac{t-\xi_k}{\xi_{k+1}-\xi_k}N_{k,1}(t).
\end{aligned}$$

Several important properties of B-splines are established in Exercise 6. In particular we note that

$$N_{i,m}(t) = 0 \text{ if } t \notin [\xi_i, \xi_{i+m}] . \qquad (6.16)$$

This local support characteristic is one of the key reasons for the popularity of B-splines.

Since any element of $S^m(\xi_1, \ldots, \xi_k)$ can be expressed as a linear combination of the $N_{im}, i = -(m-1), \ldots, k$, estimation of μ can be conducted using either B-splines or the truncated power basis. To compute the B-spline form of the estimator, define the matrix

$$\mathbf{N}_\lambda = \{N_{j,m}(t_i)\}_{i=1,n,j=-(m-1),k} . \tag{6.17}$$

Then, using the B-spline basis in our least-squares criterion produces the estimator $\mu_\lambda = \sum_{j=1}^{m+k} b_{\lambda j} N_{j-m,m}$, where $\mathbf{b}_\lambda = (b_{\lambda 1}, \ldots, b_{\lambda(m+k)})^T$ is any solution of the normal equations

$$\mathbf{N}_\lambda^T \mathbf{N}_\lambda \mathbf{c} = \mathbf{N}_\lambda^T \mathbf{y} . \tag{6.18}$$

We can now see the utility of the B-spline basis for estimation with least-squares splines. The matrix $\mathbf{N}_\lambda^T \mathbf{N}_\lambda$ has typical element

$$\sum_{r=1}^{n} N_{i,m}(t_r) N_{j,m}(t_r) .$$

This sum vanishes if $|i-j| \geq m$ as a result of (6.16). Thus, $\mathbf{N}_\lambda^T \mathbf{N}_\lambda$ is $2m-1$ banded and $\mathbf{b}_\lambda$ can be obtained by solving the normal equations (6.18) in $O(m^2(m+k))$ arithmetic operations. This is to be compared to the $O((m+k)^3)$ operations required for computing the coefficients under the truncated power basis. Notice that m will usually be small while k can frequently be quite large. Thus, B-splines have a definite computational advantage over the truncated power basis and appear to be the basis of choice from this standpoint. FORTRAN computer code for both the evaluation of B-splines and computation of the $b_{\lambda j}$ (for specified λ) can be found in de Boor (1978).

6.5 Asymptotic Analysis

In this section we study some of the large sample properties of least-squares spline estimators. For this purpose it will be mathematically convenient to assume that the design points and knots are generated by positive, continuous densities p and q through the relations

$$\int_0^{t_i} w(t)dt = i/(n+1),\ i = 1,\ldots,n, \tag{6.19}$$

and

$$\int_0^{\xi_j} q(\xi)d\xi = j/(k+1),\ j = 1,\ldots,k. \tag{6.20}$$

Assumption (6.20) means that the operative smoothing parameter for the least-squares spline is just the number of knots since for any given k the knot locations $\lambda_k = \{\xi_1,\ldots,\xi_k\}$ are determined from the ***knot density*** q using relation (6.20). Thus, strictly speaking, the asymptotics that follow are for splines with "fixed" knots that do not require estimation from the data. However, we will eventually derive an (asymptotically) optimal choice for q which can be interpreted as giving an indication of how estimated knots from nonlinear least squares will be located in large samples. It is also necessary to place some restrictions on the growth of k as a function of n. A natural condition to impose is that $k/n \to 0$ which insures that the number of knots grows at a slower rate than the sample size.

Under the above assumptions, it follows from Zhou, et al (1999) that for $t \in (\xi_i, \xi_{i+1}]$

$$\mathrm{E}\mu_{\lambda_k}(t) - \mu(t) = -\frac{\mu^{(m)}(t)B_m\left(\frac{t-\xi_i}{\xi_{i+1}-\xi_i}\right)}{m!q^m(t)}k^{-m} + o(k^{-m}) \tag{6.21}$$

with $B_m(\cdot)$ the mth Bernoulli polynomial defined recursively by $B_0(u) \equiv 1, B_i(u) = i\int_0^u B_{i-1}(z)dz + B_i, i = 1,\ldots,$ and $B_i = i\int_0^1\int_0^u B_{i-1}(z)dzdu$ is the ith Bernoulli number. In particular, $B_1(u) = u - \frac{1}{2}, B_2(u) = u^2 - u + \frac{1}{6}, B_3(u) = u^3 - \frac{3}{2}u^2 + \frac{1}{2}u$ and $B_4(u) = u^4 - 2u^3 + u^2 - \frac{1}{30}$ which gives the specific form of (6.21) for estimation by piecewise constants up to cubic splines.

Equation (6.21) gives the expected conclusion that the pointwise bias for least-squares splines will be largest in magnitude in regions where $|\mu^{(m)}(t)|$ is large. The level of bias can be reduced at a particular point t by either increasing the overall number of knots or by locating more knots in the neighborhood of t. The asymptotic form of the bias is somewhat similar to that of a variable bandwidth kernel estimator with "bandwidth" $1/kq(\cdot)$. However, there is an additional factor involving $B_m(\frac{t-\xi_i}{\xi_{i+1}-\xi_i})$ which can change sign once or twice over an interval between knots depending on whether m is odd or even. In particular, for the case of linear least-squares splines $B_2(\cdot)$ changes signs twice on (0, 1). Thus, unlike the comparable second order kernel estimator, we would not expect linear least-squares splines to fill in the valleys and round off the peaks of a regression function.

Now the variance of $\mu_{\lambda_k}(t)$ is

$$\mathrm{Var}(\mu_{\lambda_k}(t)) = \frac{\sigma^2}{n}\mathbf{N}^T(t)(n^{-1}\mathbf{N}_{\lambda_k}^T\mathbf{N}_{\lambda_k})^{-1}\mathbf{N}(t), \tag{6.22}$$

where $\mathbf{N}_{\lambda_k}$ is defined in (6.17) and $\mathbf{N}(t) = (N_{-(m-1),m}(t), \ldots, N_{k,m}(t))^T$. Zhou, et al (1999) show that there are constants $0 < C_1 \le C_2 < \infty$ such that the maximum and minimum eigenvalues of the matrix $(n^{-1}\mathbf{N}_{\lambda_k}^T\mathbf{N}_{\lambda_k})^{-1}$ both fall in the interval $[C_1k, C_2k]$ for sufficiently large n. It follows from this (Exercise 8) that

$$C_1\sigma^2k(mn)^{-1} \le \mathrm{Var}(\mu_{\lambda_k}(t)) \le C_2\sigma^2kn^{-1} \tag{6.23}$$

uniformly in $t \in [0,1]$.

Combining (6.21) and (6.23) reveals that a choice of k growing at a rate of $n^{1/(2m+1)}$ will make the mth order least-squares spline a pointwise mth order estimator. This result is actually true globally because (6.21) and (6.23) hold uniformly in t thereby allowing us to sum or integrate to assess global performance. More exact global approximations can be obtained from the work of Agarwal and Studden (1980). In the case of the average risk $R_n(\lambda_k) = n^{-1}\sum_{i=1}^n \mathrm{E}(\mu_{\lambda_k}(t_i) - \mu(t_i))^2$ their work has the consequence that

$$R_n(\lambda_k) \sim \frac{k\sigma^2}{n} + \frac{|B_{2m}|}{(2m)!}k^{-2m}\int_0^1 \frac{(\mu^{(m)}(t))^2}{q(t)^{2m}}w(t)dt, \tag{6.24}$$

where B_{2m} is the $2m$th Bernoulli number. Thus, the risk decays at the optimal $n^{-2m/(2m+1)}$ rate if the number of knots grows like $n^{1/(2m+1)}$. It is interesting to note that this rate is obtained using splines of order m that require no boundary modifications. This is in contrast to smoothing splines, where a spline of order $2m$ is required to attain this rate, and kernel estimators, which require special modifications at the boundary of $[0, 1]$.

The choices of q and k which minimize (6.24) are

$$q_{opt}(t) = \frac{\left((\mu^{(m)}(t))^2 w(t)\right)^{1/(2m+1)}}{\int_0^1 \left((\mu^{(m)}(s))^2 w(s)\right)^{1/(2m+1)} ds} \tag{6.25}$$

and

$$k_{opt} = \left[\frac{n|B_{2m}|}{(2m-1)!\sigma^2}\right]^{1/(2m+1)} \int_0^1 [\left(\mu^{(m)}(s)\right)^2 w(s)]^{1/(2m+1)} ds. \tag{6.26}$$

Note that q_{opt} is proportional to $\left((\mu^{(m)})^2 w\right)^{1/(2m+1)}$ suggesting that i) knots should be more heavily concentrated where the regression function appears to be more wiggly and ii) fewer knots should be placed in regions where the design is sparse than in regions where it is dense. The optimal number of knots k_{opt} is of exact order $n^{1/(2m+1)}$ which has the practical implication that the number of knots should be kept small and, in any case, should be much smaller than n.

Under assumption (6.20), μ_{λ_k} is a linear estimator that can be written as

$$\mu_{\lambda_k}(t) = n^{-1} \sum_{i=1}^{n} K_{\lambda_k}(t, t_i) y_i \tag{6.27}$$

for weights $K_{\lambda_k}(t, t_i), i = 1, \ldots, n$, that depend on the design and knots but not on the responses. The form of the weight function in (6.27) has been characterized asymptotically by Huang and Studden (1993) in the special case of cubic splines ($m = 4$) and uniform design and knot placement ($w = q = 1$). They show that

$$K_{\lambda_k}(t, t_i) = \sum_{j=1}^{4} u_{ji} k K\left(k(t - \xi_{ij})\right) + O\left(\frac{k^2}{n} + ke^{-Ck}\right)$$

for C a positive constant, $u_{j1}, \ldots, u_{j4}$ nonnegative weights that sum to one, K a function on the line that integrates to one and $\xi_{ij}, j = 1, \ldots, 4$, the four knots that are closest to t_i. Huang and Studden (1993) also show there are positive constants C_1 and C_2 such that

$$|K_{\lambda_k}(t, s)| \leq C_1 \exp\{-C_2 k|t - s|\}.$$

Since we can think of $k|t-s|$ as being essentially the number of knots between t and s, this means that the weights in (6.27) decay rapidly to zero for observations whose corresponding design points are close to knots that are far from the point of estimation t. Thus, least-squares splines have a local character similar to that of kernel and smoothing spline estimators. However, the weight applied to the ith response by a least-squares spline depends on factors involving distances between the point of estimation and knots that are near to t_i rather than the distance between t and t_i as we would expect from a kernel smoother. This makes it clear that least-squares splines differ in a fundamental way from kernel type estimators.

The linear form (6.27) leads one to anticipate a limiting standard normal distribution for $(\mu_{\lambda_k}(t) - \mathrm{E}\mu_{\lambda_k}(t))/\sqrt{\mathrm{Var}\,(\mu_{\lambda_k}(t))}$ under suitable restrictions on k and n. Results of this kind have been established by Zhou, et al (1999) who explore how to translate them into practical methods for interval estimation and also develop simultaneous confidence bands for μ using least-squares splines.

6.6 Extensions

There are several extensions of the basic least-squares spline estimator which can be useful in practice. A particularly useful generalization can be motivated by examination of the data on the natural logarithm of specific retention volume of methylene chloride in polyethylene terephthalate plotted as a function of the reciprocal of temperature (times 10^3) in Figure 6.3. The data appears to be roughly quadratic in nature on each side of a cusp at around $t = 2.87$. However, as seen from the figure, a quadratic spline (with one knot at $\xi_1 = 2.87$) is a dreadfully poor estimator. The problem is that it is too smooth to satisfactorily resolve the apparent cusp in the data.

The methylene chloride data suggests that it may be useful in some cases to use "splines" with reduced continuity constraints at the knots. This can be easily accomplished by appending new basis functions to the elements of the truncated power basis. To be precise, suppose it is desired to have

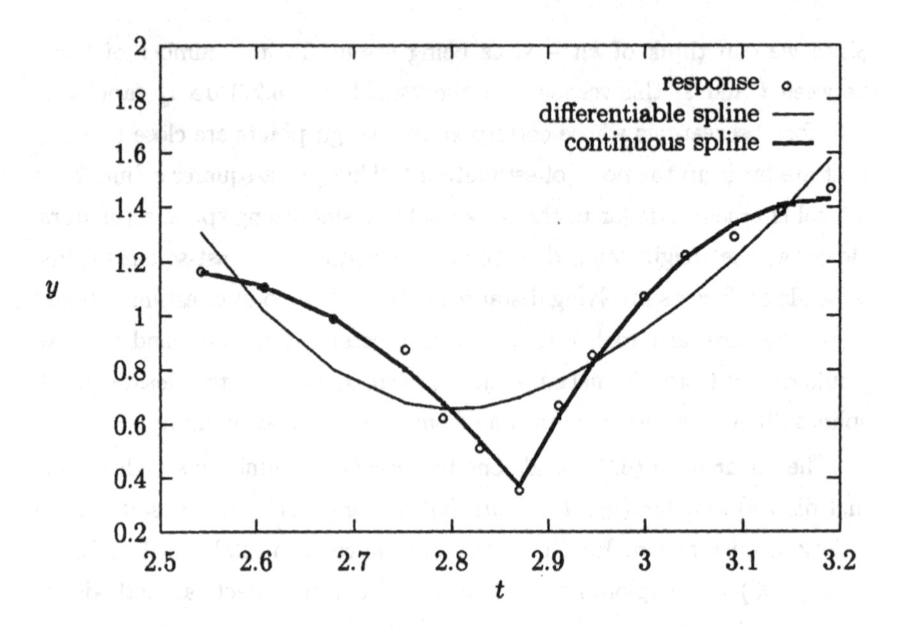

Figure 6.3: Spline Fits to Methelyene Chloride Data.

$\nu_j - 1$ continuous derivatives at ξ_j, $j = 1, \ldots, k$. Then an approximation to μ which incorporates these characteristics is the modified "spline" function

$$s(t) = \sum_{j=1}^{k} \theta_j t^{j-1} + \sum_{j=1}^{k} \sum_{r=\nu_j}^{m-1} \delta_{jr}(t - \xi_j)_+^r. \tag{6.28}$$

Approximations of this type can also be motivated as in Section 6.2 through the use of integration by parts on Rem(t) and the approximation of each term in the resulting sum by a quadrature formula. Given a choice for $\nu_1, \ldots, \nu_k$, an estimator for μ can be obtained essentially as before by minimizing $\sum_{i=1}^{n}(y_i - s(t_i))^2$ with respect to the basis function coefficients and possibly the knot locations.

To demonstrate the utility of using reduced continuity constraints return to the methelyene chloride data in Figure 6.3. In addition to our original quadratic spline fit we have also fitted a quadratic spline with one knot at $\xi_1 = 2.87$ and $\nu_1 = 1$. The resulting function is continuous, but not differentiable, and has the form $s(t) = \sum_{j=1}^{3} \theta_j t^{j-1} + \delta_{11}(t - \xi_1)_+ + \delta_{12}(t - \xi_1)_+^2$. As can be seen from the figure, this function does a much more satisfactory job of resolving the cusp in the data.

Another possible extension of least-squares spline estimators is suggested by our analysis of the weight-height ratio data in Section 6.2. We found in that example that the fitted curve was a type of spline with different order segments. Estimation of μ using splines with different order segments has been considered by Fuller (1969), Gallant (1974), Gallant and Fuller (1973) and Smith (1979). Such functions can be constructed by constraining the coefficients for the spline basis functions appropriately. To illustrate the idea, suppose we wish to fit a spline function with two knots that consists of three segments which are linear, quadratic and linear, respectively. The general form of a quadratic spline with two knots ξ_1 and ξ_2 with continuity levels $\nu_1 = \nu_2 = 1$ is

$$\theta_0 + \theta_1 t + \theta_2 t^2 + \delta_{11}(t - \xi_1)_+ + \delta_{12}(t - \xi_1)_+^2 + \delta_{21}(t - \xi_2)_+ + \delta_{22}(t - \xi_2)_+^2.$$

In order for the first and last segments to be linear it suffices to require that $\theta_2 = 0$ and $\delta_{12} = -\delta_{22}$. We can then fit a curve with the desired properties by deleting the θ_2 term from our original formulation and estimating the remaining coefficients subject to the constraint $\delta_{12} = -\delta_{22}$.

The basic method for constructing general splines with different order segments parallels the techniques used in the example, although the details can be quite tedious in certain cases. The fitting of curves of this nature will generally require the use of restricted least squares. An illustration of this can be found in Gerig and Gallant (1975).

Other extensions of least-squares spline methods include estimation with multiple predictor variables (Stone 1985, 1986, 1994 and Friedman 1991), robustified estimators (Lenth 1977 and Shi and Li 1995) and estimation in survival models (Kooperberg, Stone and Truong 1995). The application of least-squares splines for estimation in partially linear models has been considered by Chen (1988), He and Shi (1996) and Shi and Tsai (1997). Estimation with dependent errors is studied, for example, by Burman (1991).

6.7 Exercises

1. The data in Table 6.1 was simulated from the model (6.1) using normal random errors with variance $\sigma^2 = .0025$ and the regression function
$$\mu(t) = 4.26\left\{e^{-3.25t} - 4e^{-6.5t} + 3e^{-9.75t}\right\}.$$
Discuss fitting this data by a least-squares spline through an application of the guidelines discussed in Section 6.2. What order should be used and where should the knot(s) be placed. Fit the spline suggested by these considerations and discuss your results.

Table 6.1: Simulated Data

t	y	t	y	t	y	t	y
.00	-.125	.02	-.499	.04	-.756	.06	-.852
.08	-.958	.10	-.972	.12	-1.030	.14	-.898
.16	-.824	.18	-.711	.20	-.538	.22	-.484
.24	-.427	.26	-.231	.28	-.177	.30	-.034
.32	.028	.34	-.003	.36	.056	.38	.112
.40	.205	.42	.215	.44	.255	.46	.211
.48	.178	.50	.376	.52	.295	.54	.328
.56	.340	.58	.326	.60	.344	.62	.254
.64	.301	.66	.264	.68	.228	.70	.252
.72	.204	.74	.198	.76	.205	.78	.170
.80	.153	.82	.239	.84	.182	.86	.187
.88	.211	.90	.115	.92	.050	.94	.161
.96	.208	.98	072				

2. Repeat your analysis in Exercise 1 for the voltage drop data in Table 3.6.

3. Obtain a fit via nonlinear least squares for the data in Table 6.1 using a cubic spline with one knot. Compare the optimal knot location to the actual inflection point of the curve.

4. Obtain a fit via nonlinear least squares for the voltage drop data in Table 3.6 using a cubic spline with two knots.

5. Draw graphs of the linear B-splines in Section 6.3. Derive expressions for the quadratic B-splines and draw graphs of these functions as well.

6. Show that for B-splines of order m

   (a) $N_{i,m}(t) = 0$ for $t \notin [\xi_i, \xi_{i+m}]$,

   (b) $N_{i,m}(t) \geq 0$ and

   (c) $\sum_{i=-(m-1)}^{k} N_{i,m}(t) = 1$ for $t \in (0,1)$.

7. Discuss how to construct a continuous piecewise polynomial consisting of three segments (i.e., with two knots): the first piece being cubic, the second quadratic and the third linear.

8. For $\lambda_k = \{\xi_1, \ldots, \xi_k\}$ let $\mathbf{N}_{\lambda_k}$ be defined as in (6.17) and assume that there are finite, positive constants C_1 and C_2 such that the maximum and minimum eigenvalues of the matrix $(n^{-1}\mathbf{N}_{\lambda_k}^T \mathbf{N}_{\lambda_k})^{-1}$ fall in the interval $[C_1 k, C_2 k]$. Show that $C_1\sigma^2 k(mn)^{-1} \leq \mathrm{Var}(\mu_\lambda(t)) \leq C_2\sigma^2 kn^{-1}$. [Hint: Use Exercise 6.]

9. Another approach to selecting k and the knot locations for least-squares splines has been suggested by Smith (1983). Her approach utilizes a pool of knots chosen to provide a partition of $[0,1]$ which insures that no interesting characteristics of μ are overlooked. If $\boldsymbol{\alpha} = (\alpha_1, \ldots, \alpha_K)$ is such a knot pool, then Smith proposes fitting a curve of the form $\sum_{j=1}^{m} \theta_j t^{j-1} + \sum_{j=1}^{K} \beta_j (t - \alpha_j)_+^{m-1}$ with variable selection techniques used to delete those terms $(\cdot - \alpha_j)_+^{m-1}$ which do not contribute to the overall fit. When all such nonsignificant terms have been deleted, the final fit will be a spline of order m with $k \leq K$ knots at points $\xi_1, \ldots, \xi_k$ with $\xi_i = \alpha_j$ for some j.

   (a) Contrast the computational aspects of Smith's approach to nonlinear least squares for knot selection with GCV used to determine k.

Table 6.2: Titanium Heat Data

t	y	t	y	t	y	t	y
595	.644	605	.622	615	.638	625	.649
635	.652	645	.639	655	.646	665	.657
675	.652	685	.655	695	.644	705	.663
715	.663	725	.668	.735	.676	745	.676
755	.686	765	.679	775	.678	785	.683
795	.694	805	.699	815	.710	825	.730
835	.763	845	.812	855	.907	865	1.044
875	1.336	885	1.881	895	2.169	905	2.075
915	1.598	925	1.211	935	.916	945	.746
955	.672	965	.627	975	.615	985	.607
995	.606	1005	.609	1015	.603	1025	.601
1035	.603	1045	.601	1055	.611	1065	.601
1075	.608						

Source: de Boor and Rice (1968).

(b) The data in Table 6.2 represents a property of titanium as a function of heat. The least-squares estimators for the knots for a cubic spline fit to this data with $k = 5$ are (Jupp 1972, 1978) $\widehat{\lambda} = \{835.967, 876.4016, 898.1462, 916.3145, 973.9075\}$. Fit this data by Smith's approach using forward selection, cubic splines and a knot pool consisting of ten equally spaced points over [785, 1055]. (Note: You must retain the variables 1, t, t^2, t^3 in all fits and delete only terms of the form $(t - \alpha_j)^3_+$ on each step.) Also fit the data using 5 uniformly spaced points on [595, 1075]. Contrast the three fits and discuss the relative performance of optimal and uniformly placed knots.

Bibliography

G. Agarwal and W. Studden. Asymptotic integrated mean square error using least squares and bias minimizing splines. *The Annals of Statistics*, 8:1307–1325, 1980.

H. Akaike. Statistical predictor identification. *Annals of the Institute of Statistical Mathematics*, 22:203–217, 1970.

H. Akaike. Information theory and an extension of the maximum likelihood principle. In B. N. Petrov and F. Csàki, editors, *2nd International Symposium on Information Theory*, pages 267–281, Budapest, 1973. Akademia Kiadó.

H. Akaike. A new look at the statistical model identification. *I. E. E. E. Transactions on Automatic Control*, 19:716–723, 1974.

H. Akaike. A Bayesian analysis of the minimum AIC procedure. *Annals of the Institute of Statistical Mathematics*, 30:9–14, 1978.

D. Allen. The relationship between variable selection and data augmentation and a method of prediction. *Technometrics*, 16:125–127, 1974.

N. Altman. Kernel smoothing of data with correlated errors. *Journal of the American Statistical Association*, 85:749–759, 1990.

N. Altman. An introduction to kernel and nearest-neighbor nonparametric regression. *American Statistician*, 46:175–185, 1992.

P. Anselone and P. Laurent. A general method for the construction of interpolating or smoothing spline functions. *Numerische Mathematik*, 12:66–82, 1968.

C. F. Ansley and R. Kohn. Efficient generalized cross-validation for state space models. *Biometrika*, 74:139–148, 1987.

C. F. Ansley and W. E. Wecker. Extensions and examples of the signal extraction approach to regression. In *Proceedings of the ASA-CENSUS-NBER Conference on Applied Time Series Analysis of Economic Data*, pages 181–192, 1981.

R. Barham and W. Drane. An algorithm for least squares estimation of nonlinear parameters when some of the parameters are linear. *Technometrics*, 14:757–766, 1972.

D. Barry and J. Hartigan. An omnibus test for departures from constant mean. *The Annals of Statistics*, 18:1340–1357, 1990.

J. Benedetti. Kernel estimation of regression functions. In *Proceedings in Computer Science and Statistics: 8th Annual Symposium on the Interface*, pages 405–412, 1975.

J. Benedetti. On the nonparametric estimation of regression functions. *Journal of the Royal Statistical Society, Series B*, 39:248–253, 1977.

H. J. Bierens. Uniform consistency of kernel estimators of a regression function under generalized conditions. *Journal of the American Statistical Association*, 78:699–707, 1983.

S. Bjerve, K. Doksum, and B. Yandell. Uniform confidence bands for regression based on a simple moving average. *Scandinavian Journal of Statistics*, 12:159–169, 1985.

B. Blight and L. Ott. A Bayesian approach to model inadequacy for polynomial regression. *Biometrika*, 62:79–88, 1975.

S. Blyth. Optimal kernel weights under a power criterion. *Journal of the American Statistical Association*, 88:1284–1286, 1993.

C. de Boor. *A Practical Guide to Splines*. Springer-Verlag, New York, 1978.

C. de Boor and J. Rice. Least squares cubic spline approximation I: fixed knots. CSD Technical Report Number 20, 1968.

M. Brockman, T. Gasser, and E. Herrmann. Locally adaptive bandwidth choice for kernel regression estimators. *Journal of the American Statistical Association*, 88:1302–1309, 1993.

M. Buckley. Detecting a smooth signal: optimality of cusum based procedures. *Biometrika*, 78:253–262, 1991.

M. Buckley and G. Eagleson. A graphical method for estimating the residual variance in nonparametric regression. *Biometrika*, 76:203–210, 1989.

M. Buckley, G. Eagleson, and B. Silverman. The estimation of residual variance in non-parametric regression. *Biometrika*, 75:189–199, 1988.

P. Burman. Regression function estimation from dependent observations. *Journal of Multivariate Analysis*, 36:236–279, 1991.

C. Carter, G. Eagleson, and B. Silverman. A comparison of the Reinsch and Speckman splines. *Biometrika*, 79:81–91, 1992.

C. Carter and R. Kohn. Robust Bayesian nonparametric regression. In W. Härdle and M. Schimek, editors, *Statistical Theory and Computational Aspects of Smoothing*, pages 128–148. Physica-Verlag, Heidelberg, 1996.

H. Chen. Convergence rates for parametric components in a partly linear model. *The Annals of Statistics*, 16:136–146, 1988.

K. Chen. Asymptotically optimal selection of a piecewise polynomial estimator of a regression function. *Journal of Multivariate Analysis*, 22:230–244, 1987.

K. Cheng and P. Lin. Nonparametric estimation of a regression function. *Zeitschrift für Wahrscheinlichkeitstheorie und Verwandte Gebiete*, 57:223–233, 1981.

S. Chiu. On the asymptotic distributions of bandwidth estimates. *The Annals of Statistics*, 18:1696–1711, 1990.

S. Chiu. Some stabilized bandwidth selectors for nonparametric regression. *The Annals of Statistics*, 19:1528–1546, 1991.

S. Chiu and J. S. Marron. The negative correlations between data-determined bandwidths and the optimal bandwidth. *Statistics and Probability Letters*, 10:173–180, 1990.

C. Chu and J. S. Marron. Choosing a kernel regression estimator. *Statistical Science*, 6:404–436, 1991.

C. Chu and J. S. Marron. Comparison of two bandwidth selectors with dependent errors. *The Annals of Statistics*, 4:1906–1918, 1991.

R. Clark. Non-parametric estimation of a smooth regression function. *Journal of the Royal Statistical Society, Series B*, 39:107–113, 1977.

R. Clark. Calibration, cross-validation and carbon-14, II. *Journal of the Royal Statistical Society, Series A*, 143:177–194, 1980.

W. Cleveland. Robust locally weighted regression and smoothing scatterplots. *Journal of the American Statistical Association*, 74:829–836, 1979.

W. Cleveland and S. Devlin. Locally weighted regression: An approach to regression analysis by local fitting. *Journal of the American Statistical Association*, 83:596–610, 1988.

R. D. Cook and S. Weisberg. *Residuals and Influence in Regression.* Chapman and Hall, New York, 1982.

D. Cox. Asymptotics for M-type smoothing splines. *The Annals of Statistics*, 11:530–551, 1983.

D. Cox. Approximation of least squares regression on nested subspaces. *The Annals of Statistics*, 16:713–732, 1988.

D. Cox and E. Koh. A smoothing spline based test of model adequacy in polynomial regression. *Annals of the Institute of Statistical Mathematics*, 41:383–400, 1989.

P. Craven and G. Wahba. Smoothing noisy data with spline functions: estimating the correct degree of smoothing by the method of generalized cross-validation. *Numerische Mathematik*, 31:377–403, 1979.

J. Cuzick. Efficient estimates in semiparametric additive regression models with unknown error distribution. *The Annals of Statistics*, 20:1129–1136, 1992.

M. Davidian, R. Carroll, and W. Smith. Variance functions and the minimum detectable concentration in assays. *Biometrika*, 75:549–556, 1988.

P. Davis. *Interpolation and Approximation*. Dover, New York, 1975.

A. Demmler and C. Reinsch. Oscillation matrices with spline smoothing. *Numerische Mathematik*, 24:375–382, 1975.

L. Devroye. The uniform convergence of the Nadraya-Watson regression function estimate. *Canadian Journal of Statistics*, 6:179–191, 1978.

L. Devroye. On the almost everywhere convergence of nonparametric regression function estimates. *The Annals of Statistics*, 9:1310–1319, 1981.

L. Devroye and T. Wagner. Distribution-free consistency results in nonparametric discrimination and regression function estimation. *The Annals of Statistics*, 8:231–239, 1980.

P. Diggle and M. Hutchinson. On spline smoothing with autocorrelated errors. *Australian Journal of Statistics*, 31:166–182, 1989.

D. Donoho and M. Low. Renormalization exponents and optimal pointwise rates of convergence. *The Annals of Statistics*, 20:944–970, 1992.

N. Draper and I. Guttman. Incorporating overlap effects from neighboring units into response surface models. *Applied Statistics*, 29:128–134, 1980.

S. Efromovich. On nonparametric regression for iid observations in a general setting. *The Annals of Statistics*, 24:1126–1144, 1996.

S Efromovich and M. Low. On optimal adaptive estimation of a quadratic functional. *The Annals of Statistics*, 24:1106–1125, 1996.

B. Efron. *The Jackknife, the Bootstrap and Other Resampling Plans.* CBMS-NSF series. SIAM, Philadelphia, 1982.

R. Engle, C. Granger, J. Rice, and A. Weiss. Semiparametric estimates of the relation between weather and electricity sales. *Journal of the American Statistical Association*, 81:310–320, 1986.

V. Epanechnikov. Non-parametric estimation of a multi-variate probability density. *Theory of Probability and Its Applications*, 14:153–158, 1969.

E. Eppright, H. Fox, B. Fryer, G. Lamkin, V. Vivian, and E. Fuller. Nutrition of infants and preschool children in the north central region of the United States of America. *World Review of Nutrition and Dietetics*, 14:269–332, 1972.

R. Eubank. On testing for no effect in nonparametric regression. Manuscript, 1999.

R. Eubank and B. Jayasuriya. The asymptotic average squared error from polynomial regression. *Statistics*, 24:311–319, 1993.

R. Eubank and P. Speckman. Curve fitting by polynomial-trigonometric regression. *Biometrika*, 77:1–9, 1990.

R. Eubank and P. Speckman. Convergence rates for trigonometric and polynomial-trigonometric regression estimators. *Statistics and Probability Letters*, 11:119–124, 1991.

R. Eubank and P. Speckman. A bias reduction theorem with applications in nonparametric regression. *Scandinavian Journal of Statistics*, 18:211–222, 1992.

R. Eubank and P. Speckman. Confidence bands in nonparametric regression. *Journal of the American Statistical Association*, 88:1287–1301, 1993.

J. Fan. On the estimation of quadratic functionals. *The Annals of Statistics*, 19:1273–1294, 1991.

J. Fan. Design-adaptive nonparametric regression. *Journal of the American Statistical Association*, 87:998–1004, 1992.

J. Fan and I. Gijbels. Variable bandwidth and local linear regression smoothers. *The Annals of Statistics*, 20:2008–2036, 1992.

J. Fan and I. Gijbels. *Local Polynomial Modeling and Its Applications.* Chapman and Hall, New York, 1996.

J. Fan, N. Heckman, and M. Wand. Local polynomial kernel regression for generalized linear models and quasi-likelihood functions. *Journal of the American Statistical Association*, 90:141–150, 1995.

W. Feller. *An Introduction to Probability Theory and Its Applications, Volume I.* John Wiley, New York, 1968.

J. Franklin. *Matrix Theory.* Prentice-Hall, New Jersey, 1968.

J. Friedman. Multivariate adaptive regression splines (with discussion). *The Annals of Statistics*, 19:1–141, 1991.

W. A. Fuller. Grafted polynomials as approximating functions. *Australian Journal of Agricultural Economics*, 13:35–46, 1969.

W. A. Fuller. *Introduction to Statistical Time Series.* John Wiley, New York, 1976.

R. Gallant. The theory of nonlinear regression as it relates to segmented polynomial regression with estimated join points. Mimeograph Series No. 925, Institute of Statistics, North Carolina State Univiversity, Raleigh, 1974.

R. Gallant and W. Fuller. Fitting segmented polynomial regression models whose join points have to be estimated. *Journal of the American Statistical Association*, 68:144–147, 1973.

Th. Gasser, A. Kneip, and W. Köhler. A flexible and fast method for automatic smoothing. *Journal of the American Statistical Association*, 86:643–652, 1991.

Th. Gasser and H. G. Müller. Kernel estimation of regression functions. In Th. Gasser and M. Rosenblatt, editors, *Smoothing Techniques for Curve Estimation*, pages 23–68. Springer-Verlag, Heidelberg, 1979.

Th. Gasser and H. G. Müller. Estimating regression functions and their derivatives by the kernel method. *Scandinavian Journal of Statistics*, 11:171–185, 1984.

Th. Gasser, H. G. Müller, W. Köhler, L. Molinari, and A. Prader. Nonparametric regression analysis of growth curves. *The Annals of Statistics*, 12:210–229, 1984.

Th. Gasser, H. G. Müller, and V. Mammitzsch. Kernels for nonparametric curve estimation. *Journal of the Royal Statistical Society, Series B*, 47:238–252, 1985.

Th. Gasser, L. Sroka, and C. Jennen-Steinmetz. Residual variance and residual pattern in nonlinear regression. *Biometrika*, 73:625–633, 1986.

T. Gerig and A. R. Gallant. Computing methods for linear models subject to linear parametric constraints. *Journal of Statistical Computation and Simulation*, 3:283–296, 1975.

D. Girard. A fast 'Monte-Carlo cross-validation' procedure for large least squares problems with noisy data. *Numerische Mathematik*, 56:1–23, 1989.

L. Goldstein and K. Messer. Optimal plug-in estimators for nonparametric functional estimation. *The Annals of Statistics*, 20:1306–1328, 1992.

G. Golub, M. Heath, and G. Wahba. Generalized cross-validation as a method for choosing a good ridge parameter. *Technometrics*, 21:215–223, 1979.

G. Golub and V. Pereyra. The differentiation of pseudo-inverses and nonlinear least squares problems whose variables separate. *SIAM Journal of Numerical Analysis*, 10:413–432, 1973.

G. Golubev and M. Nussbaum. A risk bound in Sobolev class regression. *The Annals of Statistics*, 18:758–778, 1990.

I. S. Gradshteyn and I. M. Ryzhik. *Tables of Integrals, Series, and Products*. Academic Press, New York, 1980.

W. Greblicki, A. Krzyzak, and M. Pawlak. Distribution-free pointwise consistency of kernel regression estimate. *The Annals of Statistics*, 12:1570–1575, 1984.

P. Green. Linear models for field trials, smoothing and cross-validation. *Biometrika*, 72:527–537, 1985.

P. Green, C. Jennison, and A. Seheult. Analysis of field experiments by least squares smoothing. *Journal of the Royal Statistical Society, Series B*, 47:299–315, 1985.

P. Green and B. Silverman. *Nonparametric Regression and Generalized Linear Models*. Chapman and Hall, New York, 1994.

R. Gunst and R. Mason. *Regression Analysis and Its Application: A Data-Oriented Approach*. Marcel-Dekker, New York, 1980.

I. Guttman, V. Pereyra, and H. Scolnik. Least squares estimation for a class of non-linear models. *Technometrics*, 15:209–218, 1973.

P. Hall. On bootstrap confidence intervals in nonparametric regression. *The Annals of Statistics*, 20:695–711, 1992.

P. Hall and I. Johnstone. Empirical functionals and efficient smoothing parameter selection (with discussion). *Journal of the Royal Statistical Society, Series B*, 54:475–530, 1992.

P. Hall, J. Kay, and D. M. Titterington. Asymptotically optimal difference based estimation of variance in nonparametric regression. *Biometrika*, 77:521–528, 1990.

P. Hall and J. S. Marron. On variance estimation in nonparametric regression. *Biometrika*, 77:415–419, 1990.

P. Hall and W. Schucany. A local cross-validation algorithm. *Statistics and Probalility Letters*, 8:109–117, 1989.

P. Hall and D. M. Titterington. On confidence bands in nonparametric density estimation and regression. *Journal of Multivariate Analysis*, 27:228–254, 1988.

P. Hall and T. Wehrly. A geometrical method for removing edge effects from kernel-type nonparametric regression estimators. *Journal of the American Statistical Association*, 86:665–672, 1991.

S. Hamilton and Y. Truong. Local linear estimation in partly linear models. Manuscript, 1995.

W. Härdle. Asymptotic maximal deviation of M-smoothers. *Journal of Multivariate Analysis*, 29:163–179, 1989.

W. Härdle. *Applied Nonparametric Regression*. Cambridge University Press, New York, 1990.

W. Härdle and A. Bowman. Bootstrapping in nonparametric regression: local adaptive smoothing and confidence bands. *Journal of the American Statistical Association*, 83:102–110, 1988.

W. Härdle, P. Hall, and J. S. Marron. How far are automatically chosen regression smoothing parameters from their optimum ? (with discussion). *Journal of the American Statistical Association*, 83:86–95, 1988.

W. Härdle, P. Hall, and J. S. Marron. Regression smoothing parameters that are not far from their optimum. *Journal of the American Statistical Association*, 87:227–233, 1992.

W. Härdle and J. S. Marron. Asymptotic nonequivalence of some bandwidth selectors in nonparametric regression. *Biometrika*, 72:481–484, 1985.

W. Härdle and J. S. Marron. Bootstrap simultaneous error bars for nonparametric regression. *The Annals of Statistics*, 19:778–796, 1991.

J. Hart. Kernel regression estimation with time series errors. *Journal of the Royal Statistical Society, Series B*, 53:173–187, 1991.

J. Hart. *Nonparametric Smoothing and Lack-of-Fit Tests*. Springer-Verlag, New York, 1997.

J. Hart and T. Wehrly. Kernel regression estimation using repeated measurements data. *Journal of the American Statistical Association*, 81:1080–1088, 1986.

J. Hart and T. Wehrly. Kernel regression estimation when the boundary region is large, with an application to testing the adequacy of polynomial models. *Journal of the American Statistical Association*, 87:1018–1024, 1992.

J. Hart and T. Wehrly. Consistency of cross-validation when the data are curves. *Stochastic Processes and their Applications*, 45:351–361, 1993.

J. Hart and S. Yi. One-sided cross-validation. *Journal of the American Statistical Association*, 93:620–631, 1998.

T. Hastie and C. Loader. Local regression: Automatic kernel carpentry. *Statistical Science*, 8:120–143, 1993.

T. Hastie and R. Tibshirani. *Generalized Additive Models*. Chapman and Hall, New York, 1990.

X. He and P. Shi. Bivariate tensor product B-splines in a partly linear model. *Journal of Multivariate Analysis*, 58:162–181, 1996.

N. Heckman. Spline smoothing in a partly linear model. *Journal of the Royal Statistical Society, Series B*, 48:244–248, 1986.

N. Heckman. Minimax estimates in a semiparametric model. *Journal of the American Statistical Association*, 83:1090–1096, 1988.

E. Herrmann, Th. Gasser, and A. Kneip. Choice of bandwidth for kernel regression when residuals are correlated. *Biometrika*, 79:783–795, 1992.

R. Hocking. The analysis and selection of variables in linear regression. *Biometrics*, 32:1–49, 1976.

A. Hoerl and R. Kennard. Ridge regression: applications to nonorthogonal problems. *Technometrics*, 12:69–82, 1970.

A. Hoerl and R. Kennard. Ridge regression: biased estimation for nonorthogonal problems. *Technometrics*, 12:55–67, 1970.

S. Huang and W. Studden. An equivalent kernel method for least squares spline regression. *Statistics and Decisions*, 3:179–201, 1993.

C. Hurvich and W. Tsai. Relative rates of convergence for efficient model selection criteria in linear regression. *Biometrika*, 82:418–425, 1995.

M. Hutchinson and F. deHoog. Smoothing noisy data with spline functions. *Numerische Mathematik*, 47:99–106, 1985.

M. Hutchinson and F. deHoog. An efficient method for calculating smoothing splines using orthogonal transformations. *Numerische Mathematik*, 50:311–319, 1987.

I. Ibragimov and R. Hasminski. *Statistical Estimation: Asymptotic Theory*. Springer Verlag, New York, 1981.

C. Jennen-Steinmetz and Th. Gasser. A unifying approach to nonparametric regression estimation. *Journal of the American Statistical Association*, 83:1084–1089, 1988.

G. Johnston. Probability of maximal deviations for nonparametric regression function estimates. *Journal of Multivariate Analysis*, 12:402–414, 1982.

D. Jupp. Curve fitting by splines as an application of unconstrained optimization. In R. S. Anderssen, L. S. Jennings, and D. M. Ryan, editors, *Optimization*, pages 49–59. University of Queensland Press, St. Lucia, Queensland, 1972.

D. Jupp. The "lethargy" theorem–a property of approximation by γ-polynomials. *Journal of Approximation Theory*, 14:204–217, 1975.

D. Jupp. Approximation to data by splines with free knots. *SIAM Journal of Numerical Analysis*, 15:328–343, 1978.

W. Kallenberg and T. Ledwina. On data driven Neyman's tests. *Probability and Mathematical Statistics*, 15:409–426, 1995.

W. Kennedy and J. Gentle. *Statistical Computing*. Marcel-Dekker, New York, 1980.

G. Knafl, J. Sacks, and D. Ylvisaker. Model robust confidence intervals. *Journal of Statistical Planning and Inference*, 6:319–334, 1982.

G. Knafl, J. Sacks, and D. Ylvisaker. Confidence bands for regression functions. *Journal of the American Statistical Association*, 80:683–691, 1985.

R. Kohn and C. Ansley. A new algorithm for spline smoothing based on smoothing a stochastic process. *SIAM Journal of Scientific and Statistical Computing*, 8:33–48, 1987.

R. Kohn, C. Ansley, and D. Tharm. The performance of cross-validation and maximum likelihood estimators of spline smoothing parameters. *Journal of the American Statistical Association*, 86:1042–1050, 1991.

R. Kohn, C. Ansley, and C. Wong. Nonparametric spline regression with autoregressive moving average errors. *Biometrika*, 79:335–346, 1992.

C. Kooperberg, C. Stone, and Y. Truong. The L_2 rate of convergence for hazard regression. *Scandinavian Journal of Statistics*, 22:143–157, 1995.

S. Kotz, N. L. Johnson, and C. B. Read. *Encyclopedia of Statistical Science, Volume VIII.* John Wiley, New York, 1988.

P. J. Laurent. Construction of spline functions on a convex set. In I. J. Schoenberg, editor, *Approximation with Special Emphasis on Spline Functions*, pages 415–446, 1969.

E. L. Lehmann. *Testing Statistical Hypotheses.* John Wiley, New York, second edition, 1986.

R. Lenth. Robust splines. *Communications in Statistics, Series A*, 6:847–854, 1977.

K. C. Li. Minimaxity of the method of regularization on stochastic processes. *The Annals of Statistics*, 10:937–942, 1982.

K. C. Li. Regression models with infinitely many parameters: consistency of bounded linear functions. *The Annals of Statistics*, 12:601–611, 1984.

K. C. Li. From Stein's unbiased risk estimates to the method of generalized cross-validation. *The Annals of Statistics*, 13:1352–1377, 1985.

K. C. Li. Asymptotic optimality of C_L and generalized cross-validation in ridge regression with application to spline smoothing. *The Annals of Statistics*, 14:1101–1112, 1986.

K. C. Li. Asymptotic optimality for C_p, C_L, cross-validation and generalized cross-validation: discrete index set. *The Annals of Statistics*, 15:958–975, 1987.

A. van der Linde. A note on smoothing splines as Bayesian estimates. *Statistics and Decisions*, 11:61–67, 1993.

M. J. Lindstrom. Penalized estimation of free-knot splines. *Journal of Computational and Graphical Statistics*, 1998. To appear.

D. G. Luenberger. *Optimization by Vector Space Methods*. John Wiley, New York, 1969.

T. Lyche and L. Schumaker. Computation of smoothing and interpolating natural splines via local bases. *SIAM Journal of Numerical Analysis*, 10:1027–1038, 1973.

Y. Mack and H. G. Müller. Convolution type estimators for nonparametric regression. *Statistics and Probability Letters*, 7:229–239, 1989.

Y. Mack and B. Silverman. Weak and strong uniform consistency of kernel regression estimates. *Zeitschrift für Wahrscheinlichkeitstheorie und Verwandte Gebiete*, 60:405–415, 1982.

C. Mallows. Some comments on C_p. *Technometrics*, 15:661–675, 1973.

P. McCullagh and J. Nelder. *Generalized Linear Models*. Chapman and Hall, New York, 1983.

K. Messer. A comparison of a spline estimate to its equivalent kernel estimate. *The Annals of Statistics*, 19:817–829, 1991.

K. Messer and L. Goldstein. A new class of kernels for nonparametric curve estimation. *The Annals of Statistics*, 21:179–195, 1993.

D. Montgomery and E. Peck. *Introduction to Linear Regression Analysis*. John Wiley, New York, 1982.

H. G. Müller. Optimal designs for nonparametric kernel regression. *Statistics and Probability Letters*, 2:285–290, 1984.

H. G. Müller. Smooth optimum kernel estimators of densities, regression curves and modes. *The Annals of Statistics*, 12:766–774, 1984.

H. G. Müller. *Nonparametric Regression Analysis of Longitudinal Data*. Springer-Verlag, New York, 1988.

H. G. Müller. Adaptive nonparametric peak estimation. *The Annals of Statistics*, 17:1053–1069, 1989.

H. G. Müller. Smooth optimum kernel estimators near endpoints. *Biometrika*, 78:521–530, 1991.

H. G. Müller. On the boundary kernel method for non-parametric curve estimation near endpoints. *Scandinavian Journal of Statistics*, 20:313–328, 1993.

H. G. Müller and U. Stadtmüller. Estimation of heteroscedasticity in regression analysis. *The Annals of Statistics*, 15:610–625, 1987.

H. G. Müller and U. Stadtmüller. Variable bandwidth kernel estimators of regression curves. *Annals of Statistics*, 15:182–201, 1987.

H. G. Müller and U. Stadtmüller. Detecting dependencies in smooth regression models. *Biometrika*, 75:639–650, 1988.

H. G. Müller, U. Stadtmüller, and T. Schmitt. Bandwidth choice and confidence intervals for derivatives of noisy data. *Biometrika*, 74:743–749, 1987.

H. G. Müller and J. L. Wang. Hazard rate estimation under random censoring with varying kernels and bandwidths. *Biometrics*, 50:61–76, 1994.

E. Nadaraya. On estimating regression. *Theory of Probability and Its Applications*, 9:141–142, 1964.

E. Nadaraya. On nonparametric estimation of density functions and regression curves. *Theory of Probability and Its Applications*, 10:186–190, 1965.

E. Nadaraya. Remarks on non-parametric estimates for density functions and regression curves. *Theory of Probability and Its Applications*, 15:134–137, 1970.

S. Narula. Orthogonal polynomial regression. *International Statistical Review*, 47:31–36, 1979.

M. Neumann. Automatic bandwidth choice and confidence intervals in nonparametric regression. *The Annals of Statistics*, 23:1937–1959, 1995.

K. Noda. Estimation of a regression function by the Parzen kernel-type density estimators. *Annals of the Institute of Statistical Mathematics*, 28:221–234, 1976.

M. Nussbaum. Spline smoothing in regression models and asymptotic efficiencey in L_2. *The Annals of Statistics*, 13:984–997, 1985.

D. Nychka. Bayesian confidence intervals for smoothing splines. *Journal of the American Statistical Association*, 83:1134–1143, 1988.

D. Nychka. The average posterior variance of a smoothing spline and a consistent estimate of the average squared error. *The Annals of Statistics*, 18:415–428, 1990.

D. Nychka. Choosing a range for the amount of smoothing in nonparametric regression. *Journal of the American Statistical Association*, 86:653–664, 1991.

D. Nychka. Splines as local smoothers. *The Annals of Statistics*, 23:1175–1197, 1995.

G. Oehlert. Relaxed boundary smoothing splines. *The Annals of Statistics*, 20:146–160, 1992.

F. O'Sullivan. A statistical perspective on ill-posed inverse problems. *Statistical Science*, 1:502–527, 1986.

F. O'Sullivan, B. Yandell, and W. Raynor. Automatic smoothing of regression functions in generalized linear models. *Journal of the American Statistical Association*, 81:96–103, 1986.

B. Park and J. S. Marron. Comparison of data-driven bandwidth selectors. *Journal of the American Statistical Association*, 85:66–72, 1990.

M. Priestley and M. Chao. Non-parametric function fitting. *Journal of the Royal Statistical Society, Series B*, 34:385–392, 1972.

E. Rafajlowicz. Nonparametric orthogonal series estimators of regression: a class attaining the optimal convergence rate in L_2. *Statistics and Probability Letters*, 5:219–224, 1987.

E. Rafajlowicz. Nonparametric least squares estimation of a regression function. *Statistics*, 3:349–358, 1988.

J. Ramsay. Monotone regression splines in action. *Statistical Science*, 4:425–461, 1988.

B. L. S. Prakasa Rao. *Nonparametric Functional Estimation*. Academic Press, New York, 1983.

B. Ray and R. Tsay. Bandwidth selection for kernel regression with long-range dependent errors. *Biometrika*, 84:791–802, 1997.

C. Reinsch. Smoothing by spline functions. *Numerische Mathematik*, 10:177–183, 1967.

C. Reinsch. Smoothing by spline functions, II. *Numerische Mathematik*, 16:451–454, 1971.

P. Révész. On the nonparametric estimation of the regression function. *Problems of Control and Information Theory*, 8:297–302, 1979.

J. Rice. A approach to peak area estimation. *Journal of Research of the National Bureau of Standards*, 87:53–65, 1982.

J. Rice. Bandwidth choice for nonparametric regression. *The Annals of Statistics*, 12:1215–1230, 1984a.

J. Rice. Boundary modification for kernel regression. *Communications in Statistics, Series A*, 13:893–900, 1984b.

J. Rice. Bandwidth choice for differentiation. *Journal of Multivariate Analysis*, 19:251–264, 1986a.

J. Rice. Choice of smoothing parameter in deconvolution problems. *Contemporary Mathematics*, 59:137–151, 1986b.

J. Rice. Convergence rates for partially splined models. *Statistics and Probability Letters*, 4:203–208, 1986c.

J. Rice and M. Rosenblatt. Smoothing splines: regression, derivatives and deconvolution. *The Annals of Statistics*, 11:141–156, 1983.

M. Rosenblatt. Conditional probabililty density and regression estimators. In P. Krishnaiah, editor, *Multivariate Analysis II*, pages 25–31. Academic Press, New York, 1969.

M. Rosenblatt. Density estimates and Markov sequences. In M. L. Puri, editor, *Nonparametric Techniques in Statistical Inference*, pages 199–210. Cambridge University Press, Oxford, 1970.

G. Roussas. Nonparametric estimation of the transition distribution function of a Markov process. *The Annals of Mathematical Statistics*, 40:1386–1400, 1969.

G. Roussas and L. Tran. Asymptotic normality of the recursive kernel regression estimate under dependence conditions. *The Annals of Statistics*, 20:98–120, 1992.

D. Ruppert and M. Wand. Multivariate locally weighted least squares regression. *The Annals of Statistics*, 22:1346–1370, 1994.

L. Rutkowski. On system identification by nonparametric function fitting. *IEEE Transactions on Automatic Control*, 27:225–227, 1982.

I. J. Schoenberg. Spline functions and the problem of graduation. *Proceedings of the National Academy of Science USA*, 52:947–950, 1964.

E. Schuster. Joint asymptotic distribution of the estimated regression function at a finite number of distinct points. *The Annals of Mathematical Statistics*, 43:84–88, 1972.

E. Schuster and S. Yakowitz. Contributions to the theory of nonparametric regression, with application to system identification. *The Annals of Statistics*, 7:139–149, 1979.

G. Schwarz. Estimating the dimension of a model. *The Annals of Statistics*, 6:461–464, 1978.

D. Scott. *Multivariate Density Estimation.* John Wiley, New York, 1992.

D. Scott and G. Terrell. Biased and unbiased cross-validation in density estimation. *Journal of the American Statistical Association*, 82:1131–1146, 1987.

R. Serfling. *Approximation Theorems of Mathematical Statistics.* John Wiley, New York, 1976.

T. Severini and J. Staniswalis. Quasi-likelihood estimation in semiparametric models. *Journal of the American Statistical Association*, 89:501–511, 1994.

P. Shi and G. Li. Global convergence rates of B-spline M-estimators in nonparametric regression. *Statistica Sinica*, 5:303–318, 1995.

P. Shi and C. Tsai. Semiparametric regression model selections. Manuscript, 1997.

J. J. Shiau and G. Wahba. Rates of convergence of some estimators for a semiparametric model. *Communications in Statistics, Series B*, 17:1117–1133, 1988.

R. Shibata. Selection of the order of an autoregressive model by Akaike's information criterion. *Biometrika*, 63:117–126, 1976.

R. Shibata. An optimal selection of regression variables. *Biometrika*, 68:45–54, 1981.

G. Shorack and J. Wellner. *Empirical Processes with Applications in Statistics*. John Wiley, New York, 1986.

B. Silverman. A fast and efficient cross-validation method for smoothing parameter choice in spline regression. *Journal of the American Statistical Association*, 79:584–589, 1984a.

B. Silverman. Spline smoothing: the equivalent variable kernel method. *The Annals of Statistics*, 12:898–916, 1984b.

B. Silverman. Some aspects of the spline smoothing approach to non-parametric regression curve fitting (with discussion). *Journal of the Royal Statistical Society, Series B*, 47:1–52, 1985.

B. Silverman. *Density Estimation for Statistics and Data Analysis*. Chapman and Hall, New York, 1986.

J. Simonoff. *Smoothing Methods in Statistics*. Springer-Verlag, New York, 1995.

P. Smith. Splines as a useful and convenient statistical tool. *American Statistician*, 33:57–62, 1979.

P. Smith. Curve fitting and modeling with splines using statistical variable selection techniques. Manuscript, 1983.

P. Speckman. Spline smoothing and optimal rates of convergence in non-parametric regression models. *The Annals of Statistics*, 13:970–983, 1985.

P. Speckman. Kernel smoothing in partial linear models. *Journal of the Royal Statistical Society, Series B*, 50:413–436, 1988.

F. Spitzer. A combinatorial lemma with its applications to probability theory. *Transactions of the American Mathematical Society*, 82:323–339, 1956.

U. Stadtmüller. Asymptotic properties of nonparametric curve estimates. *Periodica Mathematica Hungarica*, 17:83–108, 1986.

J. Staniswalis. The kernel estimate of a regression function in likelihood-based models. *Journal of the American Statistical Association*, 84:276–283, 1989a.

J. Staniswalis. Local bandwidth selection for kernel estimates. *Journal of the American Statistical Association*, 84:284–288, 1989b.

C. Stein. Estimation of the mean of a multivariate normal distribution. *The Annals of Statistics*, 9:1135–1155, 1981.

C. Stone. Nearest neighbor estimators of a nonlinear regression function. In *Proceedings on Computer Science and Statistics: 8th Annual Symposium on the Interface*, pages 413–418, 1975.

C. Stone. Optimal rates of convergence for nonparametric estimators. *The Annals of Statistics*, 8:1348–1360, 1980.

C. Stone. Optimal global rates of convergence for nonparametric regression. *The Annals of Statistics*, 10:1040–1053, 1982.

C. Stone. Additive regression and other nonparametric models. *The Annals of Statistics*, 13:689–705, 1985.

C. Stone. The dimensionality reduction principle for generalized additive models. *Annals of Statistics*, 14:590–606, 1986.

C. Stone. The use of polynomial splines and their tensor products in multivariate function estimation. *The Annals of Statistics*, 22:118–184, 1994.

W. Stuetzle and Y. Mittal. Some comments on the asymptotic behavior of robust smoothers. In Th. Gasser and M Rosenblatt, editors, *Smoothing Techniques for Curve Estimation*, pages 191–195. Springer-Verlag, Heidelberg, 1979.

J. Sun and C. Loader. Simultaneous confidence bands for linear regression and smoothing. *The Annals of Statistics*, 22:1328–1345, 1994.

G. Szegö. *Orthogonal Polynomials*, volume 23 of *Colloquium Publications*. American Mathematical Society, Providence, Rhode Island, 1986.

M. Thompson. Selection of variables in multiple regression. *International Statistical Review*, 46:1–49, 1978.

M. Thompson, J. Kay, and M. Titterington. Noise estimation in signal restoration using regularization. *Biometrika*, 78:475–488, 1991.

M. Titterington. Common structure of smoothing techniques in statistics. *International Statistical Review*, 53:141–170, 1985.

Y. Truong and C. Stone. Nonparametric function estimation involving time series. *The Annals of Statistics*, 20:77–97, 1992.

J. Tukey. Curves as parameters and touch estimation. In *Proceedings of the 4th Berkeley Symposium on Mathematical Statistics*, volume 1, pages 681–694, 1961.

F. Utreras. Optimal smoothing of noisy data using spline functions. *SIAM Journal of Scientific and Statistical Computing*, 2:349–362, 1981.

F. Utreras. Smoothing noisy data under monotonicity constraints: existence, characterization and convergence rates. *Numerische Mathematik*, 47:611–625, 1985.

F. Utreras. Boundary effects on convergence rates for Tikhonov regularization. *Journal of Approximation Theory*, 54:235–249, 1987.

M. Villalobos and G. Wahba. Inequality-constrained multivariate smoothing splines with application to the estimation of posterior probabilities. *Journal of the American Statistical Association*, 82:239–248, 1987.

G. Wahba. Improper priors, spline smoothing and the problem of guarding against model errors in regression. *Journal of the Royal Statistical Society, Series B*, 40:364–372, 1978.

G. Wahba. Bayesian "confidence intervals" for the cross-validated smoothing spline. *Journal of the Royal Statistical Society, Series B*, 45:133–150, 1983.

G. Wahba. Partial spline models for the semi-parametric estimation of functions of several variables. In *Statistical Analysis of Time Series*, pages 319–329. Institute of Statistical Mathematics, Tokyo, 1984.

G. Wahba. *Spline Models for Observational Data.* CBMS-NSF series. SIAM, Philadelphia, 1990.

G. Wahba, Y. Wang, C. Gu, R. Klein, and B. Klein. Smoothing spline ANOVA for exponential families, with applications to the Wisconsin epidemiological study of diabetic retinopathy. *The Annals of Statistics*, 23:1865–1895, 1995.

G. Wahba and J. Wendelberger. Some new mathematical methods for variational objective analysis using splines and cross validation. *Monthly Weather Review*, 108:1122–1143, 1980.

M. Wand and M. C. Jones. *Kernel Smoothing.* Chapman and Hall, New York, 1995.

F. Wang and D. Scott. The L_1 method for robust nonparametric regression. *Journal of the American Statistical Association*, 89:65–76, 1994.

S. Wang. Saddlepoint methods for bootstrap confidence bands in nonparametric regression. *Australian Journal of Statistics*, 35:93–101, 1993.

Y. Wang. Smoothing spline models with correlated random errors. *Journal of the American Statistical Association*, 93:341–348, 1998.

G. Watson. Smooth regression analysis. *Sankhya, Series A*, 26:359–372, 1964.

E. T. Whittaker. On a new method of graduation. *Proceedings of the Edinburgh Mathematical Society*, 41:63–75, 1923.

S. Wold. Spline functions in data analysis. *Technometrics*, 16:1–11, 1974.

W. Wong. On the consistency of cross-validation in kernel nonparametric regression. *The Annals of Statistics*, 11:1136–1141, 1983.

S. Wood and R. Kohn. A Bayesian approach to robust binary nonparametric regression. *Journal of the American Statistical Association*, 93:203–213, 1998.

I. Wright and E. Wegman. Isotonic, convex and related splines. *The Annals of Statistics*, 8:1023–1035, 1980.

Y. Xia. Bias corrected confidence bands in nonparametric regression. *Journal of the Royal Statistical Society, Series B*, 1999. To appear.

S. Yakowitz. Nonparametric density estimation, prediction, and regression for Markov sequences. *Journal of the American Statistical Association*, 80:215–221, 1985.

Y. Yatracos. A lower bound on the error in nonparametric regression type problems. *The Annals of Statistics*, 16:1180–1187, 1988.

Y. Yatracos. On the estimation of the derivatives of a function with the derivatives of an estimate. *Journal of Multivariate Analysis*, 28:172–175, 1989.

P. Zhang. On the distributional properties of model selection criteria. *Journal of the American Statistical Association*, 87:732–737, 1992.

P. Zhang. Assessing prediction error in non-parametric regression. *Scandinavian Journal of Statistics*, 22:83–94, 1995.

S. Zhou, X. Shen, and D. Wolfe. Local asymptotics for regression splines and confidence regions. *The Annals of Statistics*, 1999. To appear.

Index

acceleration curve, 222
Akaike information criterion, 65
asymptotic normality
for kernel estimators, 195, 196
for series estimators, 105
for smoothing splines, 268, 287

bandwidth
asymptotically optimal, 169, 173, 175, 186, 187
convergence rates for estimators, 181, 183
definition of, 157
estimation by cross validation, 179
for derivative estimation, 188, 219, 220, 222
for smoothing splines, 237, 240, 249, 258
plug-in estimators, 184, 185
variable, 185, 194, 258
bias reduction lemma, 131
for boundary bias removal, 132, 178, 217, 259, 286
for estimation in partial linear models, 137, 203, 223, 278
boundary bias
for kernel estimators, 170
for smoothing splines, 254, 259
for trigonometric regression estimators, 102
removal by bias reduction lemma, 132, 178, 217
removal by the generalized jackknife, 217
boundary kernel, 171, 177, 187, 188, 190, 192, 217–220

central limit theorem, 6, 104
complete orthonormal sequence, 75
confidence bands
for kernel estimators
bias corrected, 200
Bonferroni, 200
for least squares splines, 305
for series estimators, 111
confidence intervals
for kernel estimators
bias corrected, 197
normal theory, 196
for least squares splines, 299, 305
for parameters in a partial lin-

ear model, 140, 206
for series estimators, 82, 96, 106, 128
Bonferroni bias correction, 108, 128
for smoothing splines
Bayesian, 265, 268, 287
normal theory, 268, 287
Cook's distance, 94, 125, 270
cross validation, 43

deconvolution, 274
Demmler-Reinsch representation of a smoothing spline, 234, 237
derivative estimation
using kernel estimators, 188
using smoothing splines, 274

Epanechnikov kernel, 176

Fourier coefficient
definition of, 77, 78
estimation bias, 106
estimation of, 80, 87
inference for, 82, 111, 128
rate of decay, 101, 129
Fourier series, 77, 79
Fourier series estimator
definition of, 80
relation to kenel estimators, 162
relation to smoothing splines, 234

generalized cross validation, 43
generlized linear models, 277
growth curve, 192
GSJS variance estimator, 49

Hutchinson/de Hoog algorithm, 246

interaction splines, 277

kernel
boundary, 171
for derivative estimation, 188
for least squares splines, 304
for local linear regression, 190
for smoothing splines, 237, 247, 258
higher order, 186
minimum variance, 176
optimal, 176
second order, 157

Laplacian smoothing splines, 276
Legendre polynomials, 76
loss, 28

Mallow's C_L, 39
method of regularization, 274

Nadaraya-Watson estimator, 160, 194, 211
natural boundary conditions, 258, 282

order
of a series estimator, 83
asymptotically optimal, 100, 129

estimation of, 84, 85, 128
of a spline, 281
of an estimator, 16
order selection test, 115

partial spline, 278
pseudo-residuals, 49

regression
curve, 2
function, 2
nonparametric, 3
ridge, 34
simple linear, 3
risk
definition of, 14, 28
estimation of, 39
global convergence rate
for kernel estimators, 173, 175, 187
for local linear regression, 190
for polynomial regression, 129
for polynomial-trigonometric regression, 134
for smoothing splines, 253, 258
for trigonometric regression, 100
optimal, 16
integrated, 30
pointwise convergence rate
for kernel estimators, 168, 175, 186
for smoothing splines, 252
prediction, 29
estimation of, 39, 42, 44

Schwarz criterion, 55, 56, 58, 61, 91, 114, 149
smoothing parameter
asymptotically optimal, 253, 254
definition of, 228
estimation by cross validation, 239
estimation by generalized maximum likelihood, 262, 287
relation to bandwidth, 237, 240
Speckman's minimax estimator, 259
spline
B-spline, 300
definition of, 281
interaction, 277
knots, 281
natural, 281
order of, 281
partial, 278
thin plate, 276

Taylor's Theorem, 121
thin plate splines, 276
twicing, 219

unbiased risk criterion, 39

variable selection, 31, 42, 44, 51
variance estimation

for partial linear models, 206
in nonparametric regression,
49, 81, 263
velocity curve, 192

9 780824 793371

Printed in the United States
by Baker & Taylor Publisher Services